ELECTRONICS

2ND EDITION

TOM DUNCAN, B.Sc.

JOHN MURRAY

Success Studybooks

Advertising and Promotion
Book-keeping and Accounts
Business Calculations
Chemistry
Commerce
Commerce: West African Edition
Communication
Electronics
European History 1815-1941
Information Processing
Insurance
Investment
Law
Managing People
Marketing
Politics
Principles of Accounting
Principles of Accounting: Answer Book
Psychology
Sociology
Statistics
Twentieth Century World Affairs
World History since 1945

First published 1983
by John Murray (Publishers) Ltd
50 Albemarle Street
London W1X 4BD

Reprinted 1984 and 1986 (twice) with revisions;
1987, 1989, 1990, 1991, 1993, 1995, 1996
Second edition 1997

Reprinted 1999, 2000

Typeset by Servis Filmsetting Ltd, Manchester
Printed and bound in Great Britain by
Biddles Ltd, www.biddles.co.uk

A catalogue entry for this title can be obtained from the British Library.

ISBN 0–7195–7205–3

Contents

Part Two Components

Part Four Digital circuits

Foreword

Success in Electronics is intended for anyone who wishes to gain an understanding of the basic principles of electronics as they are applied in communication, control and computer systems. Very little previous knowledge of electricity is assumed, and mathematical requirements are kept to a minimum. The treatment is practically orientated and actual devices with their uses are considered in the hope that the reader may be encouraged to 'do' some electronics. For this reason books on project work are listed in the Further Reading section at the end of Part Five of this book.

While not following any particular examination syllabus, the book is appropriate for students taking GCSE, BTEC (Electronics NII and NIII), City & Guilds and A-level courses.

Part One deals with *Basic electricity*, Part Two with *Components*, Parts Three and Four with *Linear* and *Digital circuits* respectively (in both discrete component and integrated circuit form) and in Part Five an outline is given of some *Electronic systems*. If desired, Part Four may be taken before Part Three. At the end of most Units, as an aid to checking progress, there are revision questions and numerical problems. Answers are given, where appropriate, at the end of the book.

The content has been amended for this second edition to keep pace with the rapid growth in importance of digital electronics and computers in telecommunications and in other fields. Unit 18 now gives a fuller treatment of the ever-increasing role of digital electronics in communication systems. The original Unit on *Digital systems and computers* has been replaced by two new Units. Unit 19 now deals only with *Digital systems* and gives a fuller treatment than previously of Boolean algebra and De Morgan's theorems and their role in logic circuit design. The content of the new Unit 20, *Computers and microprocessors*, has been rewritten and much enlarged to include more detailed coverage of computer peripherals and uses of computers in, for example, multimedia systems.

T.D.

Acknowledgements

I should like to thank Edward Mallory, Leslie Basford and Jim Hutton who read and criticized the book during its initial preparation. I am also much indebted to Dr. Helen Wright for editing the original text so meticulously, to my wife who typed the manuscript and to my daughter, Dr. Heather Kennett, for help in various ways. For collating the second edition and seeing it through the press in her usual efficient way, I am once again most grateful to Jane Roth.

T.D.

For permission to use copyright photographs thanks are due to:
p.2 Rosenfeld Images Ltd/Science Photo Library; **p.3** *t* David Parker/Science Photo Library, *b* Hank Morgan/Science Photo Library; **p.4** *t & b* Geoff Tompkinson/Science Photo Library; **p.5** Geoff Tompkinson/Science Photo Library; **p.16** Avo Ltd; **p.37** *t* RS Components; **p.47** *t, c & br* RS Components, *bl* Maplin Electronics plc; **p.73** *a, b & c* RS Components, *d* Maplin Electronics plc; **p.90** RS Components; **p.107** RS Components; **p.109** RS Components; **p.110** Maplin Electronics plc; **p.111** Maplin Electronics plc; **p.112** RS Components; **p.113** Maplin Electronics plc; **p.120** RS Components; **p.153** RS Components; **p.178** Ferranti Electronics Ltd; **p.180** Brierley Photo Library; **p.181** *l* David Parker/Seagate Microelectronics Ltd/Science Photo Library, *r* Ray Ellis/Science Photo Library; **p.313** STC Components Ltd; **p.341** Hank Morgan/Science Photo Library; **p.346** © Patrick Llewelyn-Davies, courtesy of Hi-Grade Computers plc; **p.349** Compaq Presario 8700 Series; **p.350** Geoff Tompkinson/Science Photo Library; **p.354** STC/A. Sternberg/Science Photo Library; **p.364** RS Components; **p.373** Unilab; **p.375** Sheila Terry, Lecroy/Science Photo Library

PART ONE

Basic electricity

UNIT 1

Introduction

1.1 Electronics today

The development of electronics initially gave us pocket calculators, digital watches, heart pacemakers, computers for industry, commerce and scientific research, electronically controlled washing machines, traffic lights and manufacturing processes, and 'instant' viewing on our televisions of events on the other side of the world.

Further advances have given us microwave ovens, video cassette recorders, teletext, personal computers (PCs), electronic games and 'multimedia' applications, computer-aided design (CAD), electronic limbs, 'keyhole' surgery, data processing, electronic cash dispensers and point-of-sale facilities, digital telephone links, fax, e-mail and the World Wide Web. Some of these many applications of electronics are shown in Fig. 1.01(*a*)–(*f*).

Fig. 1.01 (*a*) Computer control of the machining of turbine blades

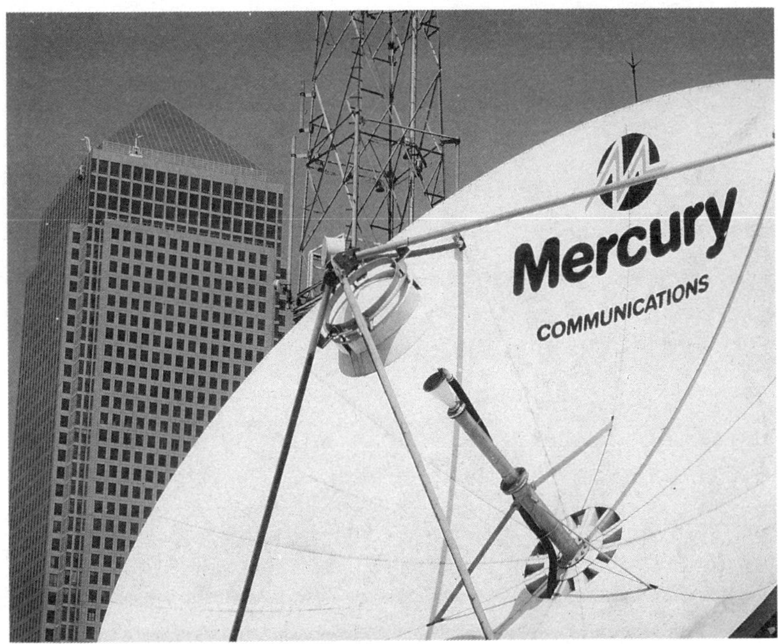

Fig. 1.01 (*b*) Satellite communications dish aerial near Canary Wharf tower, London

Fig. 1.01 (*c*) An IBM engineer testing a computer speech-recognition system. Interpretation of voice patterns is a criterion for development of fifth generation computer systems, where user-friendliness would be a major feature

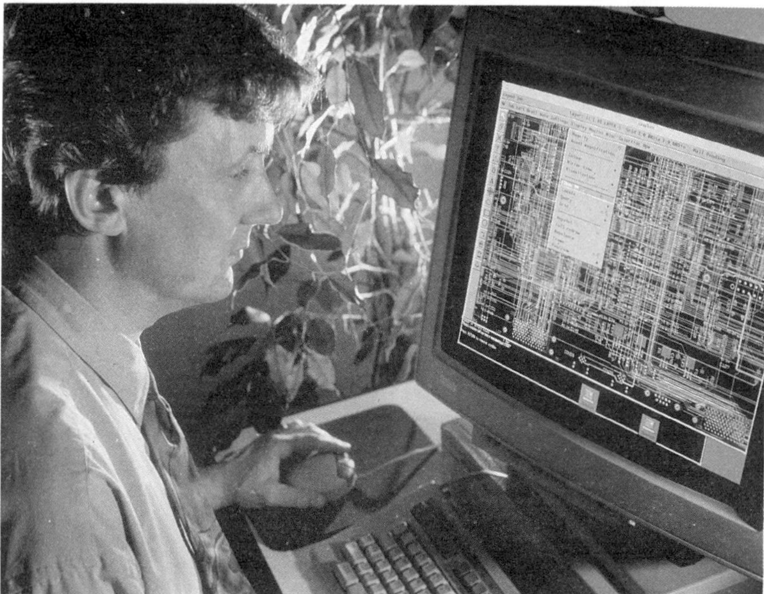

Fig. 1.01 (*d*) Computer-aided design (CAD) of electronic circuits. This is now a crucial aspect of modern electronic engineering. Once the design is fixed the computer can generate master plans to manufacture the circuits

Fig. 1.01 (*e*) Use of virtual reality in a computer game. Through his goggles the player sees a virtual battlefield and can give orders to military units. The display also includes 'floating' screens showing real-time views of the elements of the force available

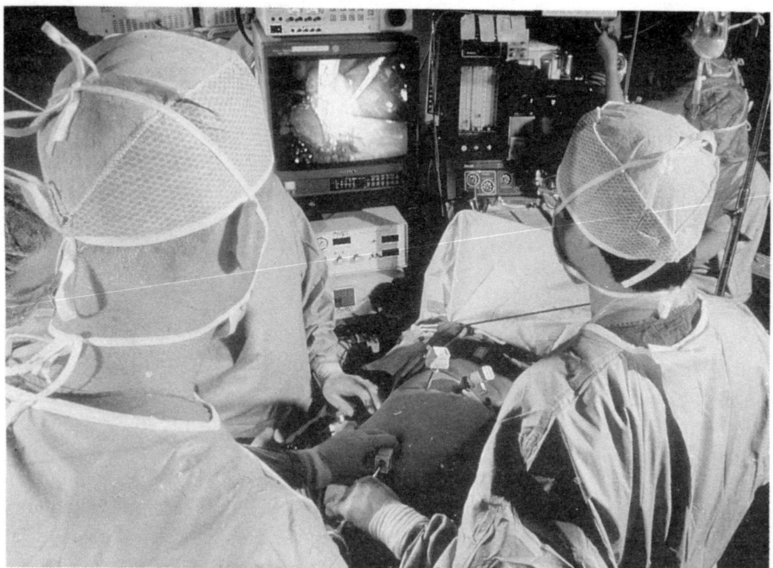

Fig. 1.01 (*f*) Use of a laparoscope in the removal of a gall bladder. The laparoscope contains a camera which provides a view of the gall bladder to guide the surgical instruments

All these advanced applications have become possible largely because we have learned how to build complete circuits, containing thousands of electronic parts, on a tiny wafer of silicon no more than 5 mm square and 0.5 mm thick. Microelectronics is concerned with these 'densely populated', miniaturized integrated circuits (ICs) or 'chips', which are changing the way we live and work and challenging us to see that the changes are for the better.

1.2 Electronic systems

After glancing inside an electronic system such as a radio or a television set or a computer, it would not be surprising if you felt slightly discouraged from studying electronics. Fortunately things are really less daunting than they seem, for which there are two reasons.

First, while an electronic system may have a large number of *components* (parts), there are only a few types of these. The main ones are resistors, capacitors, inductors, diodes, transistors, switches and transducers. Those used in integrated circuits (transistors, diodes, resistors and capacitors) have the same action as their discrete (separate) counterparts; the difference is mainly one of size.

Second, different electronic *systems* are made up from a fairly small number of basic *circuits* or building blocks. Each basic circuit consists of components connected in a certain way so that it does a particular job such as amplifying or counting (Fig. 1.02).

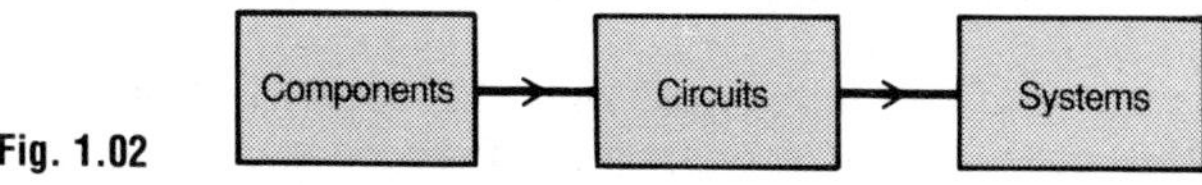

Fig. 1.02

Before the days of microelectronics, components were made separately and then wired together to give the required circuit. Today they are also produced in integrated circuit form, complete with interconnections, more or less all at the same time. The extent of the integration is now so great that the distinction between a circuit and a system is often less clear-cut. In fact, some chips qualify to be at least subsystems if not quite complete systems. In many cases, systems are built from a mixture of discrete components and integrated circuits.

1.3 Linear and digital circuits

Most electronic systems are designed to receive an electrical *input* and then 'process' it so that it produces an electrical *output* capable of doing the required task (which the input could not do without the aid of the system). For example, in a CD player, the signal from the pick-up (which cannot operate the loudspeaker directly) supplies the input to the amplifier which produces an output capable of driving the loudspeaker, Fig. 1.03(*a*). Or again, in Fig. 1.03(*b*), showing an arrangement for counting the number of packets on a moving conveyor belt in a factory, the electrical signals from the photocell (produced when the light from the lamp opposite is blocked by the passing packet) provide the input to the counter which generates an output that enables the display to record the total count.

The electronic circuits used in systems fall into two main groups—*linear* (or analogue) and *digital.*

Linear circuits are *amplifier-type* circuits handling signals which are frequently electrical representations (i.e. analogues) of quantities, such as speech and music sounds, that change smoothly over a range of values. This would be so in, for instance, the amplifier of Fig. 1.03(*a*), where the output will also vary continuously and be more or less an exact but amplified copy of the input, Fig. 1.04(*a*). The output changes in step with the input, or in mathematical terms, there is a linear connection between them, so doubling the input doubles the output. Many linear circuits use transistors as amplifiers.

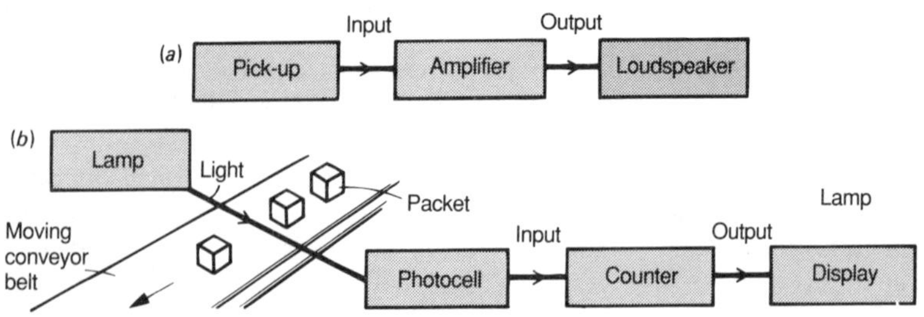

Fig. 1.03

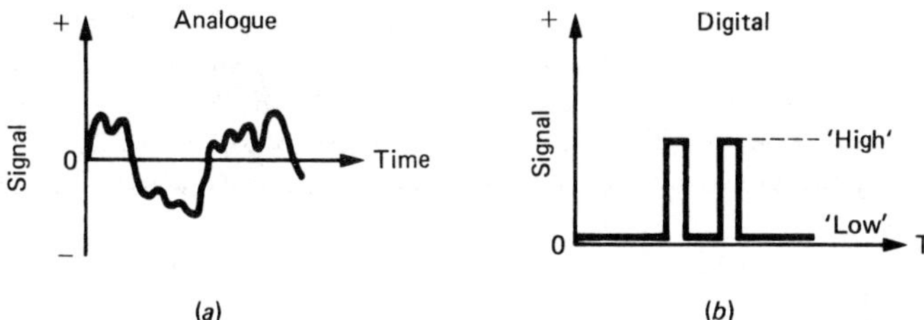

Fig. 1.04

Digital circuits are *switching-type* circuits handling signals which have only one of two values. When there is a change from one value to the other, it happens suddenly, Fig. 1.04(*b*). While linear circuits are continuous-state circuits, digital circuits are two-state ones, their inputs and outputs are either 'high', i.e. near the value of the supply, or 'low', i.e. near zero. They use transistors as switches. The counter in Fig. 1.03(*b*) is a digital circuit in which the input from the photocell is either 'low' or 'high' depending on whether or not light is interrupted. Digital circuits carry electrical pulses.

A lamp controlled by a dimmer allows a wide range of light levels and is a continuous-state system. One controlled by an on-off switch is a two-state system; it is either fully lit or it is not lit at all.

Some electronic systems contain both linear and digital circuits.

1.4 Electronic diagrams

Diagrams are the language of electronics. The 'letters' of this language are the signs or *symbols* which represent components, and the 'words' are the groups of symbols which form *circuit diagrams*. In Fig. 1.05(*a*) the symbols for a resistor, a capacitor and a transistor are shown. In Fig. 1.05(*b*) they have been combined to make the circuit diagram of a simple amplifier. With practice circuit diagrams can be 'read' and understood like any other language.

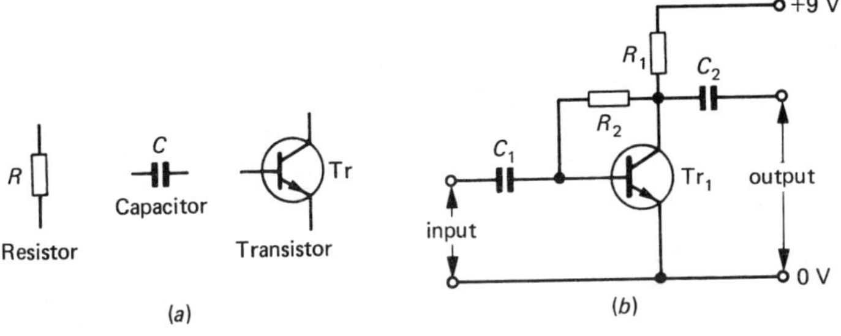

Fig. 1.05

The 'sentences' of the language are the diagrams of whole systems. Frequently, to simplify matters these are drawn as *block diagrams* in which the circuits making up the system are represented by boxes, as in Fig. 1.03. In the block diagram or systems way of looking at things it is often unnecessary to know exactly how a circuit works but sufficient simply to regard it as a 'black box' and to know what it does.

1.5 Questions

1. How is electronics helping
   a) the doctor,
   b) handicapped people,
   c) the shopkeeper,
   d) the policeman,
   e) the weather forecaster,
   f) the astronomer?

2. In what ways could electronic systems intrude on private matters and reveal personal 'secrets'?

3. a) What is the difference between linear and digital circuits?
   b) Name an electronic system that uses
      (i) linear circuits,
      (ii) digital circuits.

UNIT 2

Direct current

2.1 Electric current

All atoms contain tiny particles called *electrons* which have a negative electric charge. In metals some electrons are held loosely by their atoms and can move from one atom to another. The motion of these 'free' electrons is normally haphazard with as many electrons moving in one direction as in the opposite direction, Fig. 2.01(*a*).

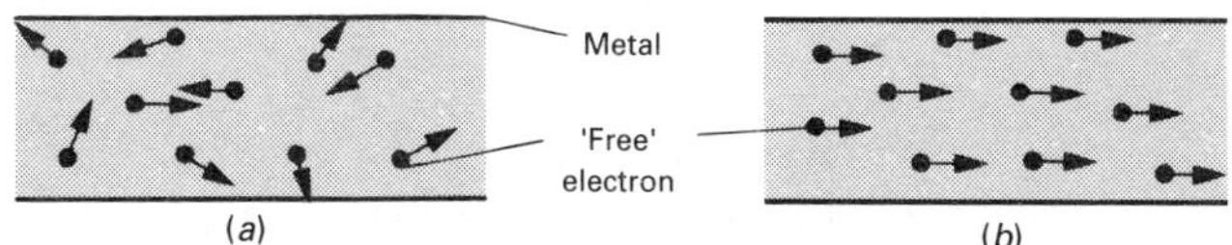

Fig. 2.01

If a force acts on the 'free' electrons in a metal wire so that they all 'flow' along the wire in the same direction, an electric current is produced, Fig. 2.01(*b*). We know that something different is happening in the wire because it can get warm and a magnetic compass near it can be deflected, i.e. a magnetic field is created in the surrounding space. A one-way flow of electrons is called a *direct current* (shortened to d.c.).

Current is measured in *amperes* (shortened to A) by connecting an *ammeter* so that the current flows through it, as illustrated in Fig. 2.02. In section 2.9 we will see that when about six million million million electrons pass any point in a wire every second, the current is 1A.

Two smaller units of current useful in electronics are:

$$1 \text{ milliampere } (1 \text{ mA}) = \frac{1}{1\,000}\text{A} = \frac{1}{10^3}\text{ A} = 10^{-3}\text{ A} = 0.001\text{ A}$$

$$1 \text{ microampere } (1\ \mu\text{A}) = \frac{1}{1\,000\,000}\text{ A} = \frac{1}{10^6}\text{ A} = 10^{-6}\text{ A} = 0.000\,001\text{ A}$$

2.2 Electromotive force

The force needed to cause a current flow in a wire can be supplied by a battery. We say the battery produces an *electromotive force* (shortened to e.m.f.) because of the chemical action which occurs inside it. A generator is another source of e.m.f. but it creates the e.m.f. magnetically.

E.m.f. is measured in *volts* (V) by connecting a *voltmeter* across the battery (or generator) as shown in Fig. 2.03.

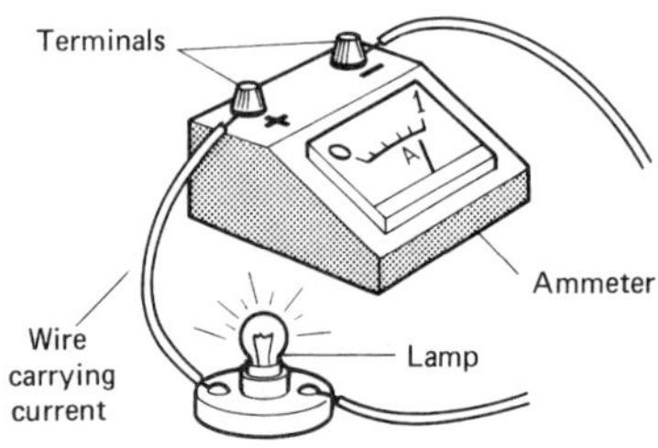

Fig. 2.02

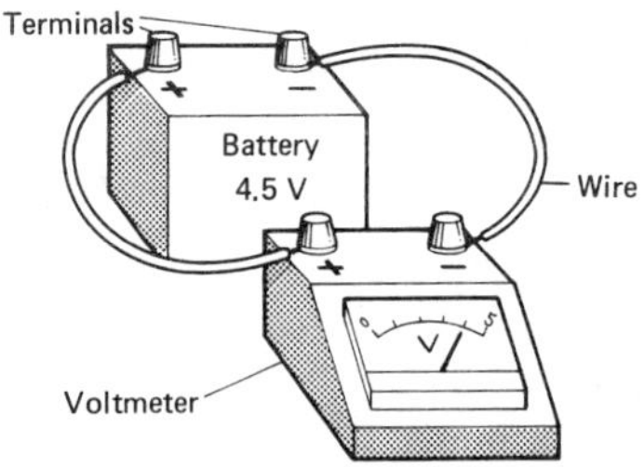

Fig. 2.03

Batteries are made up from cells—see section 21.6. A cell has two terminals, one is marked + (positive) and the other − (negative). A carbon-zinc (dry) cell has an e.m.f. of 1.5 V; if three are joined in series, i.e. + of one to − of next, the e.m.f. of the battery so formed is $3 \times 1.5 = 4.5$ V. Transistor radio and torch batteries consist of carbon-zinc cells. Cells and batteries produce direct current.

The e.m.f. acts along a wire at nearly the speed of light (300 million metres per second) and sets all the 'free' electrons drifting in the same direction at more or less the same instant. However, you may be surprised to learn that they drift quite slowly, often travelling less than one millimetre in a second.

2.3 Circuits and diagrams

An electric current requires a complete path (a circuit) before it can flow. The symbols used to represent various parts of a circuit in a circuit diagram are shown in Fig. 2.04.

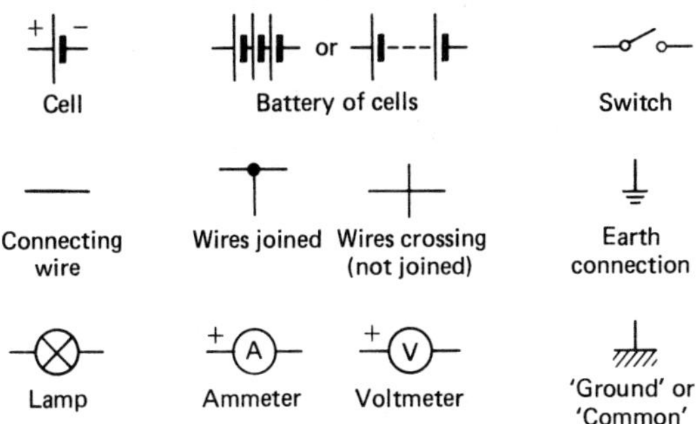

Fig. 2.04

Before the electron was discovered scientists decided that a current consisted of positive electric charges moving round a circuit in the direction from + to − of a battery. We still keep to this rule so when an arrow is put on a circuit diagram it shows the direction of what is called the *conventional current*, i.e. the direction in which positive charges would flow. This is in the opposite direction to the actual flow of the negatively charged electrons.

In a *series* circuit in which the different parts follow one after the other, the current is the same all the way round the circuit, it is not used up. In the series circuit of Fig. 2.05 the readings on ammeters A_1 and A_2 will be equal.

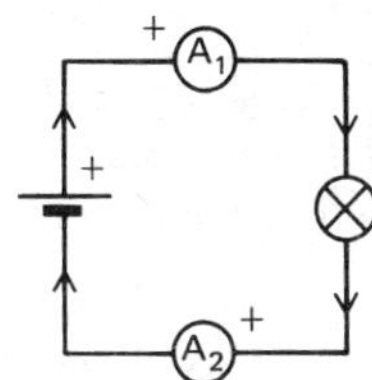

Fig. 2.05

2.4 Potential difference

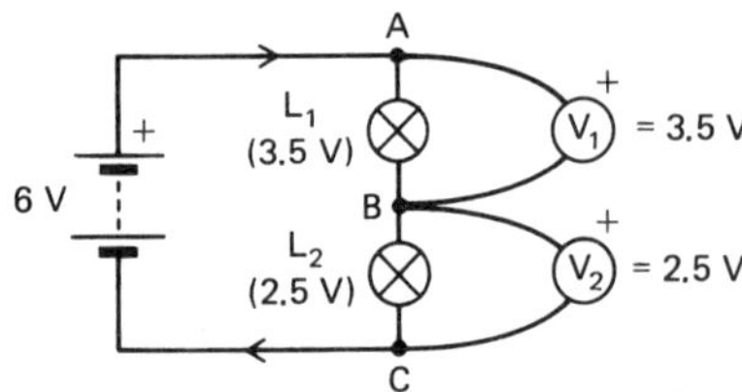

Fig. 2.06

In the circuit diagram of Fig. 2.06 two lamps L_1 and L_2 are connected in series with a 6 V battery. L_1 is designed to be lit by a 3.5 V supply and L_2 by a 2.5 V supply. The e.m.f. of the battery divides between the lamps so that each gets the correct supply. A voltmeter, V_1, connected across L_1 would read 3.5 V and another voltmeter, V_2, across L_2 would read 2.5 V.

The portion of the e.m.f. appearing across each lamp is called the *potential difference* (shortened to p.d.) across that lamp. The p.d. across L_1 is 3.5 V and across L_2 is 2.5 V. In general, using letters instead of numbers, if E is the e.m.f. of a battery and V_1 and V_2 are the p.ds across two appliances (e.g. lamps) in series with it, then:

$$E = V_1 + V_2$$

Although it is usually the p.d. between two points in a circuit we have to consider, there are occasions when it is a help to deal with what we call the *potential at a point*. To do this we have to choose one point in the circuit and say it has zero potential, i.e. 0 V. The potentials of all other points are then stated with reference to it, that is, the potential at any point is the p.d. between the point and the point of zero potential.

For example, if we take point C in Fig. 2.06 (in effect the negative terminal of the battery) as our zero, the potential at B is +2.5 V and at A (the positive terminal of the battery) it is +6 V. In electronic circuits a point of zero potential is called *ground* or *common*, or *earth* if it is connected to earth (for example via the earthing pin on a 3-pin mains plug); the signs used are shown in Fig. 2.04.

The terms e.m.f., p.d. and current have been devised by scientists to help us to understand electric circuits. What you should remember about them is that:

(i) e.m.f. is something a *battery* (or generator) has;
(ii) p.d. appears across *appliances in a circuit*;
(iii) current (conventional flows from points of *higher* potential to points of *lower* potential in a circuit.

E.m.fs and p.ds are often called *voltages* (since both are measured in volts) and we will follow this practice.

2.5 Resistance

Electrons move more easily through some materials than others when a voltage is applied. Opposition to direct current flow is called *resistance*.

A *conductor* has low resistance. Copper is a very good conducting material and is used as connecting wire in circuits. (Usually it has a coating of tin to protect it from the atmosphere.) Short thick wires have less resistance than long thin ones. Conductors especially made to have resistance are termed *resistors*—see section 4.1.

An *insulator* has very high resistance and is a poor conductor. Plastics such as polythene and PVC (polyvinyl chloride) are good insulators and only pass tiny currents even with high voltages; they are used as insulation to cover connecting wire.

A *semiconductor* falls between a conductor and an insulator. Silicon is one, and is used to make transistors and integrated circuits (chips).

Resistance is stated in *ohms* (shortened to Ω, the Greek letter omega). It is found by dividing the voltage by the current. For example, if the current through a lamp is 2 A when the voltage across it is 12 V, its resistance is 12/2 = 6 Ω. In general, if I is the current through a resistor when the voltage across it is V (Fig. 2.07), its resistance R is given by:

$$R = \frac{V}{I} \tag{1}$$

This is a reasonable way to quantify resistance because it makes R large when I is small and vice versa. Note that V must be in volts and I in amperes to get R in ohms. If the current is in mA, say 60 mA, we can write $I = 60/1\,000$ A, then if $V =$ 6 V we have:

$$R = \frac{V}{I} = \frac{6}{60/1\,000} = 6 \times \frac{1\,000}{60} = \frac{600}{6} = 100\ \Omega$$

Alternatively, using decimals, we can say $I = 60/1\,000 = 0.06$ A giving:

$$R = \frac{V}{I} = \frac{6}{0.06} = \frac{6 \times 100}{0.06 \times 100} = \frac{600}{6} = 100\ \Omega$$

Larger units of resistance are:

$$1 \text{ kilohm } (1\ \text{k}\Omega) = 1\,000\ \Omega = 10^3\ \Omega$$
$$1 \text{ megohm } (1\ \text{M}\Omega) = 1\,000\,000\ \Omega = 10^6\ \Omega$$

In electronics, R is often in kΩ and I in mA. Using these units V is still obtained in V.

Equation (1) can also be written:

$$V = I \times R \tag{2}$$

and

$$I = \frac{V}{R} \tag{3}$$

The equations (1), (2) and (3) are useful in calculations. The triangle in Fig. 2.08 is an aid to remembering them. If you cover the quantity you want with your finger, e.g. I, it equals what you can still see, i.e. V/R. To find V, cover V and you get $V = I \times R$. To find R, cover R and $R = V/I$.

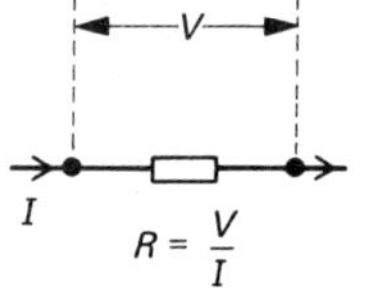

Fig. 2.07

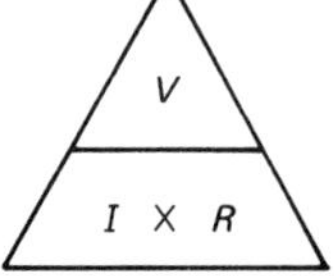

Fig. 2.08

2.6 Worked examples

Example 1

What is the current flowing through resistors of
a) 10 Ω, b) 1 kΩ, when the voltage across them is 9 V?

a) $R = 10\ \Omega$ and $V = 9$ V.

$$\therefore \quad I = \frac{V}{R} = \frac{9}{10}\text{A} = 0.9\text{ A}$$

b) R = 1 kΩ = 1 000 Ω and V = 9 V.

$$\therefore \quad I = \frac{V}{R} = \frac{9}{1\,000}\text{A} = 0.009\text{ A} = 9\text{ mA}$$

Example 2

What is the voltage across a resistor of
a) 100 Ω, b) 1.5 kΩ, when a current of 2 mA flows?

a) $R = 100\ \Omega$ and $I = 2$ mA = 2/1 000 A.

$$\therefore \quad V = I \times R = \frac{2}{1\,000} \times 100 = \frac{2}{10} = 0.2\text{ V} = 200\text{ mV}$$

b) $R = 1.5$ kΩ = 1.5 × 1 000 Ω and $I = 2/1\,000$ A.

$$\therefore \quad V = I \times R = \frac{2}{1\,000} \times 1.5 \times 1\,000 = 2 \times 1.5 = 3\text{ V}$$

Alternatively, if R is left in kΩ and I in mA, then V is obtained more quickly in V:

$$V = I \times R = 2\text{ mA} \times 1.5\text{ k}\Omega = 3\text{ V}$$

Example 3

Calculate the resistance of a resistor through which a current of 5 mA passes when the voltage across it is 1.5 V.

$I = 5$ mA = 5/1 000 A and $V = 1.5$ V.

$$\therefore \quad R = \frac{V}{I} = \frac{1.5}{5/1\,000} = \frac{1.5 \times 1\,000}{5} = \frac{1\,500}{5} = 300\ \Omega$$

2.7 Ammeters, voltmeters and multimeters

(a) Ammeters

To use an ammeter the circuit must be broken at some point and the ammeter connected in *series* so that current flows through it. In d.c. circuits the ammeter must have the terminal marked + (or coloured red) leading to the + terminal of the battery, as shown in Fig. 2.05. In that way conventional current enters the + terminal of the ammeter and leaves by the − terminal; otherwise the pointer on the ammeter is deflected in the wrong direction and damage may occur. Note that with cells it is different, when they are connected in series the + terminal of one cell is joined to the − terminal of the next.

An ammeter should have a very *small* resistance compared with the resistance of the rest of the circuit so that it does not reduce the current it has to measure when connected in the circuit.

(b) Voltmeters

A voltmeter is connected in *parallel*, that is, side-by-side with the component across which the voltage is to be measured. As with an ammeter the terminal marked + (or coloured red) should lead to the + terminal of the battery, as shown in Fig. 2.06, to ensure that conventional current deflects the pointer in the correct direction.

A voltmeter should have a very *large* resistance compared with the resistance of the component across which it is connected. Otherwise the current drawn from the main circuit by the voltmeter (which is needed to make it give a reading) becomes an appreciable fraction of the main current and the voltage across the component changes—see section 4.7.

(c) Multimeters

A multimeter measures a wide range of currents and voltages for both direct current and alternating current—see section 3.1, and also resistances. It is a combined ammeter, voltmeter and ohmmeter and is an essential test instrument in electronics. The analogue type will be considered here and the digital type will be discussed in section 21.7.

Like many ammeters and voltmeters, an analogue multimeter is basically a moving-coil microammeter having resistors connected internally. It consists of a coil of wire pivoted so that it can turn between the poles of a U-shaped magnet, as shown in Fig. 2.09. The coil has a pointer fixed to it which moves over a scale. When current passes through the coil, it turns and winds up two small springs until the force caused by the current in the coil balances the force due to the wound-up springs. The greater the current the more the coil turns. The deflection of the pointer therefore continuously represents the value of the current, i.e. it is an analogue of it.

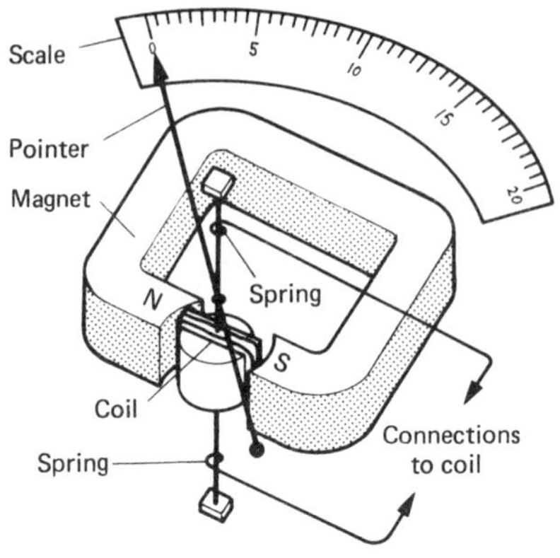

Fig. 2.09

In the multimeter of Fig. 2.10, the rotary switch allows different ranges to be chosen. On any resistance range there is a battery B (e.g. 1.5 V for the lower ranges and 15 V for the highest range) and a variable resistor R in series with the meter, Fig. 2.11. To measure resistance the terminals are joined (by the test leads or a piece of connecting wire) and R (often labelled 'zero Ω') adjusted until the pointer gives a full-scale deflection, i.e. is on zero of the ohms scale. The 'short-circuit' is then removed from across the terminals and the unknown resistor connected instead. The pointer reading falls owing to the smaller current passing through the meter and indicates the resistance in ohms. The resistance scale is a non-linear, 'reverse' one, i.e. the zero is on the right.

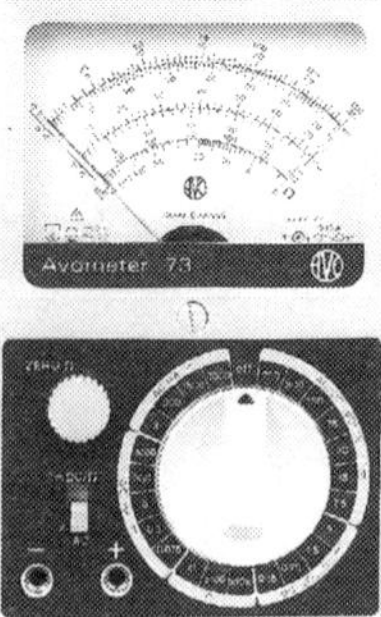

Fig. 2.10

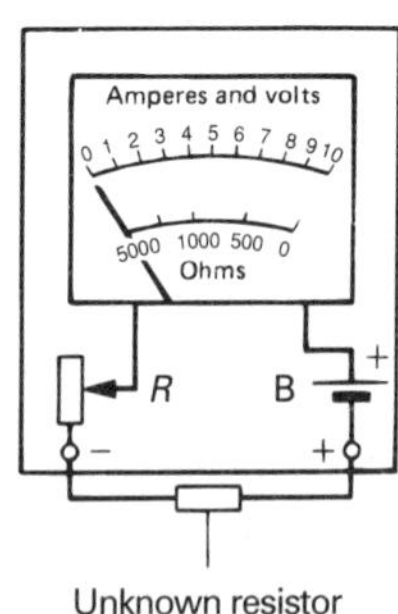

Fig. 2.11

Note In most analogue multimeters the terminal marked − (often coloured black) is connected to the positive of the internal battery (B in Fig. 2.11) when on a resistance range. The + terminal (often red) goes to the negative of the internal battery.

2.8 Ohm's law

The measurements in Table 2.1 show how the current I through a certain conductor varies when there are different voltages V across it.

Table 2.1 Relationship between current and voltage

V/volts	1	2	3	4	5
I/amperes	0.1	0.2	0.3	0.4	0.5
V/I/ohms	10	10	10	10	10

We see that the ratio V/I has a constant value, 10, and since V/I = resistance R, it follows that R for this conductor does not change when V and I change. *Ohm's law* sums up the behaviour of this and similar conductors.

It states that the current through a conductor is directly proportional to the voltage across it if the temperature and other physical conditions do not change.

Directly proportional imples that doubling V doubles I, trebling V trebles I, or halving V halves I, and so on. This is obviously true from the table. In mathematical terms Ohm's law can be written as $I \propto V$.

Another way of stating Ohm's law would be to say that the resistance of a conductor does not change so long as the physical conditions remain the same.

Conductors which obey Ohm's law are made of metals, carbon and some alloys. They are called *ohmic* conductors. Conductors not obeying Ohm's law are non-ohmic—their resistance changes with the voltage and current. Semiconductor devices such as transistors (see Unit 8) are non-ohmic.

It is often easier to understand the behaviour of an electronic component from a graph. A useful one for a conductor is obtained by plotting I along the y-axis (vertical) and V along the x-axis (horizontal). This is called a *characteristic curve*. The characteristic curve for an ohmic conductor is given in Fig. 2.12. It is a straight line passing through the origin 0 of the graph (which shows that $I \propto V$) and for this reason ohmic conductors are also called *linear* conductors. The axes of the graph are marked I/A and V/V to indicate that I is measured in amperes and V in volts.

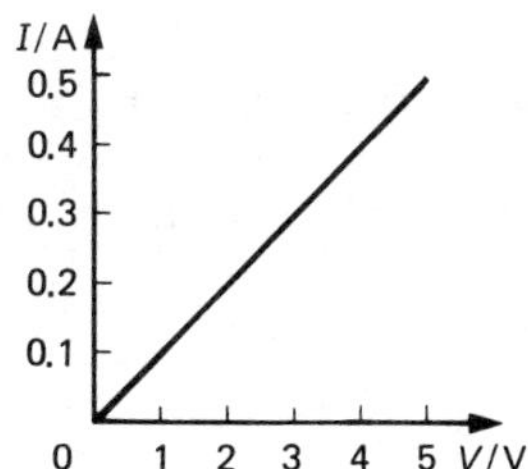

Fig. 2.12

2.9 Electric charge

An atom consists of a small central nucleus containing positively charged particles called *protons*, surrounded by an equal number of negatively charged *electrons*. The magnitude of the charge on a proton equals that on an electron, making the atom as a whole electrically neutral. Uncharged particles called *neutrons* are also present in the nucleus of every atom except hydrogen.

Hydrogen has the simplest atom with one proton and one electron as represented in Fig. 2.13, whereas a copper atom has 29 protons in its nucleus and 29 surrounding electrons.

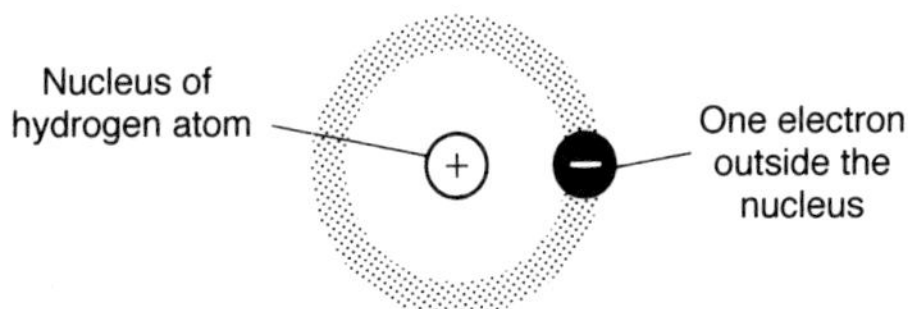

Fig. 2.13

If an atom loses one or more electrons, it becomes positively charged (because it now has more protons than electrons); if it gains one or more electrons, it becomes negatively charged. A charged atom is called an *ion*. Two positive charges repel one another, as do two negative charges. A positive charge attracts a negative charge and vice versa. Summing up:

like charges (+ and + or − and −) *repel*,
unlike charges (+ and −) *attract*.

In insulators all the electrons are bound to their atoms but they can be charged by rubbing because electrons are transferred from or to them by whatever does the rubbing. The charge produced cannot move from where the rubbing occurs, i.e. it is static. A nylon garment often becomes charged with static electricity when it is rubbed by being taken off; small sparks cause a crackling noise.

When electrons move in an orderly way, electric charge is moving and we have an electric current.

The unit of charge is the *coulomb* (C) which is the charge on 6.3 million million million electrons. The charge on one electron is therefore very small.

The current in a circuit is *one ampere* if charge passes any point in the circuit at the rate of *one coulomb per second*, i.e. $1\text{ A} = 1\text{ C s}^{-1}$. If 2 C pass in 1 s, the current is 2 A. If 6 C pass in 2 s, the current is $6\text{ C}/2\text{ s} = 3\text{ C s}^{-1} = 3\text{ A}$. In general if Q coulombs pass in t seconds, the current I amperes is given by:

$$I = \frac{Q}{t}$$

We can also write this equation as:

$$Q = I \times t$$

2.10 Electrical energy

In a circuit, electrical energy is supplied to the charges from, for example, a battery and is changed into other forms of energy by appliances in the circuit which have resistance. (Copper connecting wires have small resistance and 'use' very little electrical energy.) For example, a lamp produces heat and light. Note that it is not charge or current that is 'used up' in a circuit.

We can picture the charges (electrons) as 'picking up' electrical energy in the battery and, after 'delivering' most of it to the appliance in the circuit, they return to the battery for more. Electrical energy, like all energy, is measured in *joules* (J). If one joule of electrical energy is changed into other kinds of energy when one colomb passes through an appliance, the voltage across the appliance is one volt. That is, 1 volt = 1 joule per coulomb (1 V = 1 J C^{-1}).

If 2 J are given up by each coulomb, the voltage is 2 V. If 6 J are changed when 2 C passes, the voltage is 6 J/2 C = 3 V. In general if W joules is the electrical energy changed when charge Q coulombs passes through an appliance, the voltage V volts across it is given by:

$$V = \frac{W}{Q}$$

2.11 Power

The *power* of an appliance is the rate at which it changes energy from one form into another, i.e. the energy change per second.

The unit of power is the *watt* (W); it equals an energy change rate of one joule per second, i.e. 1 W = 1 J s^{-1}.

An electric lamp with a power of 100 W changes 100 J of electrical energy into heat and light energy each second. A larger unit of power is the kilowatt (kW): 1 kW = 1 000 W.

The power P watts of an electrical appliance, through which a current I amperes flows when there is a voltage V volts across it, is given by:

$$P = I \times V$$

In units:

$$\text{watts} = \text{amperes} \times \text{volts}$$

For example, if a 12 V battery causes a current of 3 A to flow through a car head-lamp bulb, as shown in Fig. 2.14, the power of the bulb is 3 × 12 = 36 W. This means that the bulb changes 36 J of the electrical energy supplied to it by the battery into heat and light energy, every second.

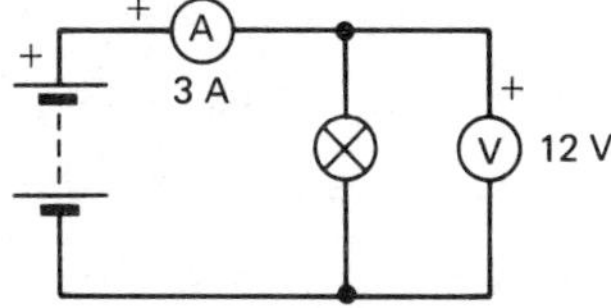

Fig. 2.14

In the case of a conductor of resistance R which changes *all* the electrical energy supplied to it into heat only, we have $V = I \times R$ and the rate of production of heat is given by:

$$P = I \times V = I \times I \times R = I^2 \times R$$

From this it follows that if $R = 3\ \Omega$ and I increases from 1 A to 2 A, then P increases from 3 W (i.e. $1^2 \times 3 = 1 \times 3 = 3$) to 12 W (i.e. $2^2 \times 3 = 4 \times 3 = 12$). Thus, if the current is doubled, four times as much heat is produced.

2.12 Internal resistance

A source of e.m.f. such as a cell has resistance, called *internal* or *source resistance*. When supplying current, the cell 'wastes' some electrical energy in driving current through itself against the internal resistance. Less energy is therefore available for any appliance connected to it.

Suppose a voltmeter connected across a dry cell on 'open' circuit, i.e. not supplying current, reads 1.5 V, Fig. 2.15(*a*). This is the e.m.f. E of the cell (provided the resistance of the voltmeter is sufficiently high not to draw much current from the cell).

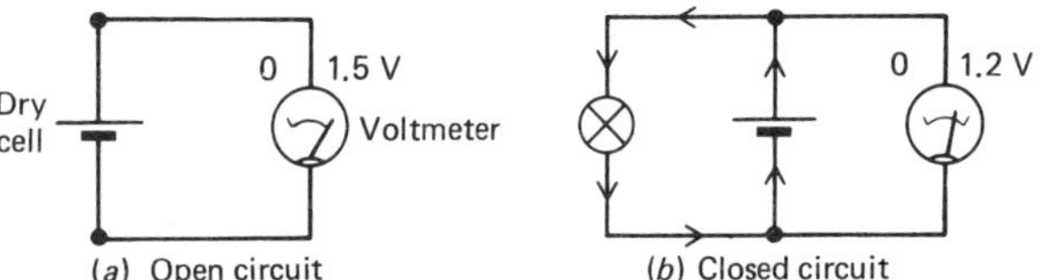

Fig. 2.15 (*a*) Open circuit (*b*) Closed circuit

If a lamp is now connected across the cell so putting it on 'closed' circuit, i.e. supplying current, the voltmeter reading falls, say to 1.2 V, Fig. 2.15(*b*). The 'lost' voltage, v is 0.3 V and the 'useful' or *terminal voltage*, V, available to the lamp is 1.2 V. We can say:

$$\text{e.m.f.} = \text{'useful' voltage} + \text{'lost' voltage}$$

or:

$$E = V + v$$

If the current in the circuit is I and the internal resistance of the cell r, then since $v = I \times r$, we have:

$$E = V + I \times r \qquad (4)$$

In the above example $E = 1.5$ V and $V = 1.2$ V and, if we take $I = 0.30$ A, then substituting in (4), we get:

$$1.5 = 1.2 + 0.30 \times r$$

$$1.5 - 1.2 = 0.30 \times r$$

$$0.30r = 0.30$$

$$r = 0.30/0.30 = 1\ \Omega$$

Check for yourself that the resistance of the lamp is 4 Ω.

The internal resistance of a dry battery increases with age and reduces its ability to supply current. The advantage of a car battery (lead-acid type) is that it has a very low internal resistance and can supply much larger currents than a dry battery without its terminal voltage falling.

A knowledge of internal resistance is important when considering how to get a source of electrical energy to deliver maximum power to an appliance (called the *load*) connected to it. It can be shown that this occurs when the resistance r of the source equals the resistance R of the load—a statement known as the *maximum power transfer theorem*. However, the power P_1 wasted in the source then equals the power P_2 developed in the load. (The same current I flows in each, therefore $P_1 = I^2 \times r = I^2 \times R = P_2$.) Thus the *efficiency* of the *power* transfer process is only 50%.

In many electronic circuits we are more concerned with transferring maximum *voltage* from one circuit or source to another circuit or load. In these cases the best 'matching' occurs when the load resistance is large compared with that of the source—see section 11.5.

2.13 Revision questions

1. What do the following letters mean: d.c.; e.m.f.; p.d.?

2. Link the following terms with their units:

a) current	(i) joule
b) e.m.f.	(ii) coulomb
c) p.d.	(iii) watt
d) resistance	(iv) volt
e) charge	(v) ohm
f) energy	(vi) ampere
g) power	(vii) volt

3. Draw the circuit diagram for a battery of two 1.5 V cells connected in series with a switch, a lamp and an ammeter. Show how a voltmeter would be connected to measure the voltage across the lamp. Mark with a +, the 'positive' terminals of the ammeter and voltmeter for correct connection.

4. Why should an ammeter have a very low resistance and a voltmeter a very high resistance?

5. What is a characteristic curve?

6. Why does a lamp change much more electrical energy per second into heat and light than the wires connecting it to the supply? Use the term 'resistance' in your answer.

7. What energy changes occur in

a) a cell,
b) a loudspeaker,
c) a microphone,
d) a television tube?

8. Does the internal resistance of a battery increase or decrease its terminal voltage when it supplies current?

2.14 Problems

1. What are the e.m.fs of the batteries of 1.5 V cells connected as in Fig. 2.16(*a*),(*b*)?

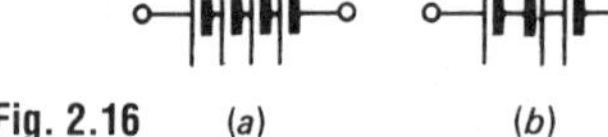

Fig. 2.16 (*a*) (*b*)

2. The lamps and the cells in all the circuits of Fig. 2.17 are the same. If ammeter A_1 reads 0.3 A, will the readings on ammeters A_2, A_3 and A_4 be less than, equal to or more than 0.3 A? State which in each case.

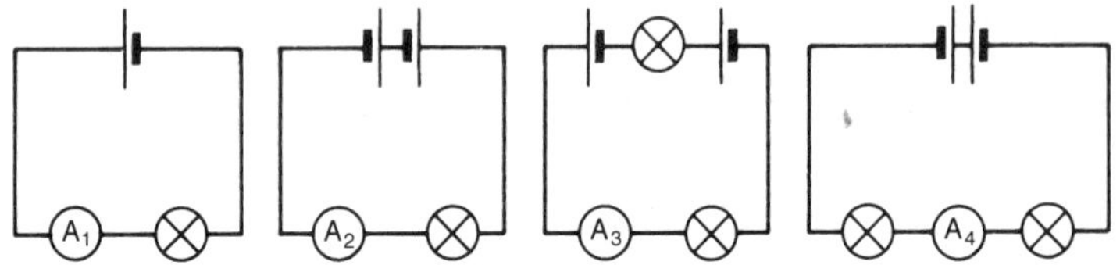

Fig. 2.17

3. Three voltmeters V, V_1 and V_2 are connected as in Fig. 2.18.
   a) Copy and complete the table of voltmeter readings shown below which were obtained with three different batteries.

V/V	V_1/V	V_2/V
. .	12	6
6	4	. .
12	. .	4

   b) How will the readings compare on ammeters A_1 and A_2 for a certain battery voltage?

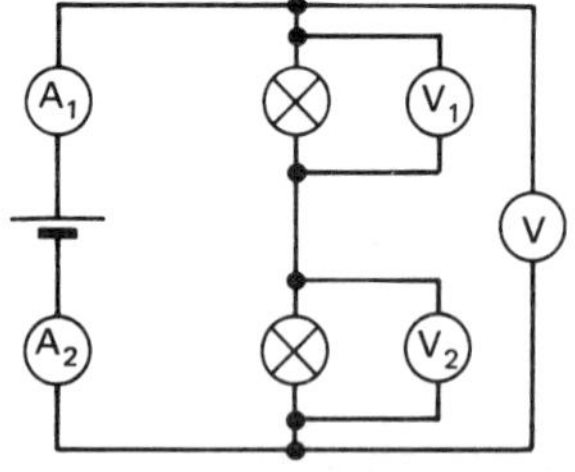

Fig. 2.18

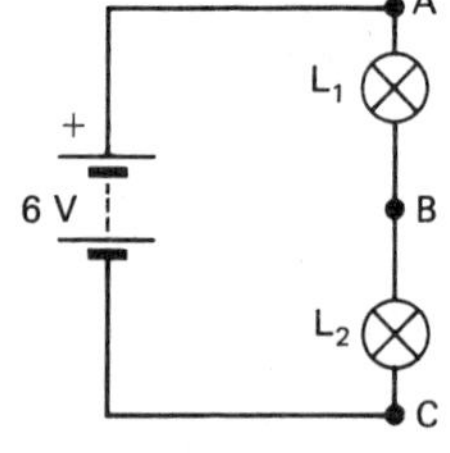

Fig. 2.19

4. In Fig. 2.19 L_1 and L_2 are identical lamps. What are the potentials at points A, B and C if the point of zero potential (i.e. 'ground') is taken as
   (i) C (i.e. negative terminal of battery),
   (ii) A (i.e. positive terminal of battery,
   (iii) B?

5. a) What is the resistance of a lamp when a voltage of 12 V across it causes a current of 4 A?
   b) Calculate the voltage across a wire of resistance 10 Ω carrying a current of 2 A.
   c) The voltage across a wire of resistance 2 Ω is 4 V. What current flows?

6. a) What is the voltage across a 220 Ω resistor when a current of 3 mA flows through it?
b) The voltage across a 2.7 kΩ resistor is 5.4 V. What current flows?
c) Calculate the resistance of a resistor if a voltage of 9 V across it causes a current of 1.5 mA to flow through it.
d) A current of 0.4 mA flows through a 10 kΩ resistor. What is the voltage across its ends?

7. a) What is the current in a circuit if the charge passing each point is (i) 10 C in 2 s, (ii) 20 C in 40 s, (iii) 240 C in 2 minutes?
b) If the current through a lamp is 5 A, what charge passes in (i) 1 s, (ii) 10 s, (iii) 1 minute?

8. The voltage across a lamp is 12 V. How many joules of electrical energy are changed into light and heat when
a) a charge of 1 C passes through it,
b) a charge of 5 C passes through it,
c) a current of 2 A flows through it for 10 s?

9. How much electrical energy in joules does a 20 W soldering iron change in
a) 1 s, b) 5 s, c) 1 minute?

10. a) What is the power of a lamp rated at 12 V 2 A?
b) How many joules of electrical energy are changed per second by a lamp rated at 6 V 0.5 A?

11. A current of 2 A passes through a resistance of 4 Ω. Calculate
a) the voltage across it, b) the power used.

12. A lamp is labelled 12 V 36 W. When used on a 12 V supply,
a) what current will it take? b) what is its resistance?

13. A high resistance voltmeter reads 3 V when connected across the terminals of a battery on open circuit and 2.6 V when the battery sends a current of 0.2 A through a lamp. What is
a) the e.m.f. of the battery,
b) the terminal voltage of the battery when supplying 0.2 A,
c) the 'lost' voltage,
d) the internal resistance of the battery, and
e) the resistance of the lamp?

14. In Fig. 2.20, V is a high resistance voltmeter.
a) It reads 6 V when the switch is open and 4.8 V when it is closed. Calculate the e.m.f. and internal resistance of the battery.
b) What value of resistor must replace the 12 Ω resistor to give maximum power output? What will be that power output?

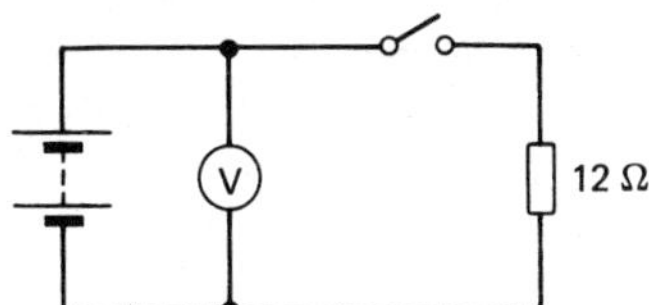

Fig. 2.20

UNIT 3

Alternating current

3.1 Direct and alternating currents

In a *direct current* (d.c.) electrons flow in one direction only. Cells and batteries produce d.c. voltages and currents. Graphs of steady and varying direct currents are given in Fig. 3.01. They show how the size of the current changes during a certain time: the shape of the graph is called the *waveform* of the current.

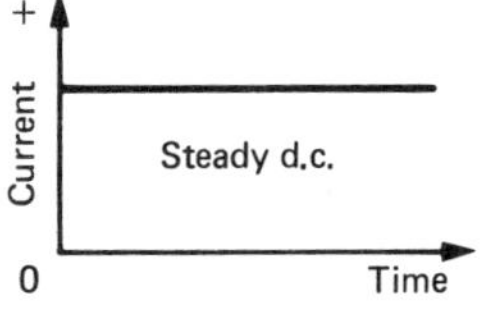

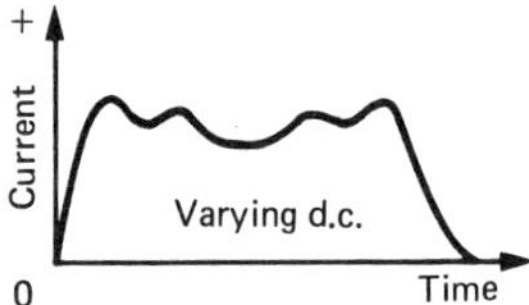

Fig. 3.01

In an *alternating current* (a.c.) the electrons flow first in one direction and then in the other, i.e. they alternate. The current starts from zero, rises to a maximum in one direction, falls to zero again before becoming a maximum in the opposite direction and then rises to zero once more. The alternating current with the simplest waveform (mathematically speaking) is given in Fig. 3.02; it has a sine wave (or sinusoidal) shape. The + and − signs show the two possible directions of current flow in the circuit. The electricity mains supply is a sine wave a.c.

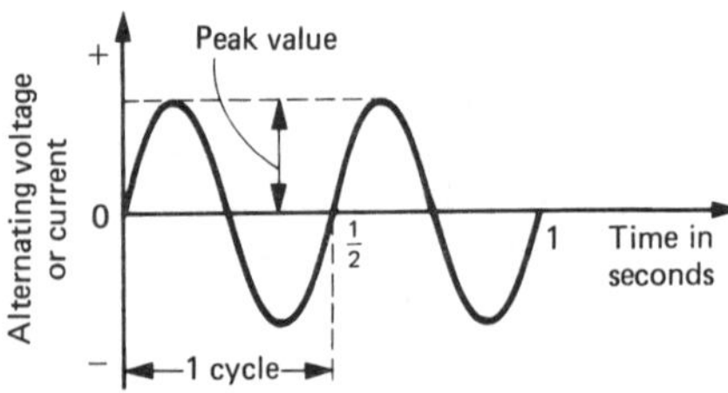

Fig. 3.02

Electric heaters and lamps will work off either a.c. or d.c. but radio and television sets require d.c. as do processes such as battery charging. When necessary, d.c. can be obtained from a.c. by rectification—see section 21.2. The pointer of a moving-coil ammeter (the commonest type of meter for measuring d.c.) is deflected one way by d.c.; a.c. makes it move to and fro about the zero of the scale if the changes are slow enough, otherwise there is no deflection. The symbol for a.c. is $\sim$.

3.2 Frequency

Many alternating currents and voltages change *periodically* with time, that is, their waveforms are repeated exactly over and over again as time proceeds. They give a regular pattern.

One complete alternation is called a cycle and the number of cycles occurring in one second is called the *frequency* (f) of the a.c. The unit of frequency is the *hertz* (Hz); previously it was the cycle per second (cs^{-1}). The frequency of the a.c. represented in Fig. 3.02 is 2 Hz; that of the mains supply in many countries, including Britain, is 50 Hz, i.e. the waveform is repeated fifty times every second.

The time for which one cycle lasts is called the *period* (T) of the a.c. If the frequency is 50 Hz, then:

$$T = \frac{1}{50} = 0.02 \text{ s}$$

In general $T = 1/f$.

The *amplitude* or *peak value* of an alternating quantity is its maximum positive or negative value.

Alternating currents with frequencies from 20 Hz or so to about 20 000 Hz (i.e. 20 kilohertz = 20 kHz) are called *audio frequency* (a.f.) currents because they produce a note we can hear when they are fed into a loudspeaker. In a microphone, speech, music and other sounds are changed into a.f. currents. For example, middle C played on a musical intstrument causes an a.f. current of 256 Hz in a microphone.

Currents with frequencies above 20 kHz are called *radio frequency* (r.f.) currents. They do not produce audible sounds in a loudspeaker but when they flow in the aerial of a radio transmitter they cause radio waves to be sent out into space. This does not happen to any extent with a.f. currents. A medium wave radio station working on a wavelength of 300 metres would have currents of frequency 1 000 000 Hz (i.e. 1 000 kHz = 1 megahertz = 1 MHz) in its aerial. Note that 1 gigahertz (1 GHz) = 1 000 MHz.

3.3 Waveforms

Alternating currents and voltages occur in many electronic systems and it is often helpful, e.g. in fault-finding, to study their waveforms. This can be done using a cathode ray oscilloscope which automatically displays them on a television-type screen. We will see in section 21.8 how this instrument works.

(a) Types of waveform

In addition to the sine wave of Fig. 3.02, some other waveforms that are met in electronics are shown in Fig. 3.03. Square waves and pulses are used in digital systems, sawtooth waveforms in oscilloscopes, spiky ones in analogue systems, and ripple ones occur in power supplies. Amplifiers have to deal with complex speech waveforms.

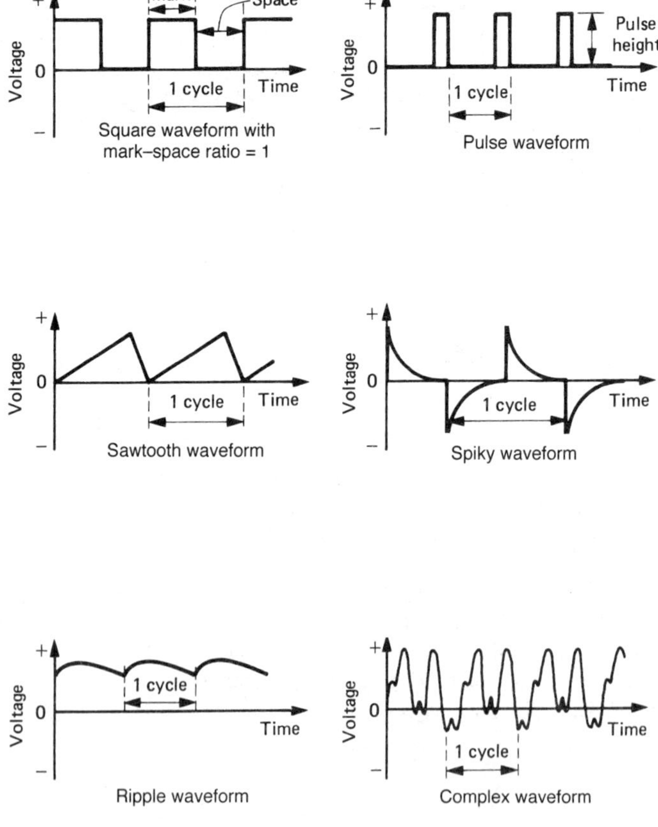

Fig. 3.03

(b) Uses of square waves

A voltage with a square waveform is sometimes used as a test signal to be applied to the input of a system. A square wave can be built up by combining a very large number of sine waves of different frequencies and amplitudes. Fig. 3.04 shows that a roughly square wave is obtained by adding sine waves with frequencies of 1, 3, 5, and 7 Hz. As more sine waves of higher frequencies are added the squarer the combined wave becomes. All periodic waves are combinations of sine waves.

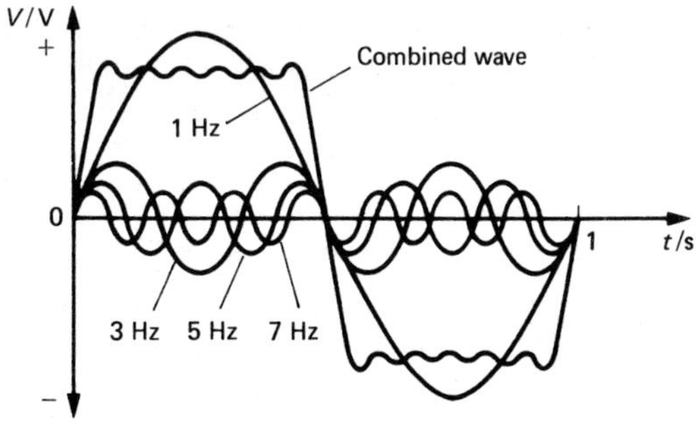

Fig. 3.04

A square wave is therefore a source of many frequencies. And, for example, by observing what an amplifier does to the shape of one, deductions can be made about how the amplifier will respond to low and high frequencies.

(c) Varying d.c. waveforms

It is sometimes useful to consider that a varying d.c. waveform is built up from a steady d.c. waveform to which an a.c. waveform has been added, as shown in Fig. 3.05.

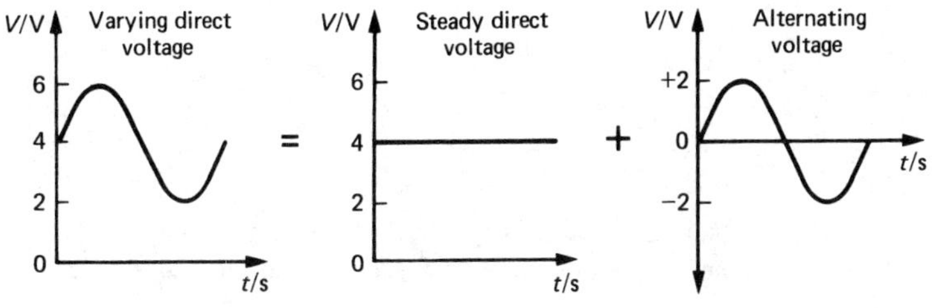

Fig. 3.05

3.4 Root mean square values

The value of an alternating current or voltage changes from one instant to the next and the problem arises of what value we should take to measure it. The average value over a complete cycle is zero; the peak value is a possibility. However, the *root mean square* (r.m.s.) value is chosen because, by using it, many calculations can be done as they would for direct currents.

The r.m.s. value of an alternating current or voltage (also called the *effective* value) is defined as the steady direct current or voltage which would give the same heating effect.

For example, if the lamp in the circuit of Fig. 3.06(*a*) is lit from a.c. (by moving the two-way switch to the left) and its brightness noted, then if 0.3 A d.c. produces the same brightness (when the switch is moved to the right and the variable resistor adjusted), the r.m.s. value of the a.c. is 0.3 A. A lamp designed to be fully lit by a current of 0.3 A d.c. will therefore be fully lit by an a.c. of r.m.s. value 0.3 A.

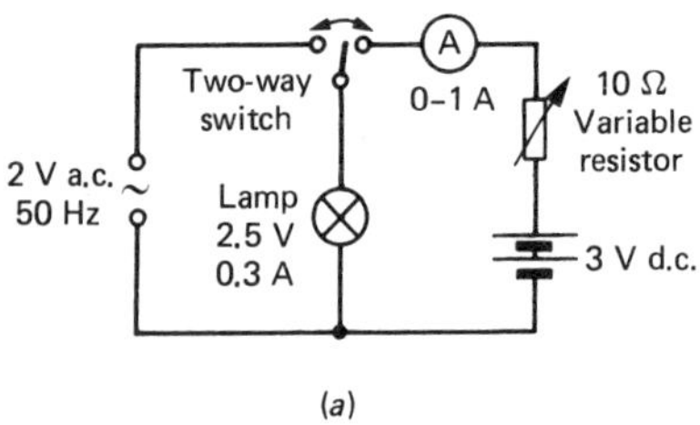

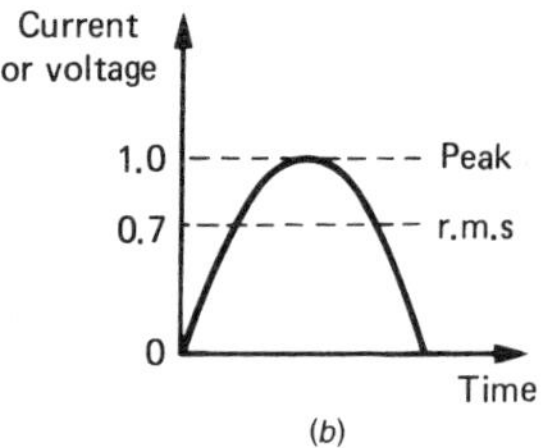

Fig. 3.06

It can be shown (Fig. 3.06(*b*)) that for sine waves:

$$\text{r.m.s. value} = \frac{\text{peak value}}{\sqrt{2}}$$

Thus, for all practical purposes, we have:

$$\text{r.m.s. value} = 0.7 \times \text{peak value}$$

The r.m.s. voltage of the mains supply in Britain is 230 V; the peak value is much higher and is given by:

$$\text{peak value} = \frac{\text{r.m.s. value}}{0.7} = \frac{230}{0.7} = 330 \text{ V approximately}$$

If an alternating voltage is said to be 12 V, this is its r.m.s. value, unless stated otherwise. The power *P* of a device on an a.c. supply is therefore given by:

$$P = I_{\text{r.m.s.}} \times V_{\text{r.m.s.}}$$

3.5 Meters for alternating current

A meter for measuring alternating currents and voltages must give a reading whichever way the current is flowing.

Rectifier meters are the commonest type of a.c. meter. They consist of a rectifier in series with a moving-coil meter, Fig. 3.07(*a*), and have scales marked to read r.m.s. values of currents or voltages that have sine waveforms.

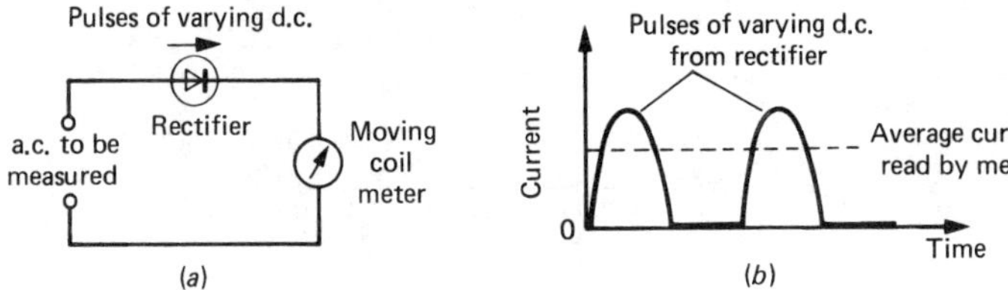

Fig. 3.07

A rectifier, e.g. a semiconductor diode (see section 7.5), has a low resistance to current flow in one direction and a very high resistance for the opposite direction. It is a one-way current conductor and when connected to an a.c. supply it produces pulses of varying d.c., Fig. 3.07(*b*). The moving-coil meter responds to the average value of the pulses.

Rectifier instruments, being based on the moving-coil meter, are much more sensitive (i.e. can read smaller currents and voltages) than other types of a.c. meter. They are used in analogue multimeters—see section 2.7.

3.6 Revision questions

1. a) How does a.c. differ from d.c.?
   b) Name a source of
      (i) d.c.,
      (ii) a.c.
   c) Name a device which works
      (i) on d.c. only,
      (ii) equally well on a.c. or d.c.

2. What do the following terms mean when applied to a.c.:
   a) frequency,
   b) peak voltage,
   c) r.m.s. voltage?

3. Which of the waveforms in Fig. 3.03 represent alternating voltages?

4. Draw a graph of current against time for two complete cycles of alternating current of peak value 1 A and frequency 10 Hz. Label your axes and mark a correct scale of values on them. (You must calculate the time for one cycle in order to mark the x-axis.)

5. Describe what the pointer of a moving-coil meter would do if the meter was supplied with a.c. of frequency
   (i) $\frac{1}{2}$ Hz,
   (ii) 50 Hz.

3.7 Problems

1. The waveform of an alternating voltage is shown in Fig. 3.08. What is
a) the period,
b) the frequency,
c) the peak voltage,
d) the r.m.s. voltage?

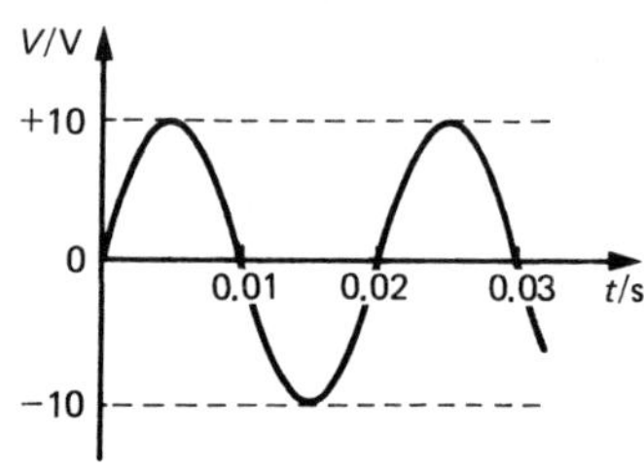

Fig. 3.08

2. What are the mark–space ratios of the square waves in Figs 3.09(*a*) and (*b*)? (Hint—see Fig. 3.03.)

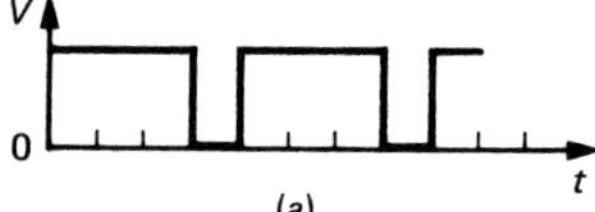

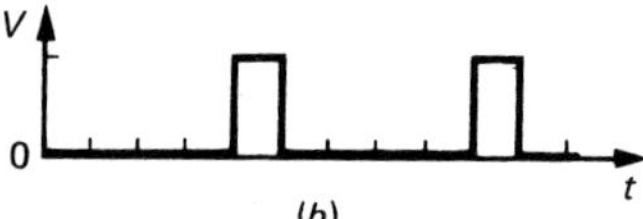

Fig. 3.09

3. What values of *steady* direct voltage and *peak* alternating voltage when added together would give the varying direct voltage represented in Fig. 3.10?

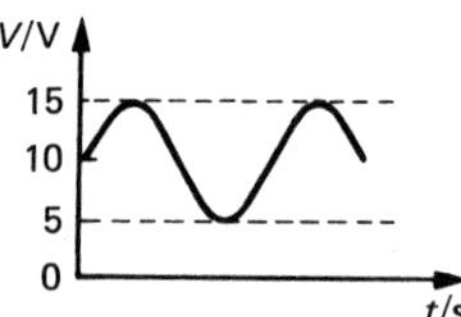

Fig. 3.10

4. An a.c. supply lights a lamp with the same brightness as does a 12 V battery. What is
a) the r.m.s. voltage,
b) the peak voltage of the a.c. supply,
c) the power of the lamp,
if it takes a current of 2 A on the a.c. supply (as read by a rectifier meter)?

PART TWO

Components

UNIT 4

Resistors

4.1 About resistors

The jobs done by resistors include directing and controlling current, making changing currents produce changing voltages (as in a voltage amplifier—see section 10.2) and obtaining variable voltages from fixed ones (as in a potential divider—see section 4.7). There are two main types of resistor—those with fixed values and those that are variable; their symbols are given in Fig. 4.01.

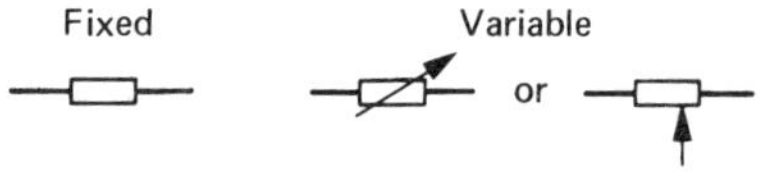

Fixed

Variable

or

Fig. 4.01

When choosing a resistor there are three factors that have to be considered, apart from the stated value.

(i) The tolerance Exact values cannot be guaranteed by mass-production methods but this is not a great disadvantage because in most electronic circuits the values of resistors are not critical. The tolerance tells us the minimum and maximum values a resistor might have, e.g. one with a stated (called nominal) value of 100 Ω and a tolerance of ±10% could have any value between 90 and 110 Ω.

(ii) The power rating If the rate at which a resistor changes electrical energy into heat exceeds its power rating, it will overheat and be damaged or destroyed. For most electronic circuits 0.25 W or 0.5 W power ratings are adequate (but see Example 1 in section 4.8 for how to make a check). The greater the physical size of a resistor the greater is its rating.

(iii) The stability This is the ability of a component to keep the same value as it 'ages' despite changes of temperature and other physical conditions. In some circuits this is an important factor.

4.2 Fixed resistors

(i) Carbon film A film of carbon is deposited on a ceramic rod and protected by a tough insulating coating, Fig. 4.02(*a*). Values range from a few ohms to 10 MΩ, a typical tolerance is ±5%, ratings are from 0.125 W to 1 W, and stability is very good.

Fig. 4.02

(ii) Metal film These offer high stability over a long period of time. Tolerance is ±1% and rating typically 0.5 W. Their construction and appearance is similar to the carbon film type, a metal film replacing carbon.

(iii) Wire-wound Low tolerance (i.e. high accuracy), high stability resistors such as are used in good quality multimeters are of this type. Those with large power ratings (e.g. over 2 W) are also wire-wound. They are made by winding nichrome, constantan or manganin wire on a tube and giving it a protective coating, Fig. 4.02(*b*). These three materials are alloys with a higher resistance than copper. Values range from a fraction of an ohm up to about 25 kΩ depending on the length and diameter of wire used. Physically, they tend to be large in size.

4.3 Resistor markings

The value and tolerance of a fixed resistor is marked on it using one of two codes.

(a) Band colour codes

In this method the resistance value and tolerance are shown by either *four* or *five* coloured bands round the resistor, the latter giving the value more accurately. The way both systems work is shown by the examples in Fig. 4.03.

The first band to read is the one at the end of the resistor where the bands are closer together—its colour gives the first digit. The second band from that end gives the second digit but while the third band in the *four band code* gives the multiplier (or the number of 0s to be added), it gives the third digit (often 0, i.e. black) in the *five band code*. In the latter system the multiplier is given by the fourth band.

In both systems the colour of the band on its own at the other end gives the *tolerance*, typically ±5% or ±10%. If this band is missing on the resistor, the tolerance is ±20%. Resistors with tolerances of ±1% and ±2% cost more; they are usually marked using the five band code.

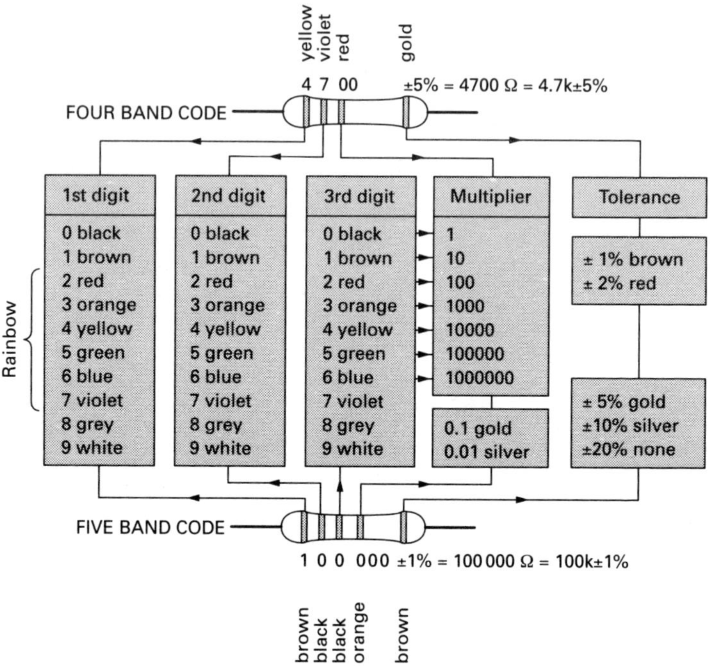

Fig. 4.03

(b) Printed code

The code is printed on the resistor and consists of letters and numbers. It is also used on variable resistors and on circuit diagrams. The examples in Table 4.1 show how it works. R means ×1, K means ×1 000, M means ×1 000 000, and the position of the letter gives the decimal point.

Table 4.1 Examples of printed code values

Value	0.27 Ω	3.3 Ω	10 Ω	220 Ω	1 kΩ	68 kΩ	100 kΩ	4.7 MΩ
Mark	R27	3R3	10R	220R	1K0	68K	100K	4M7

Tolerances are indicated by adding a letter at the end: F = ±1%, G = ±2%, J = ±5%, K = ±10%, M = ±20%. For example, 5K6K = 5.6 kΩ ±10%.

(c) Preferred values

Since exact values of fixed resistors are unnecessary in most electronic circuits, only certain *preferred values* are made. This also reduces the number of different values that have to be manufactured and stocked. The values chosen for the E12 series (with ±10% tolerance), are 1, 1.2, 1.5, 1.8, 2.2, 2.7, 3.3, 3.9, 4.7, 5.6, 6.8, 8.2 and multiples that are ten times greater.

The above values give maximum coverage with minimum overlap. For example, a 22 Ω ±10% resistor may have any value between $22 + \frac{10}{100} \times 22 \approx 22 + 2 = 24\ \Omega$ and $22 - \frac{10}{100} \times 22 \approx 22 - 2 = 20\ \Omega$, whilst the next highest value, i.e. 27 Ω ±10%, covers almost all the range 24 Ω to 30 Ω.

Resistors in the E24 series have a ±5% tolerance and 24 basic values; those in addition to the E12 series being 1.1, 1.3, 1.6, 2.0, 2.4, 3.0, 3.6, 4.3, 5.1, 6.2, 7.5 and 9.1.

4.4 Resistors in series

In Fig. 4.04 resistors of resistances R_1 and R_2 are in series. Therefore:

(i) the same current I flows through each one; and
(ii) the total voltage V across both equals the sum of the separate voltages V_1 and V_2 across each resistor.

Hence:

$$V = V_1 + V_2$$

And since $V_1 = IR_1$ and $V_2 = IR_2$ we have

$$V = IR_1 + IR_2 = I(R_1 + R_2)$$

If R is the combined resistance then $V = IR$ and

$$IR = I(R_1 + R_2)$$

Dividing both sides by I we get

$$R = R_1 + R_2$$

For example if $R_1 = 10\ \text{k}\Omega$ and $R_2 = 15\ \text{k}\Omega$ then $R = 25\ \text{k}\Omega$.

In general the combined resistance of several resistors in series equals the sum of their separate resistances and they could be replaced by a single equivalent resistor which would draw the same current from the supply.

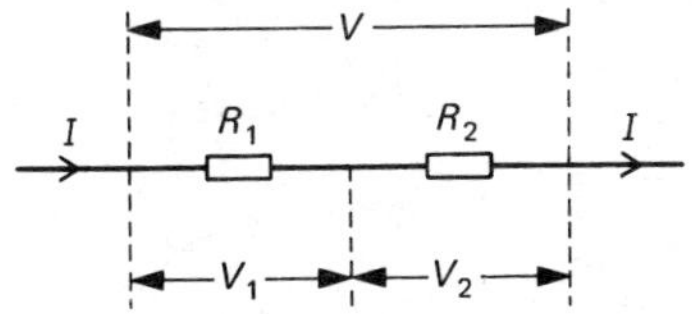

Fig. 4.04

4.5 Resistors in parallel

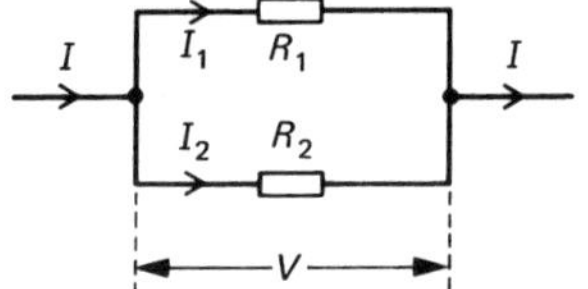

Fig. 4.05

In Fig. 4.05 resistors of resistances R_1 and R_2 are in parallel. Therefore:

(i) the same voltage V acts across each one; and
(ii) the total current I from the supply equals the sum of the separate currents I_1 and I_2 through each resistor.

Hence:

$$I = I_1 + I_2$$

Since $I_1 = V/R_1$ and $I_2 = V/R_2$ we have

$$I = \frac{V}{R_1} + \frac{V}{R_2} = V\left(\frac{1}{R_1} + \frac{1}{R_2}\right)$$

If R is the combined resistance, then $I = V/R$ and

$$\frac{V}{R} = V\left(\frac{1}{R_1} + \frac{1}{R_2}\right)$$

Dividing both sides by V we get

$$\frac{1}{R} = \frac{1}{R_1} + \frac{1}{R_2}$$

This can be written:

$$\frac{1}{R} = \frac{R_2}{R_1 \times R_2} + \frac{R_1}{R_2 \times R_1} = \frac{R_2 + R_1}{R_1 \times R_2}$$

$$\therefore \quad R = \frac{R_1 \times R_2}{R_1 + R_2} = \frac{\text{product of resistances}}{\text{sum of resistances}} \qquad (1)$$

We will call (1) the *two resistors in parallel equation*. As an example, if $R_1 = 2\ \text{k}\Omega$ and $R_2 = 4\ \text{k}\Omega$, then:

$$R = \frac{2 \times 4}{2 + 4} = \frac{8}{6} \approx 1.3\ \text{k}\Omega$$

Note that the combined resistance is less than either resistance: putting resistors in parallel provides alternative paths for the current and lowers the resistance.

4.6 Variable resistors

(a) Description

Variable resistors used as volume and other controls in radio and TV sets are usually called 'pots' (short for potential divider—see below). They consist of an incomplete circular track of either a fixed carbon resistor for high values and low power (up to 2 W) or of a fixed wire-wound resistor for high powers, Fig. 4.06(*a*). Connections to each end of the track are brought out to two terminal tags. A wiper makes contact with the track and is connected to a third terminal tag, between the other two. Rotation of the spindle (Fig. 4.06(*b*)) moves the wiper over the track and changes the resistance between the centre tag and the end ones. 'Slide' type variable resistors have a straight track.

In a *linear* track, equal changes of resistance occur when the spindle is rotated through equal angles. In a *log* track, the change of resistance at one end of the track is less than at the other for equal angular rotations.

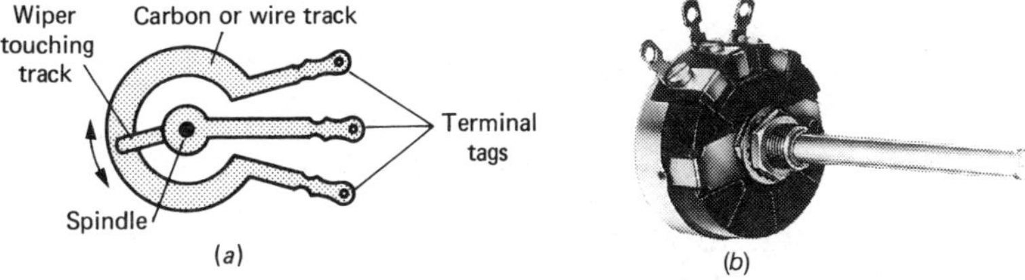

Fig. 4.06

Maximum values range from a few ohms to several megohms; common values are 10 kΩ, 50 kΩ, 100 kΩ, 500 kΩ and 1 MΩ.

Some circuits use small *preset* types, the symbol and form of which are shown in Figs. 4.07(*a*) and (*b*); these are adjusted with a screwdriver when necessary and have tracks of carbon or cermet (*cer*amic and *met*al oxide).

Fig. 4.07

(b) Uses

There are two ways of using a variable resistor. It may be used as a *rheostat* to control the current in a circuit. For this purpose only one end tag and the wiper tag are needed. In Fig. 4.08 rotating the wiper clockwise increases the resistance in the circuit and decreases the current.

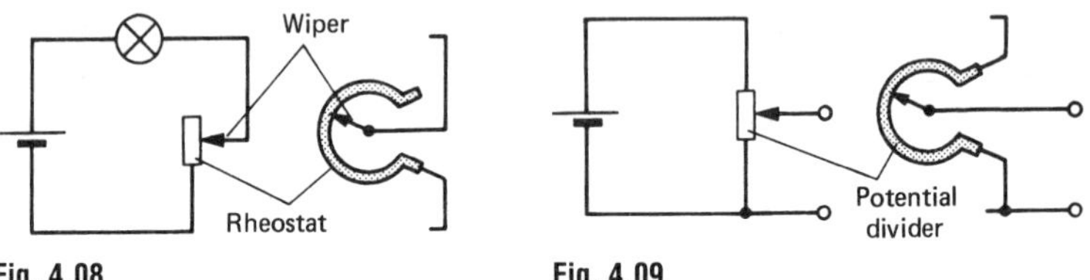

Fig. 4.08 **Fig. 4.09**

It can also act as a *potential* or *voltage divider* to obtain any voltage from zero to the maximum voltage of the supply by rotating the spindle (clockwise in Fig. 4.09). All three terminal tags are used in this case. Potential dividers will be considered in more detail in the next section.

4.7 Potential dividers

A potential or voltage divider provides a convenient way of getting a variable voltage from a fixed voltage supply.

To see how this is done, consider first two fixed resistors $R_1 = 10\ \Omega$ and $R_2 = 20\ \Omega$ connected in series across a 6.0 V supply, Fig. 4.10. The same current I passes through each resistor. Therefore:

$$I = \frac{\text{supply voltage}}{R_1 + R_2} = \frac{6.0}{30} = \frac{1}{5} = 0.2\ \text{A}$$

Thus:

$$V_1 = I \times R_1 = 0.2 \times 10 = 2.0\ \text{V}$$

and

$$V_2 = I \times R_2 = 0.2 \times 20 = 4.0\ \text{V}$$

The voltages appearing across the resistors are therefore in the ratio of their resistances, i.e. $V_1/V_2 = R_1/R_2 = 1/2$.

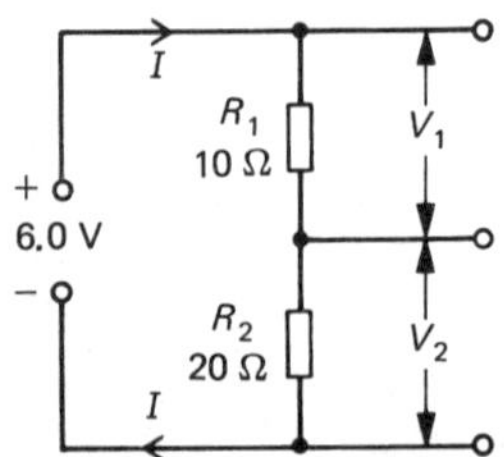

Fig. 4.10

If the resistors were changed so that $R_1 = 5\ \Omega$ and $R_2 = 25\ \Omega$, then $R_1/R_2 = 5/25 = 1/5$ and V_1/V_2 would also be equal 1/5. The new voltages across the resistors would be:

$$V_1 = \tfrac{1}{6} \times 6.0 = 1.0\text{ V and } V_2 = \tfrac{5}{6} \times 6.0 = 5.0\text{ V}$$

since $V_1 + V_2 = 6.0$ V. Different voltages are therefore obtained across R_1 and R_2 by changing their ratio.

In general, if two resistors with values R_1 and R_2 are connected in series across a supply voltage V and the voltages developed across each are V_1 and V_2 respectively, then, if I is the current flowing, we can say:

$$V_1 = I \times R_1 \qquad (2)$$

$$V_2 = I \times R_2 \qquad (3)$$

$$V = V_1 + V_2 = I(R_1 + R_2) \qquad (4)$$

Dividing (2) by (4) we obtain:

$$\frac{V_1}{V} = \frac{I \times R_1}{I(R_1 + R_2)}$$

Multiplying both sides by V gives:

$$V_1 = \frac{R_1}{R_1 + R_2} \times V$$

Similarly from (3) and (4) we get:

$$V_2 = \frac{R_2}{R_1 + R_2} \times V \qquad (5)$$

We will call (5) the *potential divider equation.*

A variable resistor connected as in Fig. 4.11 provides an easier way of changing the ratio R_1/R_2. The resistance between tags A and B represents R_1 and that between B and C represents R_2. A variable output voltage (from 0 to 6.0 V) is available between X and Y as the wiper is rotated. For example, it will be 3 V when the wiper is halfway round the track (if it is linear).

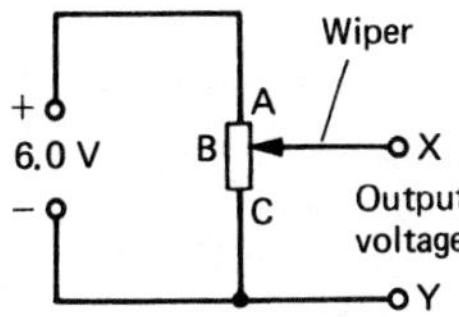

Fig. 4.11

If the output voltage is applied to, say, a lamp, the lamp draws current from the potential divider and 'loads' it. If the resistance of the load changes, say a second lamp is connected in parallel with the first, we would not want the output voltage to change or the lamps might not be fully lit. It can be shown (see Example 3 in section 4.8) that the output voltage does not alter much when the load changes provided the resistance of the load is always at least ten times greater than the resistance of the part of the potential divider across which it is connected, i.e. R_2. Putting it another way, the current drawn by the load should not exceed 10% of the current flowing through the potential divider.

4.8 Worked examples

Example 1

What electrical power is dissipated (i.e. changed to heat) in a 10 kΩ resistor carrying a current of 5 mA?

We have $I = 5$ mA $= 5/1\,000$ A and $R = 10$ kΩ $= 10\,000$ Ω. The power P in watts is given by:

$$P = I^2 \times R = \left(\frac{5}{1\,000}\right)^2 \times 10\,000 = \frac{25 \times 10\,000}{1\,000 \times 1\,000}\text{W}$$

$$= \frac{25}{100} = 0.25\text{ W}$$

Alternatively, leaving I in mA and R in kΩ, P will be in milliwatts (mW), and is given by:

$$P = I^2 \times R = 5^2 \times 10 = 25 \times 10 = 250\text{ mW (0.25 W)}$$

Example 2

What is the combined resistance of the resistor network in Fig. 4.12?

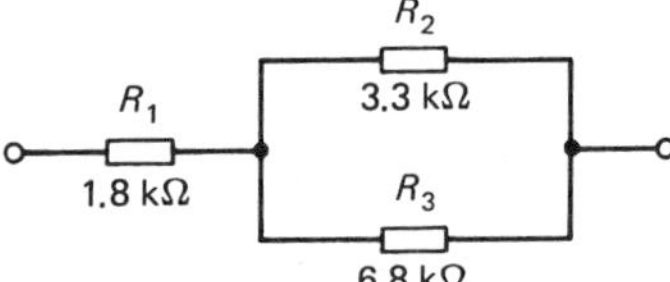

Fig. 4.12

First we find the combined resistance R of R_2 and R_3 in parallel. We have from the two resistors in parallel equation:

$$R = \frac{R_2 \times R_3}{R_2 + R_3} = \frac{3.3 \times 6.8}{3.3 + 6.8} = \frac{22}{10}\text{ kΩ (approx.)}$$

$$= 2.2\text{ kΩ}$$

We can now consider the network to consist of R and $R1$ in series. Therefore:

$$\text{combined resistance} = R + R_1 = 2.2 + 1.8\text{ kΩ}$$

$$= 4.0\text{ kΩ}$$

Example 3

A 2.5 kΩ linear 'pot' is used as a potential divider for a 9 V supply as shown in Fig. 4.13(*a*). The wiper is set at B, four-fifths of the way round from end C of the track.

a) What is the resistance of length BC of the track?
b) What is the voltage across BC?
c) If a resistor is connected as a 'load' to X and Y, Fig. 4.13(*b*), what does the voltage become across BC when the load resistance is
   (i) 20 kΩ,
   (ii) 2.0 kΩ?
   Comment on the results.

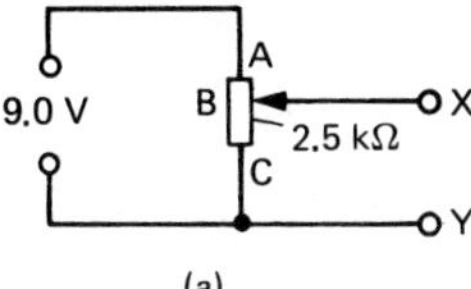

(*a*)

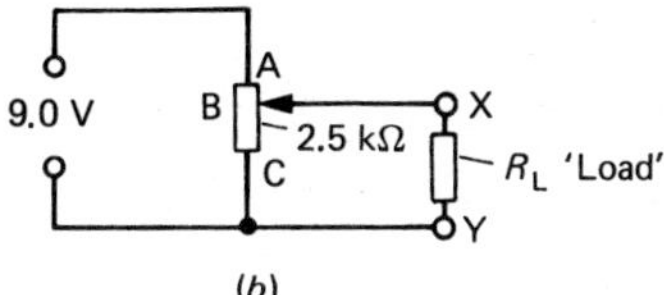

(*b*)

Fig. 4.13

a) Resistance of BC = $R_2 = \frac{4}{5} \times 2.5 = 2.0$ kΩ
b) Resistance of AB = $R_1 = 2.5 - 2.0 = 0.5$ kΩ
If V_2 is the voltage across BC (i.e. R_2) then from the potential divider equation we get:

$$V_2 = \frac{R_2}{R_1 + R_2} \times V \quad \text{where } V = 9\text{V}$$

$$\therefore \quad V_2 = \frac{2.0}{(2.0 + 0.5)} \times 9.0 = \frac{2.0}{2.5} \times 9.0 = \frac{18}{2.5}\text{V}$$

$$= 7.2 \text{ V}$$

c) (i) Resistance of 'load' = R_L = 20 kΩ
Let R' = combined resistance of R_L and R_2 in parallel. From the two resistors in parallel equation we get:

$$R' = \frac{R_2 \times R_L}{R_2 + R_L} = \frac{2 \times 20}{2 + 20} = \frac{40}{22} \approx 1.8 \text{ k}\Omega$$

We can look upon the potential divider and the load as two resistors in series, i.e. $R_1 + R' = 0.5 + 1.8 = 2.3$ kΩ. The new voltage V_2' across BC (i.e. R_2) will then be given as in (*b*) by:

$$V_2' = \frac{R'}{R_1 + R'} \times V = \frac{1.8}{2.3} \times 9.0 \approx 7.0 \text{ V}$$

(ii) Proceeding as in (i), the combined resistance R'' of the new load (R_l = 2 kΩ) and BC (R_2 = 2 kΩ) in parallel is given by:

$$R'' = \frac{R_2 \times R_L}{R_2 + R_L} = \frac{2.0 \times 2.0}{2.0 + 2.0} = \frac{4.0}{4.0} = 1.0 \text{ k}\Omega$$

If V_2'' is the new voltage across BC, then:

$$V_2'' = \frac{R''}{R_1 + R''} \times V = \frac{1.0}{0.5 \times 1.0} \times 9.0$$

$$= \frac{1.0}{1.5} \times 9.0 = 6.0 \text{ V}$$

The results show that when the resistance of the load is ten times greater than the resistance of the part of the potential divider across which it is connected (i.e. BC), the output voltage applied to the load (across XY) drops from its 'no load' value by only 0.2 V (from 7.2 to 7.0 V: a 3% drop). When the load equals the resistance across which it is connected, the drop is 1.2 V (from 7.2 to 6.0 V: a 17% drop).

4.9 Revision questions

1. Name four factors to be considered when choosing a resistor.
2. Name three types of fixed resistor, giving one good feature of each.
3. Write in shorter form:
   a) 5 600 Ω,
   b) 15 000 Ω,
   c) 100 000 Ω,
   d) 2 200 000 Ω.
4. a) What is the value and tolerance of each of the resistors R_1, R_2, R_3 and R_4 colour-coded in the table below using the *four band* code?

Band:	1	2	3	4
R_1	orange	orange	brown	silver
R_2	brown	black	orange	gold
R_3	green	blue	yellow	none
R_4	red	red	green	gold

   b) What are the values and tolerances of R_5, R_6, R_7 and R_8 shown by the *five band* code in the table below?

Band:	1	2	3	4	5
R_5	brown	black	black	brown	brown
R_6	orange	orange	black	black	gold
R_7	green	brown	black	red	red
R_8	yellow	violet	black	orange	silver

5. a) What is the four band colour code for each of the following resistors?
   (i) 150 Ω ± 1%
   (ii) 10 Ω ± 5%
   (iii) 470 kΩ ± 20%
   (iv) 8.2 kΩ ± 5%
   (v) 39 kΩ ± 2%
   (vi) 1 MΩ ± 10%

   b) What is the five band colour code for the following resistors?
   (i) 160 Ω ± 2%
   (ii) 2.4 kΩ ± 5%
   (iii) 750 kΩ ± 1%

6. What are the values and tolerances of resistors marked with the following printed codes?
   a) 2K2M
   b) 270KJ
   c) 1M0K
   d) 15RF
   e) 680RM
   f) 33KJ

7. What are the printed codes for the following resistors?

a) 100 Ω ± 5% d) 390 kΩ ± 20%
b) 4.7 kΩ ± 2% e) 10 MΩ ± 10%
c) 18 kΩ ± 1% f) 68 Ω ± 5%

8. Write equations for the combined resistance R of two resistors R_1 and R_2
a) in series,
b) in parallel.

9. When a resistor has another one connected in parallel with it, is the combined resistance greater, the same or smaller?

10. State two ways of using a variable resistor.

11. What E12 preferred values would you use if you calculated that resistors of the following values were needed for a circuit?
a) 29 Ω
b) 5.0 kΩ
c) 72 kΩ
d) 350 kΩ

12. What are the maximum and minimum values which resistors with the following markings will have?

a) 10 Ω±10% d) 820KJ
b) 4.7 kΩ±10% e) 180RK
c) 10 kΩ±20% f) 3M3M

4.10 Problems

1. What is the electrical power dissipated in a 100 Ω resistor carrying a current of 50 mA?

2. Calculate the wattage of a 6 V 60 mA lamp.

3. Would a 0.5 W rating be suitable for a 10 kΩ resistor through which a current of 10 mA flows? If it is not, what rating would be suitable?

4. a) What single resistor with a preferred value could replace a 3.3 kΩ resistor connected in series with an 8.2 kΩ resistor?
b) The combined resistance of a 10 kΩ resistor and an unknown resistor in series with it is 25 kΩ. What is the value of the unknown resistor?

5. Calculate the combined resistance between A and B in Figs. 4.14(*a*), (*b*) and (*c*).

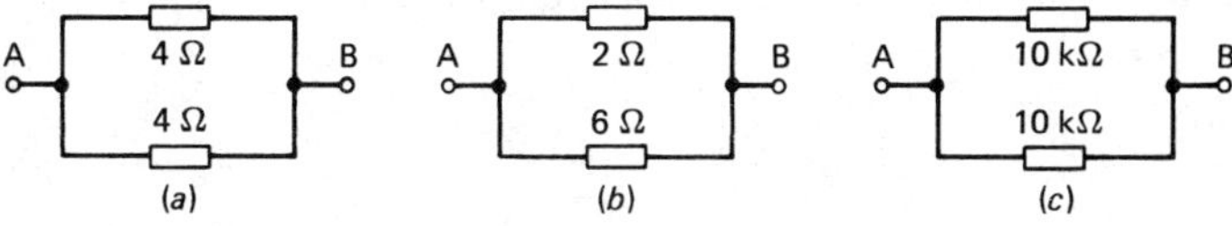

Fig. 4.14

6. What is the combined resistance of a 10 kΩ and a 22 kΩ resistor in parallel?

7. What preferred value of resistor should be placed in parallel with a 100 kΩ resistor to reduce its effective value to 80 kΩ?

8. Calculate the combined resistance of the networks in Figs. 4.15(*a*), (*b*) and (*c*).

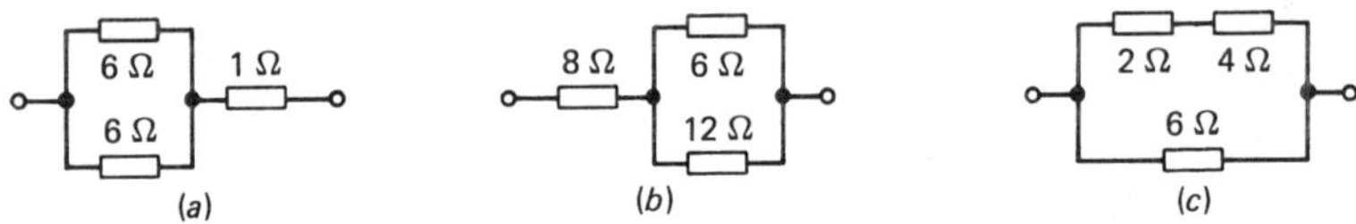

Fig. 4.15

9. What preferred value of single resistor could replace the network in Fig. 4.16?

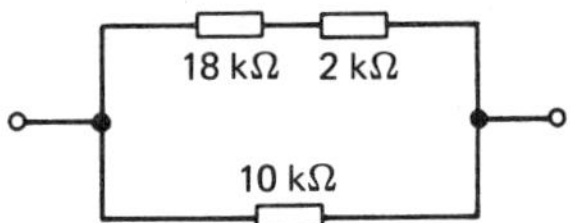

Fig. 4.16

10. What are the voltages across R_2 in Figs. 4.17(*a*), (*b*) and (*c*)?

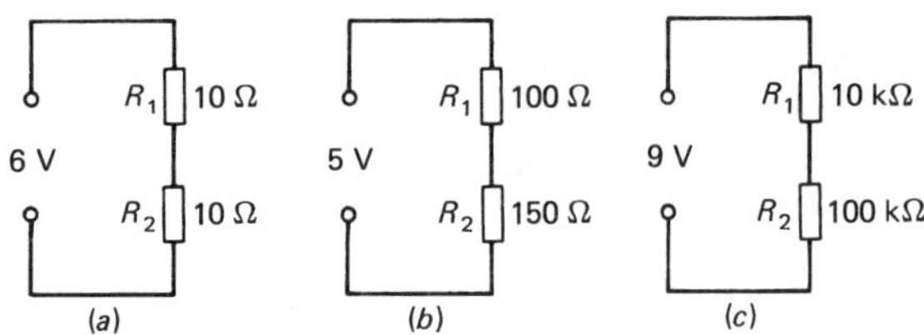

Fig. 4.17

11. What is the voltage between B and C when the wiper on the linear 'pot' in Fig. 4.18 is
a) one-third of the way round from C,
b) halfway round from C?

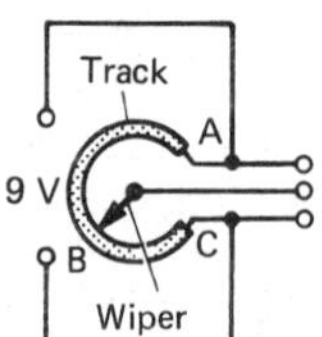

Fig. 4.18

12. A 20 kΩ linear 'pot' is used as a potential divider for a 9.0 V supply. If the wiper is set halfway round the track, what is the output voltage when the 'load' resistance is
a) very large,
b) 100 kΩ,
c) 10 kΩ?

UNIT 5

Capacitors

5.1 About capacitors

A capacitor stores electric charge. It does not allow direct current to flow through it and it behaves as if alternating current does flow through. In its simplest form it consists of two parallel metal plates separated by an insulator called the *dielectric*, Fig. 5.01(*a*). The symbols for fixed and variable capacitors are given in Fig. 5.01(*b*). Polarized types must be connected so that conventional current enters their positive terminal. Non-polarized types can be connected either way round.

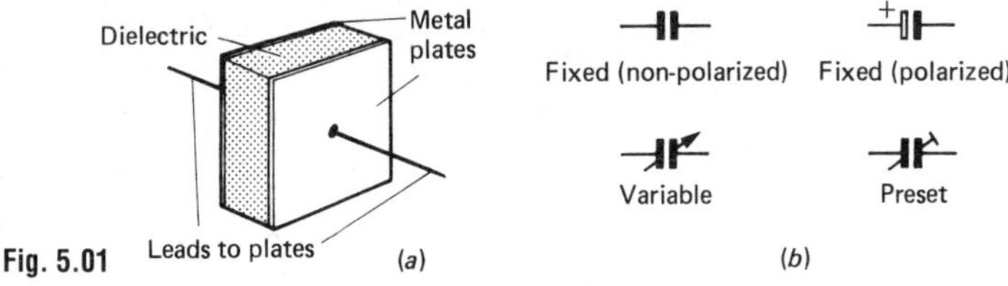

Fig. 5.01

The *capacitance* (C) of a capacitor measures its ability to store charge and is stated in *farads* (F). The farad is subdivided into smaller, more convenient units.

$$1 \text{ microfarad } (1\ \mu\text{F}) = 1 \text{ millionth of a farad} = 10^{-6}\text{F}$$
$$1 \text{ nanofarad } (1\ \text{nF}) = 1 \text{ thousand-millionth of a farad} = 10^{-9}\text{F}$$
$$1 \text{ picofarad } (1\ \text{pF}) = 1 \text{ million-millionth of a farad} = 10^{-12}\text{F}$$

In practice, capacitances range from 1 pF to about 150 000 μF: they depend on the area A of the plates (large A gives large C), the separation d of the plates (small d gives large C) and the material of the dielectric (e.g. certain plastics give large C).

When selecting a capacitor for a particular job, the factors to be considered are (as with resistors—see section 4.1) the *value* (again this is not critical in many electronic circuits), the *tolerance* and the *stability*. There are two additional factors.

(i) The working voltage This is the largest voltage (d.c. or peak a.c.) which can be applied across the capacitor and is often marked on it, e.g. 30 V wkg. If it is exceeded, the dielectric breaks down and permanent damage may result.

(ii) The leakage current No dielectric is a perfect insulator but the loss of charge through it as 'leakage current' should be small.

5.2 Capacitance codes

As with resistors, only certain *preferred values* of capacitors are made—those in the E12 range (see section 4.3). They give maximum coverage with minimum overlapping of values. Various methods are used to show capacitance values.

In one method it is marked on the capacitor in μF, nF or pF, with the submultiple abbreviation being used to indicate any decimal point. For example, 2.2 nF is shown as 2n2 and 4.7 pF as 4p7.

In another method a *three digit code* is used, like the resistor colour code but the numbers are printed on the capacitor rather than encoded in colours. The first two digits are the first two numbers of the value and the third gives the number of 0s to be added to the first two digits to give the value in *picofarads*. For example, a capacitor marked '103' has a value of 10 plus 3 zeros, which is 10 000 pF or 10 nF or 0.01 μF. Table 5.1 gives more examples.

Table 5.1 Examples of capacitor code values

Code	Value (pF)	Value (nF)	Value (μF)
101	100	0.1	0.0001
222	2 200	2.2	0.0022
333	33 000	33	0.033
474	470 000	470	0.47

Tolerances are shown as in the resistor printed code.

5.3 Types of capacitor

Capacitors can be classified according to the dielectric used; their properties depend on this.

(i) Polyester (Fig. 5.02(*a*)) Good all-round properties and small size make these suitable for general use. Values range from 0.01 μF to 10 μF or so.

(ii) Mica (Fig. 5.02(*b*)) These have low tolerance ($\pm$1%), high stability and working voltage, and low leakage. Values range from 1 pF to 0.01 μF. They are used in radio frequency tuned circuits. *Polystyrene types* are similar and cheaper.

(iii) Ceramic (Fig. 5.02(*c*)) These offer values up to 1 μF with small size but poor stability and tolerance. They are useful where exact values are not too important.

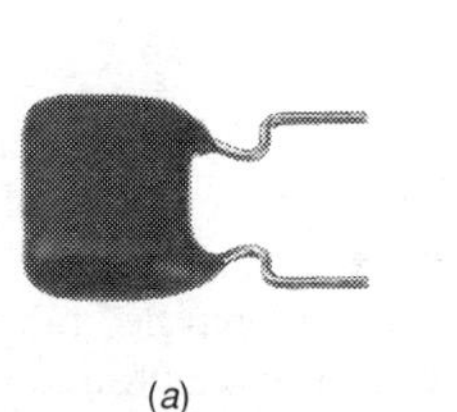

(*a*)

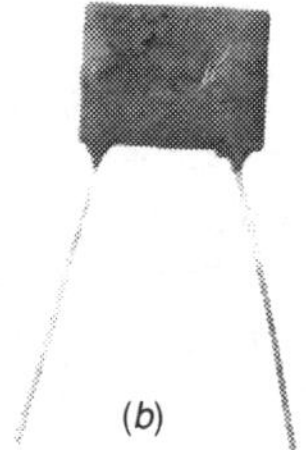

(*b*)

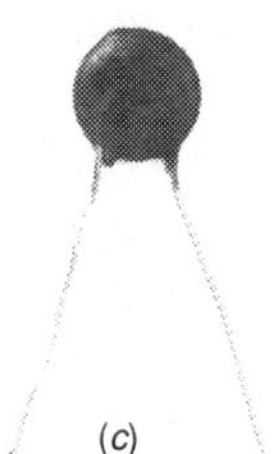

(*c*)

Fig. 5.02

(iv) Electrolytic (polarized) The *aluminium* type, Fig. 5.03(*a*), is used where very large, fixed values (up to 100 000 μF or so) are required. They are compact but have a wide tolerance range (-10 to $+50\%$), high leakage, poor stability and wrong connection can destroy the aluminium oxide dielectric. They are used in power supplies.

Tantalum electrolytics, Fig. 5.03(*b*), can be used instead of aluminium in low voltage circuits where values need not exceed about 100 μF. They have lower leakage currents and better tolerance ($\pm 20\%$).

(*a*)

(*b*)

Fig. 5.03

(v) Variable (Fig. 5.04) Variable capacitors are used mainly to tune radio receivers and consist of two sets of parallel metal plates; one set is fixed and the other moves on a spindle within the fixed set, with air as the dielectric. Often two or more are 'ganged' together so that the capacitance in several circuits changes simultaneously. The maximum capacitances for the plates fully interleaved may be 500 pF for a single unit.

Preset capacitors or *trimmers*, Fig. 5.05, are small variable types used to make fine, infrequent adjustments to the capacitance of a circuit.

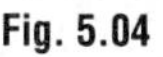

Fig. 5.04

Fig. 5.05

5.4 Charging a capacitor

(a) Action using a battery

A capacitor can be charged by connecting a battery across it. In Fig. 5.06, when the switch is closed, electrons from the upper plate X of the capacitor are attracted to the positive terminal of the battery and an equal number are repelled to the lower plate Y by the negative terminal. Positive charge builds up on X and an equal negative charge on Y. While this charging action is occurring, electrons are moving along the connecting wires (but not through the dielectric if it is a perfect capacitor) and a brief current would be detected at any point in the circuit by a sensitive meter.

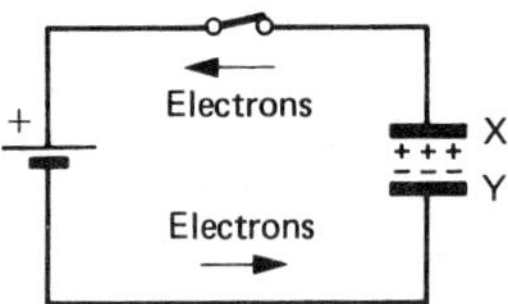

Fig. 5.06

In a short time, depending on the capacitance of the capacitor and the resistance of the battery and wires, the voltage between X and Y is equal and opposite to that of the battery. Electron flow then stops, i.e. the current is zero. If plate X has charge $+Q$, then plate Y has charge $-Q$ and we sum this up by saying that the capacitor has charge Q.

(b) Charge stored

Doubling the charging voltage doubles the charge stored by a capacitor. The equation which enables us to calculate the charge Q which is stored in a capacitor of capacitance C when it is charged to voltage V follows from the way we define capacitance (see below) and is:

$$Q = V \times C$$

For example, if $C = 1\,000\ \mu\text{F} = 1\,000/1\,000\,000\text{F} = 1/1\,000\text{F} = 0.001\text{F}$ and $V =$ 10 V, then Q is given in coulombs (C) by:

$$Q = V \times C = 10 \times 0.001 = 0.01\ \text{C}$$

Alternatively, leaving C in μF gives Q in microcoulombs (μC):

$$Q = V \times C = 10 \times 1\,000 = 10\,000\ \mu\text{C}$$

The equation is a result of the fact that we define *capacitance in farads* as the charge on a capacitor when the voltage across it is one volt. If the charge is 6 C when the voltage is 2 V, the capacitance is 6 C/2 V = 3 F. In general for charge Q and voltage V, capacitance $C = Q/V$ (giving $Q = V \times C$).

(c) Energy stored

A charged capacitor stores the electrical energy it received from whatever charged it, e.g. a battery. If a lamp (in a photographic electronic flash unit for example) or an electric motor is connected across it, current flows, and the capacitor loses all its charge, i.e. it is discharged. Some other form of energy is produced.

It can be shown that the energy W in joules (J) stored by a capacitor is given by:

$$W = \tfrac{1}{2}(Q \times V) \qquad (1)$$

Since $Q = V \times C$ we also have:

$$W = \tfrac{1}{2}(V \times C) \times V = \tfrac{1}{2}(V^2 \times C) \qquad (2)$$

Also since $V = Q/C$:

$$W = \tfrac{1}{2}(Q \times V) = \tfrac{1}{2}\left(Q \times \frac{Q}{C}\right) = \tfrac{1}{2}\frac{Q^2}{C}$$

For example if $C = 1\,000\ \mu\text{F} = 0.001$ F and $V = 10$ V, using equation (2), we get:

$$W = \tfrac{1}{2}(10^2 \times 0.001) = \tfrac{1}{2}(0.1) = 0.5 \times 0.1$$

$$= 0.05 \text{ J}$$

This is a very small amount of energy compared with that stored in a battery.

5.5 Capacitor networks

Capacitors have sometimes to be connected to give a larger or smaller capacitance.

(a) Parallel

In Fig. 5.07(*a*) capacitors of capacitance C_1 and C_2 are in parallel. The voltage V is the *same* across each but their charges are *different* and are given by:

$$Q_1 = V \times C_1 \text{ and } Q_2 = V \times C_2$$

The total charge Q on both capacitors is:

$$Q = Q_1 + Q_2$$

$$\therefore \quad Q = V \times C_1 + V \times C_2 = V(C_1 + C_2)$$

If C is the capacitance of a capacitor which has a charge Q when the voltage across it is V, Fig. 5.07(*b*), then:

$$Q = V \times C$$

$$\therefore \quad V \times C = V(C_1 + C_2)$$

Dividing both sides by V, we get:

$$C = C_1 + C_2$$

For example, if $C_1 = 500\ \mu$F and $C_2 = 250\ \mu$F, their combined capacitance C in parallel is $C = 500 + 250 = 750\ \mu$F: they behave like a capacitor with larger plates.

The equation for capacitors in parallel is similar to that for resistors in *series*—see section 4.4.

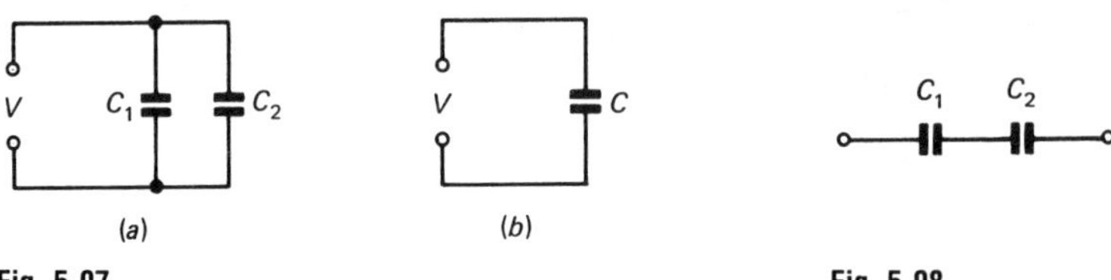

Fig. 5.07 **Fig. 5.08**

(b) Series

The equation for calculating the combined capacitance C of two capacitances C_1 and C_2 in series (Fig. 5.08) can be shown to be:

$$\frac{1}{C} = \frac{1}{C_1} + \frac{1}{C_2}$$

or

$$C = \frac{C_1 \times C_2}{C_1 + C_2} = \frac{\text{product of capacitances}}{\text{sum of capacitances}}$$

This is similar to the equation for resistors in *parallel*—see section 4.5. As an example, if $C_1 = 500\ \mu F$ and $C_2 = 250\ \mu F$, their combined capacitance C in series is:

$$C = \frac{500 \times 250}{500 + 250} = \frac{500 \times 250}{750} = \frac{500}{3} \approx 170\ \mu F$$

Note that it is less than the capacitance of either capacitor.

(c) Worked example

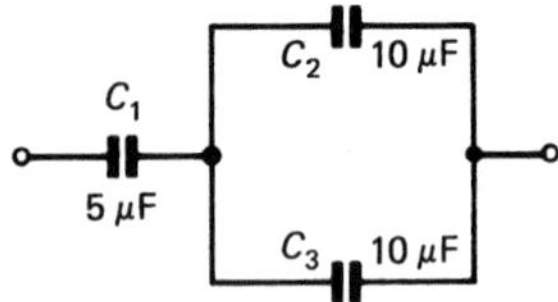

Fig. 5.09

What is the combined capacitance of the network in Fig. 5.09?
The combined capacitance of C_2 and C_3 in parallel is:

$$C_2 + C_3 = 10 + 10 = 20\ \mu F$$

This is in series with $C_1 = 5\ \mu F$ and their combined capacitance C is:

$$C = \frac{5 \times 20}{5 + 20} = \frac{100}{25} = 4\ \mu F$$

5.6 Time constant

You will meet many circuits in which a capacitor charges or discharges through a resistor. Such processes do not occur instantaneously but take time and it often helps us to understand how a circuit works if we know what is called its *time constant.*

(a) Charge

In the circuit of Fig. 5.10, when the switch is in position 1, the capacitor, capacitance C, charges through the resistor, resistance R, from a 9 V battery. The microammeter records the charging current I and the voltmeter reads the voltage V_C across the capacitor at different times t.

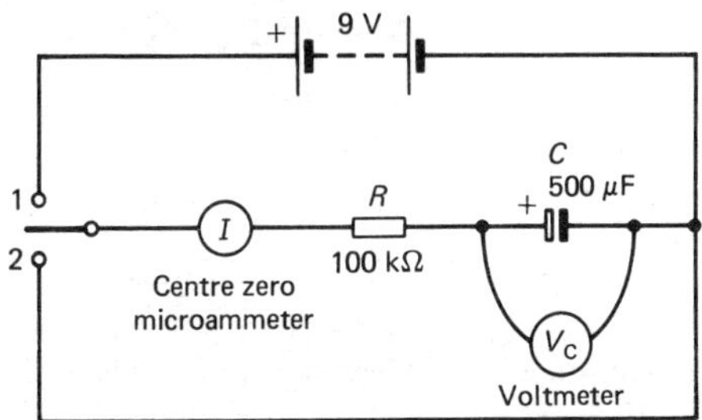

Fig. 5.10

If you do the experiment and plot graphs of I against t and V_C against t from the readings, the graphs will look like those in Figs. 5.11(*a*) and (*b*). These graphs show that:

(i) I has its maximum value at the start when the capacitor begins to charge and then decreases more and more slowly until it becomes zero;

(ii) V_C rises rapidly from zero and slowly approaches its maximum value (9 V) which it reaches when the capacitor is fully charged and $I = 0$.

If readings of the voltage V_R across the resistor were also taken, a graph of V_R against t would have the same shape as that of I against t (since $V_R = I \times R$ at all times) and is shown dashed in Fig. 5.11(*b*). All three graphs are called *exponential* curves, I and V_R being 'decaying' ones and V_C a 'growing' one.

The sum of the voltages across the resistor and the capacitor equals the battery voltage V, that is $V = V_C + V_R$.

Initially $V_C = 0$, therefore $V = V_R$. Finally $I = 0$, therefore $V_R = 0$ and $V = V_C$.

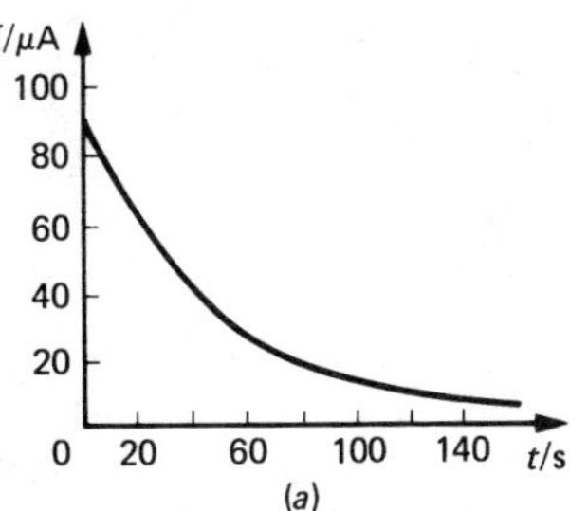

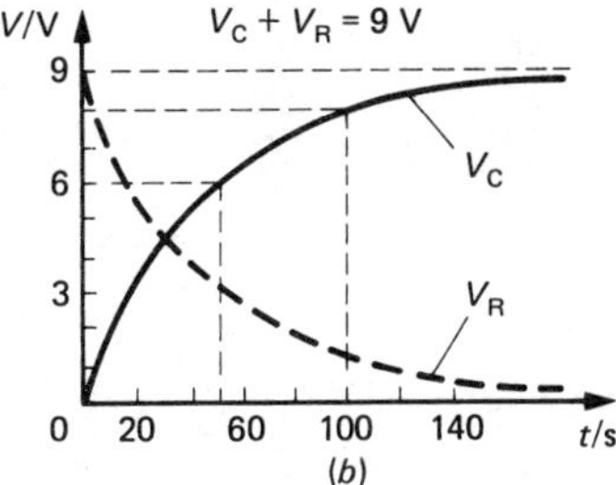

Fig. 5.11

(b) Time constant

It can be shown that if the charging current I remained steady at its starting value, a capacitor would be fully charged after a time of $C \times R$ seconds (if C is in farads and R is in ohms). In fact I decreases with time, as Fig. 5.11(*a*) shows, and the capacitor has only 0.63 of its full charge after time $C \times R$. Nevertheless $C \times R$, called the *time constant* T, is a useful measure of how long it takes a capacitor to charge through a resistor. The greater the values of C and R the greater is T and the more slowly does the capacitor charge; also the more slowly does the voltage V_C across it rise.

To calculate T for the circuit of Fig. 5.10 we have:

$$C = 500\ \mu\text{F} = 500 \times 10^{-6}\text{F}\ (0.000\,5\ \text{F})$$

$$R = 100\ \text{k}\Omega = 100 \times 10^{3}\ \Omega\ (100\,000\ \Omega)$$

Therefore:

$$T = C \times R = 500 \times 10^{-6} \times 100 \times 10^{3} = 50\ \text{s}$$

As a check you can see from the graph of V_C against t in Fig. 5.11(*b*) that after 50 s V_C is about 6 V $= \frac{2}{3} \times 9$ V. (For most purposes we can take $0.63 = \frac{2}{3}$.) After 100 s (i.e. after the second time constant) V_C rises by two-thirds of the voltage remaining after 50 s, i.e. $\frac{2}{3} \times (9 - 6) = \frac{2}{3} \times 3 = 2$ V , so making $V_C = 6 + 2 = 8$ V. After about $5 \times C \times R$ (250 s) it is fully charged, i.e. $V_C = 9$ V. In general $C \times R$ is the time for V_C to rise to two-thirds of the charging voltage remaining at the start of the time.

(c) Discharge

In Fig. 5.10 when the switch is moved from position 1 to position 2, the capacitor discharges through the resistor. If graphs of I, V_C and V_R are plotted as before, they are again exponential curves, as shown in Figs. 5.12(*a*) and (*b*). Note that:

(i) the discharge current I, and so also V_R, is in the opposite direction to that during charge (compare them with Figs. 5.14(*a*) and (*b*));

(ii) V_C and V_R are in opposition during discharge.

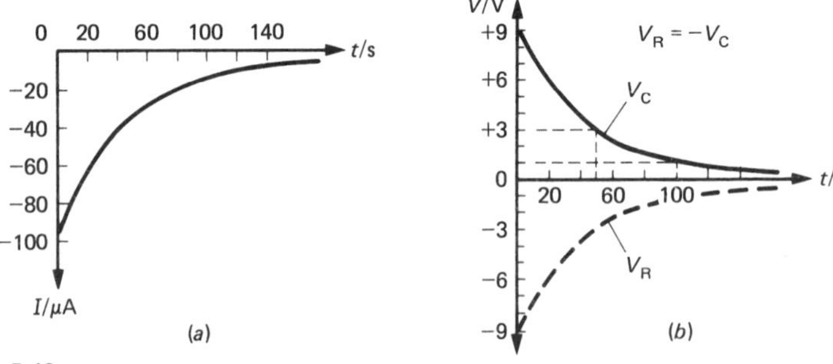

Fig. 5.12

The time constant $C \times R$ is useful here too and is the time for V_C to fall by two-thirds of its value at the start of the discharge. You can see from the V_C graph that in 50 s (T for the circuit), V_C falls from 9 V to 3 V and in the next 50 s it falls by $\frac{2}{3} \times 3$, i.e. by 2 V, making $V_C = 1$ V. After about $5 \times C \times R$ (250 s), $V_C = 0$.

5.7 Capacitive reactance

(a) Effect of a capacitor on d.c. and a.c.

If you set up the circuit of Fig. 5.13(*a*) in which the supply is d.c. from a battery, you will find the lamp does not light; in the circuit of Fig. 5.13(*b*) the supply is a.c. and the lamp lights. This suggests that a capacitor stops d.c. but lets a.c. pass, Fig. 5.13(*c*).

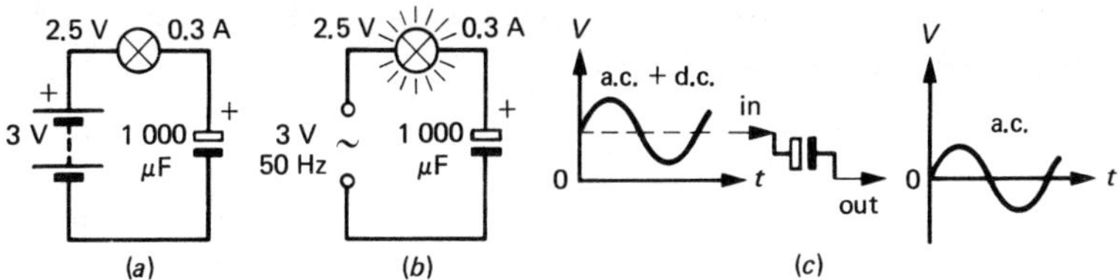

Fig. 5.13

In a d.c. circuit there is a brief current flow which charges the capacitor. In an a.c. circuit, when terminal X of the supply is +, as in Fig. 5.14(*a*), current flows and charges the capacitor with the upper plate positive and the lower plate negative. Then the direction of the current reverses, the capacitor discharges and when the a.c. supply makes terminal Y +, as in Fig. 5.14(*b*), the capacitor charges up with the lower plate positive, only to discharge again as the a.c. supply becomes zero at the end of a cycle. This process is repeated fifty times a second on a 50 Hz supply.

No current actually passes through the capacitor since its plates are separated by an insulator. Electrons, however, do flow backwards and forwards in the wires joining the a.c. supply to the capacitor, going on to and off the plates as it charges and discharges. They make it seem that current does pass through the capacitor and a current would be recorded on an a.c. ammeter.

We will meet many electronic circuits in which a capacitor is used to block d.c. but allows a.c. to pass (Fig. 5.13(*c*)).

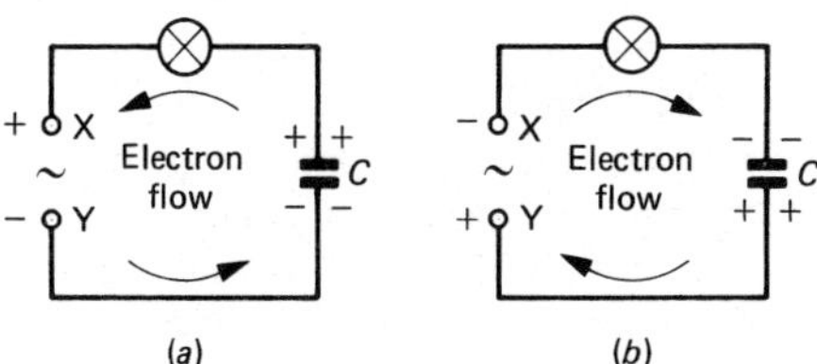

Fig. 5.14

(b) Reactance

A capacitor offers less 'opposition' to a.c. when:

(i) its capacitance C is large—so that it stores a greater charge;
(ii) the frequency f of the a.c. is high—causing the charge to flow on and off the plates in a shorter time.

Larger C and f result in larger currents appearing to flow through a capacitor.

The opposition of a resistor to current flow is measured by its resistance and is the same for d.c. as for a.c., whatever the frequency (except for very high frequencies). This is not so for capacitors.

The opposition of a capacitor to a.c. is called its *capacitive reactance*, denoted by X_C and given in ohms by the equation:

$$X_C = \frac{1}{2\pi fC}$$

C is in farads and f in hertz. The equation summarizes the fact that X_C decreases if either f or C increases. If $C = 1\,000\ \mu F = 10^3\ \mu F = 10^3/10^6\ F = 1/10^3\ F = 10^{-3}$ F and $f = 50$ Hz, then:

$$X_C = \frac{1}{2 \times 3.14 \times 50 \times 10^{-3}} = \frac{10^3}{314} = \frac{1\,000}{314} \approx 3.2\ \Omega$$

If $f = 500$ Hz and $C = 1\,000\ \mu F$, $X_C = 0.32\ \Omega$.

(c) Phase shift

When a.c. flows through a resistor, the current and voltage reach their peak values at the same instant, as shown in Fig. 5.15(*a*). They are in time with each other or as we say, they are *in phase*. A capacitor behaves differently.

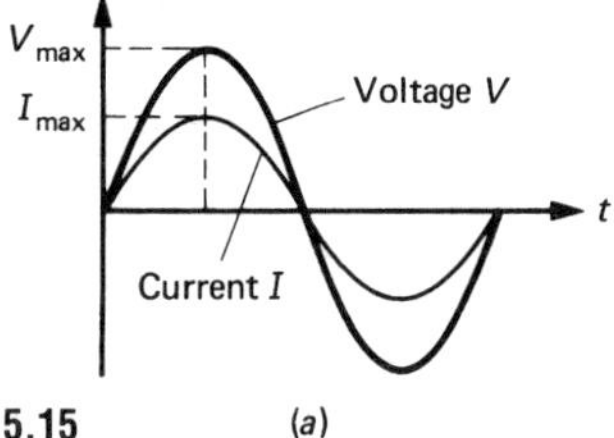

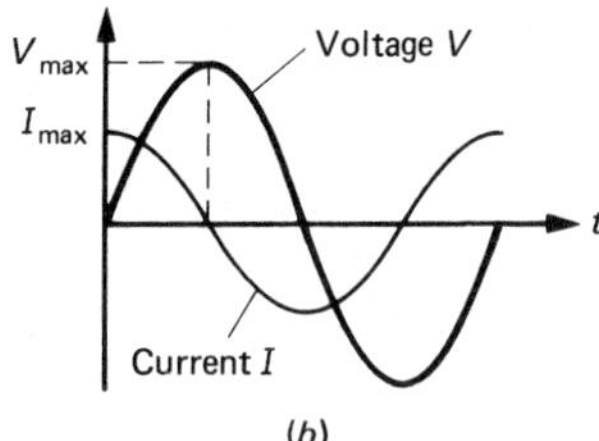

Fig. 5.15 (*a*) (*b*)

We saw in section 5.4 that the voltage and the current were not in step. Figs. 5.11(*a*) and (*b*) show that when the current I has its maximum value, the voltage V_C across the capacitor is zero and vice versa. This is also the situation when a capacitor is in an a.c. circuit.

The waveforms for the current 'through' a capacitor and the voltage across it are given for a sinusoidal a.c. input in Fig. 5.15(*b*). They show (as do Figs. 5.11(*a*) and (*b*)) that the current I is at its peak value one quarter of a cycle before the voltage V; there is a *phase shift* between them. I and V are said to be out of phase by one quarter of a cycle or 90°, with I ahead. (In an alternator—as used in a car or a power station—one complete revolution of the armature through 360° produces one cycle of a.c., so one quarter of a cycle corresponds to a 90° rotation.)

5.8 *CR* coupled circuits

Different parts of an electronic circuit are often coupled (joined) by a capacitor and a resistor. In this section we will work out what effect the time constant of the coupling circuit has on the waveform of the input. Square waves will be considered as the test waveform because they contain many frequencies.

(a) Capacitor coupling

In Fig. 5.16(*a*), a square wave input voltage (Fig. 5.16(*b*)), is obtained from a battery by moving the change-over switch from position 1 to position 2 repeatedly. The output voltage V_R appears across the resistor, resistance R.

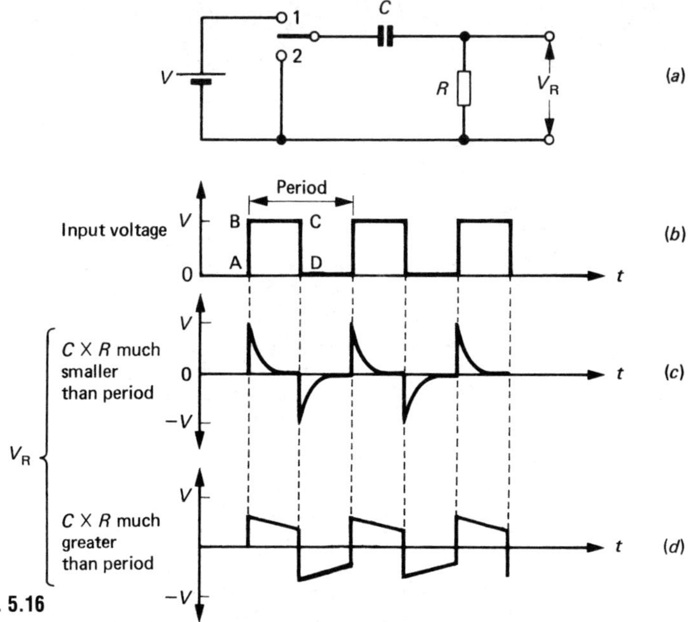

Fig. 5.16

When the input voltage rises from 0 to V volts (position 1: AB on Fig. 5.16(*b*)), V_R rises immediately to V because the voltage V_C across the capacitor cannot change suddenly (it takes time) i.e. the capacitor passes on the rapid voltage change, making $V_R = V$. The capacitor starts to charge through the resistor. V_C therefore rises and V_R falls (see the V_C and V_R graphs of Fig. 5.11(*b*)): the rate at which they do so depends on the time constant $C \times R$.

If $C \times R$ is much *smaller* than the period of the square wave, the capacitor charges quickly causing V_R to fall rapidly and making it have a spiky waveform, as in Fig. 5.16(*c*). When the input voltage falls from V volts to 0 (position 2: CD on Fig. 5.16(*b*)), the capacitor starts to discharge producing a negative spiky waveform for V_R (see V_R graph of Fig. 5.12(*b*)). In this case the waveform of the output V_R is very different from that of the input voltage and considerable distortion has occurred. The circuit is called a *differentiator* because it gives an output only when the input is *changing* (along AB and CD); the faster the rate of change, the greater the output.

If $C \times R$ is much *greater* than the period of the square wave, C charges very slowly causing a slow change in V_R. Here the waveform of V_R is much more like that of the input voltage, as is shown in Fig. 5.16(*d*), and very little distortion occurs. This kind of coupling circuit is used to connect the stages of an audio amplifier, the time constant being about ten times greater than the period of the input waveform (e.g. 1/100 s for a 1 kHz input).

Also notice that V_R is an alternating voltage (average value zero), whereas the input voltage is a direct voltage (average value greater than zero). If we think of the input voltage as being steady d.c. to which a.c. has been added, as in Fig. 3.5 of section 3.3, we see that only the a.c. part passes through the capacitor to the output.

(b) Resistor coupling

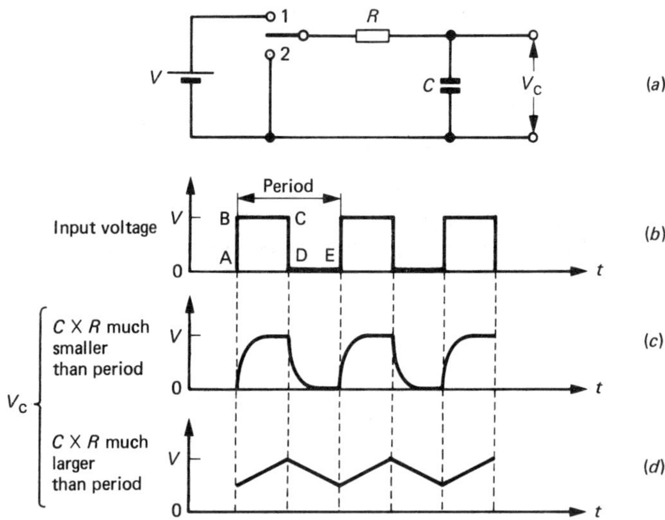

Fig. 5.17

In Fig. 5.17(*a*) the positions of the resistor and the capacitor have been changed and the output voltage V_C is taken from across the capacitor. When the time constant is relatively small, the waveform of V_C closely resembles that of the input voltage as shown in Figs. 5.17(*b*) and (*c*). When the time constant is relatively large, V_C does not change much and the voltage changes at the input are smoothed out, as shown in Fig. 5.17(*d*). The circuit is called an *integrator* because it *adds up* the input over a period of time and responds to the steady parts (e.g. BC and DE) of the input.

This kind of coupling is used for smoothing in power supplies (see section 21.3) and for demodulation in radio receivers (see section 18.3).

5.9 Testing capacitors

An analogue multimeter (see section 2.7) or a digital multimeter, on its high resistance range (see section 21.7) can be used to test a capacitor.

(i) Non-polarized types If the resistance of the capacitor is less than about 1 MΩ, it is allowing d.c. to pass (from the battery inside the multimeter), i.e. it is 'leaking', and is faulty. With large value capacitors there may be a short initial burst of current as the capacitor charges up.

(ii) Polarized types For the dielectric to form in these, a positive voltage must be applied to the positive side of the capacitor (marked with a + or a groove or both). In most analogue multimeters the terminal marked − (black) is connected to the positive of the internal battery; with digital meters the maker's instructions should be consulted for the polarities of the terminals. When the capacitor is first connected to the multimeter, its resistance is low but rises as the dielectric forms—otherwise it is faulty.

5.10 Worked examples

Example 1

After the switch is closed in the circuit of Fig. 5.18 what will be
a) the maximum current through the resistor,
b) the maximum charge on the capacitor?

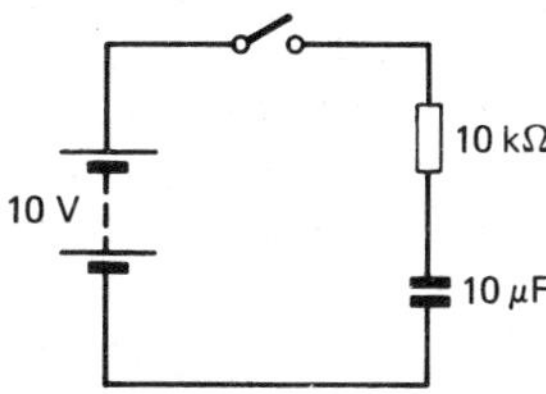

Fig. 5.18

a) The current is a maximum at the instant when the switch is closed, i.e. before the voltage starts to build up across the capacitor and opposes the battery voltage. It is given by:

$$I = V/R \text{ where } V = \text{battery voltage} = 10\text{ V and } R = 10\text{ k}\Omega$$

$$\therefore \quad I = \frac{10}{10} = 1\text{ mA}$$

b) The charge is a maximum when the charging current has fallen to zero and the voltage across the capacitor equals the battery voltage. It is given by:

$$Q = V \times C \text{ where } V = 10\text{ V and } C = 10\ \mu\text{F}$$

$$\therefore \quad Q = 10 \times 10 = 100\ \mu\text{C}$$

Example 2

In the circuit of Fig. 5.19 what are the voltages on plates A and B of the initially uncharged capacitor C and what is the voltage across R

a) immediately the switch is moved to position 1,
b) after the switch has been in position 1 for 5 s,
c) immediately the switch is moved to position 2,
d) after the switch has been in position 2 for 5 s,
e) immediately the switch is in position 1 again?

Draw graphs showing waveforms of one period of the input and output voltages.

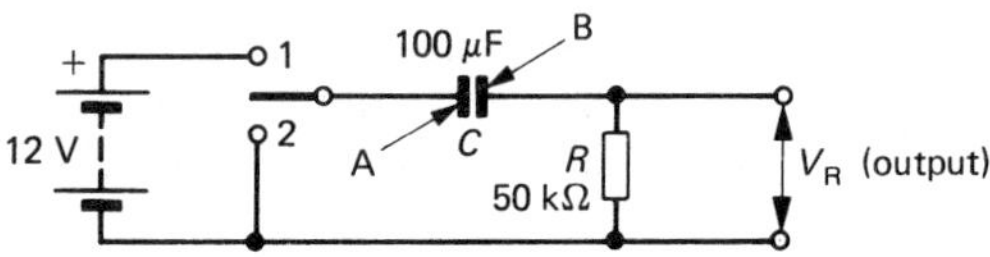

Fig. 5.19

a) Voltage on A rises from 0 V to +12 V causing the voltage on B to rise to +12 V because the capacitor has not had time to charge up and so the voltage V_C across it is zero. Hence V_R = +12 V since the lower end of the resistor is connected to battery negative, i.e. 0 V. Therefore:

$$A = +12\text{ V} \qquad B = +12\text{ V} \qquad V_R = +12\text{ V}$$

b) Time constant $C \times R = 100 \times 10^{-6} \times 50 \times 10^3 = 5$ s (since $C = 100\ \mu F = 100 \times 10^{-6}$F and $R = 50\ k\Omega = 50 \times 10^3\ \Omega$). In position 1, C charges through R and after 5 s V_C will have risen by two-thirds of 12 V (the charging voltage), i.e. V_C = 8 V.

Plate A is still connected to +12 V. Therefore, so that there can be a voltage of 8 V across C, the voltage on B must be (12 − 8) = +4 V. Hence V_R = +4 V since during charge $V_C + V_R = 12$ V. Therefore:

$$A = +12\text{ V} \qquad B = +4\text{ V} \qquad V_R = +4\text{ V}$$

c) Voltage on A falls to 0 V (because it is connected to battery negative), therefore since V_C = 8 V the voltage on B is now −8 V. Hence V_R = −8 V. Therefore:

$$A = 0\text{ V} \qquad B = -8\text{ V} \qquad V_R = -8\text{ V}$$

d) In position 2, C discharges through R and in 5 s V_C will fall by two-thirds of 8 V (its present voltage), i.e. $V_C = 8 - \frac{2}{3} \times 8 = 8 - 5\frac{1}{3} = 2\frac{2}{3} \approx 2.7$ V.

Hence V_R = −2.7 V since during discharge $V_R = -V_C$. Therefore:

$$A = 0\text{ V} \quad B = -2.7\text{ V} \quad V_R = -2.7\text{ V}$$

e) Voltage on A rises to +12 V (because it is concerned to battery positive). Therefore, since $V_C = 2.7$ V, the voltage on B is $(12 - 2.7) = +9.3$ V. Hence $V_R = +9.3$ V. Thus:

$$A = +12\text{ V} \qquad B = +9.3\text{ V} \qquad V_R = +9.3\text{ V}$$

The input and output voltage graphs are shown in Figs. 5.20(*a*) and (*b*) respectively.

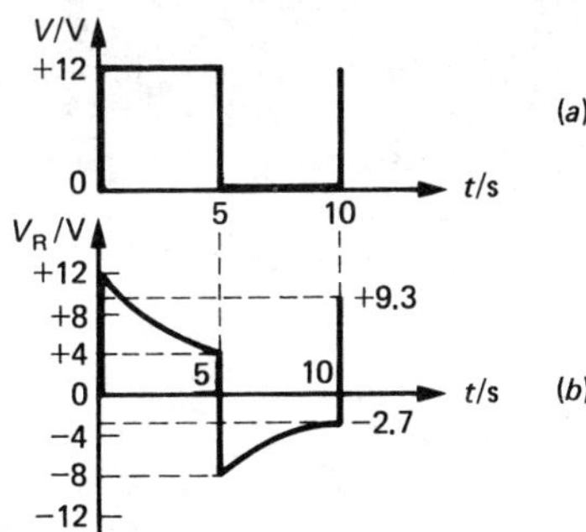

Fig. 5.20

5.11 Revision questions

1. State three useful properties of a capacitor.

2. a) Express in nF: 100 pF, 8200 pF, 10 000 pF.
b) Express in μF: 330 nF, 1 000 nF, 47 nF.
c) What is the value in pF of a capacitor with a three-digit code of: 821, 102, 563, 104?
d) What is the three-digit code for a capacitor of value: 220 pF, 82 nF, 0.1 μF?

3. What are the upper and lower values of
a) a ceramic capacitor marked 0.1 μF±20%,
b) an electrolytic capacitor marked 100 μF + 100% to −20%?

4. In the equations $Q = V \times C$ and $W = \frac{1}{2}Q \times V$, what do the symbols Q, V, C and W stand for?
What is each equation used for?
Write each equation in two other ways.

5. a) What is the combined value C of two capacitors of capacitances C_1 and C_2 which are joined
(i) in series,
(ii) in parallel?
b) Repeat (a) for three capacitances C_1, C_2, C_3.

6. a) What is meant by the *time constant* of a resistor–capacitor combination during
(i) charge,
(ii) discharge?
State the expression used to measure it, giving units.

b) Draw two graphs, one to show how the voltage V_C across a capacitor varies with time t as it charges from 0 V to 6 V and the other as it discharges from 6 V to 0 V.
What will be the value of V_C in each case after
(i) one time constant,
(ii) two time constants?

7. Graphs A and B in Fig. 5.21 are the charge curves for two *CR* circuits.
a) In which one does *C* charge quicker?
b) Which circuit has the smaller time constant?

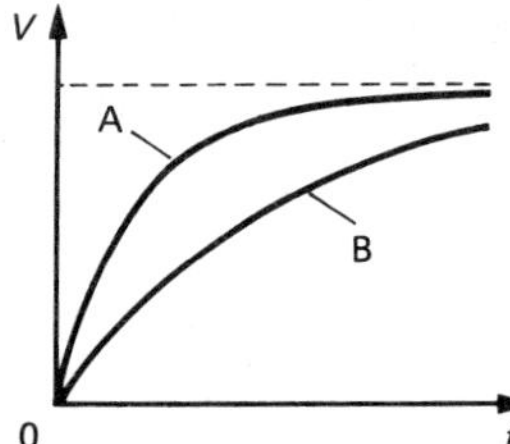

Fig. 5.21

8. a) Why does a capacitor seem to let a.c. pass through it?
b) What is meant by capacitive reactance?
c) State an equation for capacitive reactance.
d) Which meter records the larger current in
(i) Fig. 5.22,
(ii) Fig. 5.23?

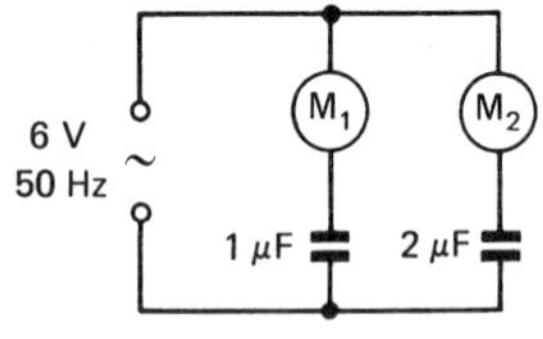

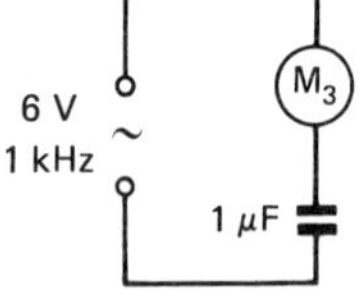

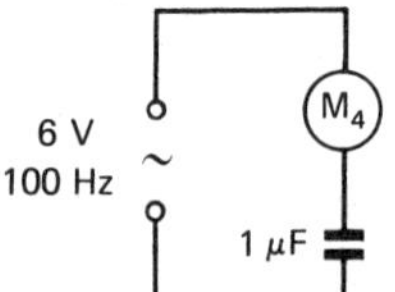

Fig. 5.22 **Fig. 5.23**

9. Waveforms for the voltage across a capacitor and the current through it in an a.c. circuit are shown in Fig. 5.24. Which one represents the current?

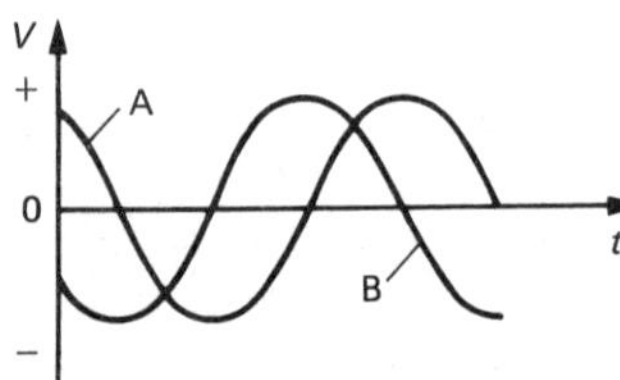

Fig. 5.24

5.12 Problems

1. What is the charge stored when the voltage across
  a) a 100 000 μF capacitor is 10 V,
  b) a 50 μF capacitor is 9 V?

2. a) What is the capacitance of a capacitor which stores 12 μC of charge when connected to a 6 V battery?
  b) Work out the voltage across the plates of a 10 μF capacitor when it has a charge of 50 μC.

3. Calculate the energy stored in
  a) a capacitor with a charge of 2C and 9V across its plates,
  b) a 1 μF capacitor charged to 500V.

4. What is the combined capacitance of
  a) 2.2 μF and 4.7 μF in parallel,
  b) 0.1 μF and 0.22 μF in parallel,
  c) two 100 μF capacitors in series?

5. Copy and complete the time constant table shown below.

C	*R*	Time constant
1 μF	1 MΩ	...
...	1 kΩ	1 s
0.22 μF	150 kΩ	...
10 μF	...	0.47 s

6. A 1.5 V battery is connected to a 1 000 μF capacitor in series with a 150 Ω resistor.
  a) What is the maximum current which flows through the resistor during charging?
  b) What is the maximum charge on the capacitor?
  c) How long does the capacitor take to charge to 1.0 V?

7. Calculate the reactance of a 1 μF capacitor at frequencies of
  a) 1 kHz,
  b) 1 MHz.

8. What value of capacitance would give a reactance of 50 Ω at 700 Hz?

9. Draw the waveforms of the voltages you would expect at the output in Fig. 5.25(*a*) if the input voltage had the waveform shown in Fig. 5.25(*b*) and
(i) $C = 1\ \mu\text{F}$ and $R = 1\ \text{k}\Omega$,
(ii) $C = 10\ \mu\text{F}$ and $R = 10\ \text{k}\Omega$.

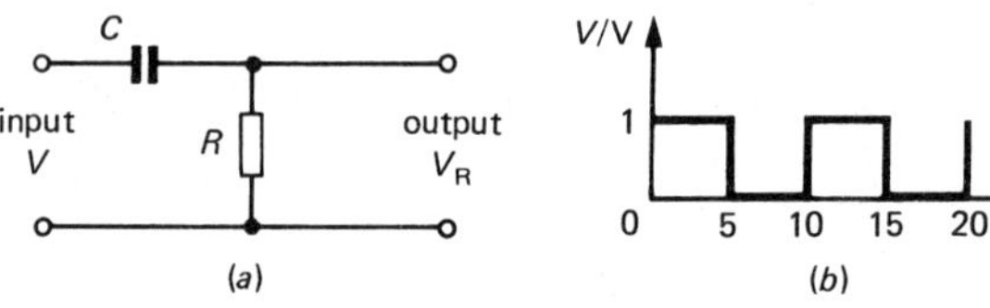

Fig. 5.25

(b) Electricity from magnetism

When a conductor is in a *changing* magnetic field, a voltage is produced in it. This can be shown by pushing a magnet into a coil, one pole first (Fig. 6.04), holding it motionless inside the coil and then withdrawing it. A microammeter shows that current flows when the magnet is *moving* but not when it is at rest. The current flows in opposite directions when the magnet enters and leaves the coil, i.e. a.c. is generated. The result is the same if the coil is moved towards and away from the stationary magnet.

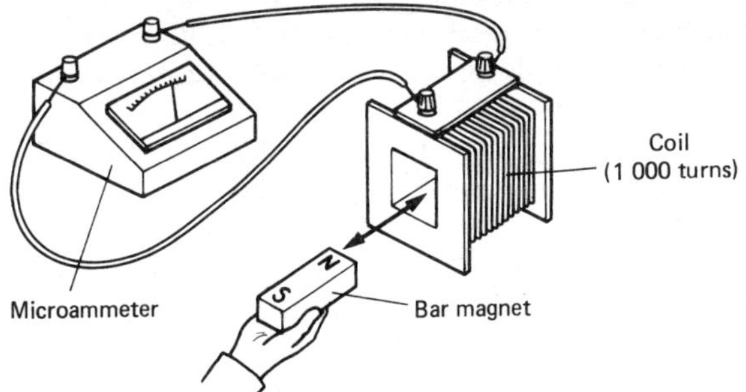

Fig. 6.04

We say that a voltage has been *induced* in the coil because the magnetic field passing through it *changed* as the magnet and coil approached or separated. If the coil is part of a complete circuit, the induced voltage causes an induced current. Experiments show that:

(i) the greater the rate at which the magnetic field changes (i.e. the faster the magnet moves), the greater is the induced voltage—a statement known as *Faraday's law* of electromagnetic induction;
(ii) the induced voltage (and current) always opposes what causes it—a statement called *Lenz's law* of electromagnetic induction (or the law of cussedness)!

Lenz's law is illustrated by Figs. 6.05(*a*) and (*b*). In Fig. 6.05(*a*), the magnet approaches the coil, north pole first. The induced current flows so that the coil behaves like a magnet with a north pole at the top. The downward motion of the magnet is then opposed since two north (like) poles repel. When the magnet is withdrawn, Fig. 6.05(*b*), the current direction reverses and makes the top of the coil a south pole. This attracts the north pole of the magnet and hinders its removal.

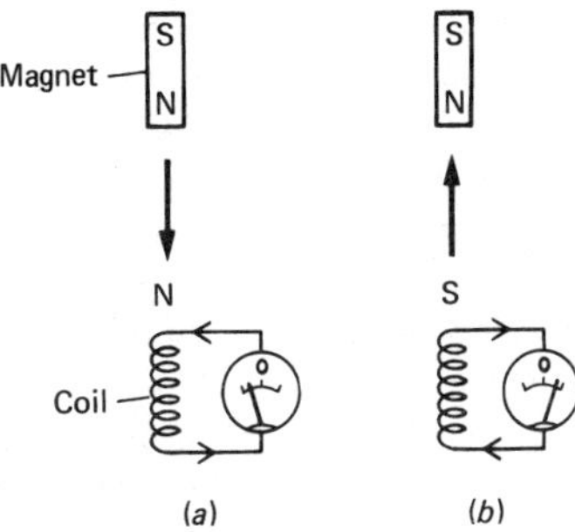

Fig. 6.05

The generation of electricity at power stations and by car alternators and dynamos is by electromagnetic induction.

6.3 How inductors work

Inductors oppose changing currents because of electromagnetic induction.

(a) Direct current

When a direct current increases from zero to its steady value, the magnetic field it produces builds up to its final shape. During the process the field passing through the coil *changes* so inducing a voltage in the coil *itself* which (as Lenz's law tells us) opposes the change that has caused it, i.e. the rising current that is trying to establish the field. This opposition delays the rise of current.

The time constant T of the circuit in Fig. 6.06(*a*) is given by L/R, where L is the inductance of the inductor and R the resistance of the resistor. T is the time for the current I to rise to 0.63 (approximately two-thirds) of its final steady value of V/R, Fig. 6.06(*b*). If L is in henries and R is in ohms, T is in seconds. In $5L/R$ seconds I will have reached its final value which depends on R but not L. The energy stored in the final magnetic field of the inductor can be shown to be $\frac{1}{2}LI^2$ joules if L is in henries and I is in amperes.

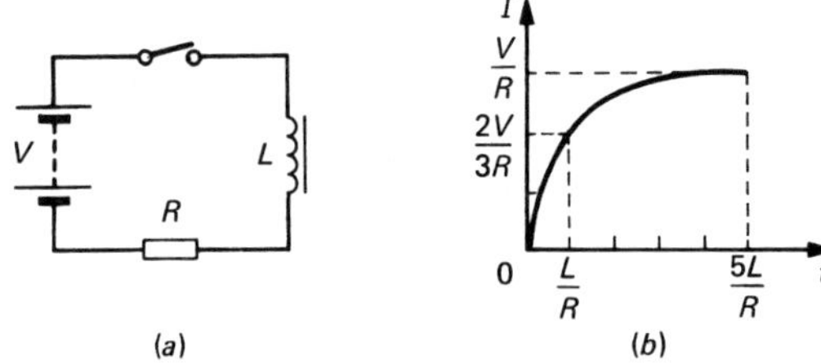

Fig. 6.06

When the current is switched off, the switch contacts open and the resistance R becomes very large, making L/R very small and causing the magnetic field to collapse rapidly. As it does so, the field passing through the coil *changes* and a large voltage is induced which opposes the change producing it, i.e. the collapsing field due to the falling current. It therefore tries to keep the current flowing longer so delaying its fall to zero and producing sparks at the switch contacts. The energy for these comes from that stored in the magnetic field.

(b) Alternating current

Since an alternating current is changing all the time, the magnetic field it produces is also changing continuously. There is therefore always an induced voltage in the coil and permanent opposition to the current. In Fig. 6.02, when the battery is replaced by an a.c. supply the lamp in series with the inductor of inductance L does not light up (unless L is reduced by removing the iron core).

You should note that the induced voltage is caused, as in all cases of electromagnetic induction, by a *change* in the strength of the magnetic field passing through the coil. But the change here arises not as a result of a magnet or a coil moving (as in section 6.2(b)), but from a changing electric current.

The unit of inductance, the *henry* (H), is defined in terms of the induced voltage. It is the inductance of a coil in which a current changing at the rate of one ampere per second induces a voltage of one volt.

6.4 Types of inductor

(a) Iron cores

Materials based on iron are used where a large inductance is required. Iron increases several hundred times the strength of the magnetic field caused by the current in the coil wound on it. Silicon steel and nickel-iron alloys such as *Mumetal* and *Stalloy* are used at audio frequencies (up to 20 kHz). Iron cores are laminated, they consist of flat sheets (stampings) coated thinly on one side by an insulating material. The laminations reduce the conversion of electrical energy to heat by making it difficult for currents in the coil to induce currents in the core. These induced currents are called *eddy currents* which flow in circles through the iron core, Fig. 6.07(*a*). If the laminations are at right angles to the plane of the coil windings, the core offers a large resistance to the eddy currents. E- and I-shaped laminations for a low frequency choke are shown in Fig. 6.07(*b*); such inductors are used as 'smoothing' chokes in mains power supply units, a typical value being 10 H.

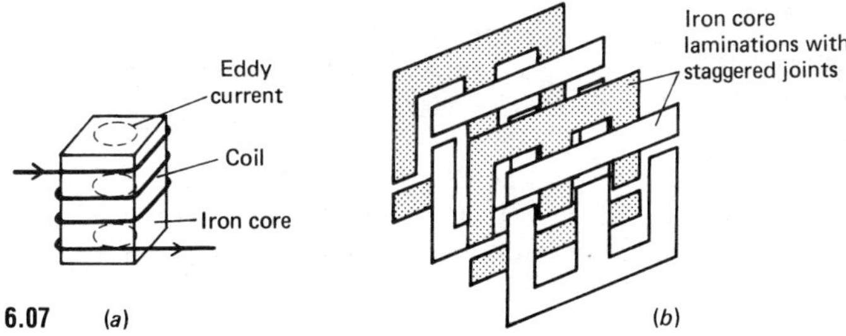

Fig. 6.07 (*a*) (*b*)

(b) Air cores

Inductors with an air core have small inductances and are used for high frequencies, either in radio tuning circuits or as 'r.f. chokes' to stop radio frequency currents (greater than 20 kHz) taking certain paths in a circuit. Coils for use at high frequencies are made of *Litz* wire which consists of several thin copper wires insulated from each other. High frequency currents tend to flow near the surface of a wire (the 'skin effect') and by providing more surface area for a given volume of copper the resistance is reduced. This reduces the loss of electrical energy to heat.

(c) Iron-dust and ferrite cores

Iron-based cores can be used at high frequencies if the material is in the form of a powder which has been coated with an insulator and pressed together. An inductor with an iron-dust core is shown (with the core removed) in Fig. 6.08(*a*) along with its symbol; it is used in radio frequency tuned circuits—see section 12.3.

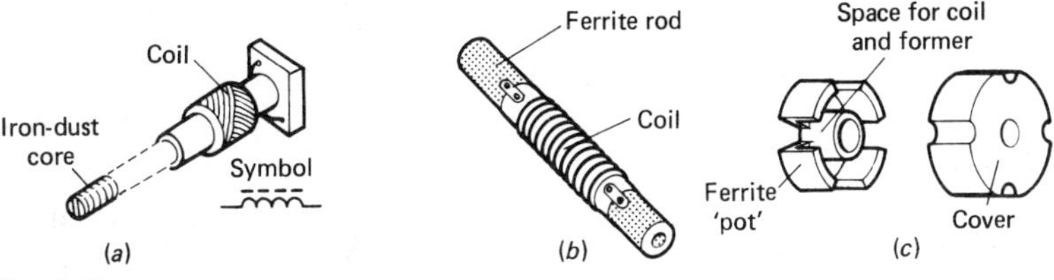

(*a*) (*b*) (*c*)

Fig. 6.08

Ferrite cores may also be used at high frequencies. They consist of ferric oxide combined with other oxides such as nickel oxide. Fig. 6.08(*b*) shows the aerial coil of a radio receiver wound on a ferrite rod as the core. A pot-type ferrite core is shown in Fig. 6.08(*c*).

Iron-dust and ferrite cores increase the inductance of a coil considerably. For example if an air-cored coil has an inductance of 1 mH, a ferrite core could increase it to about 400 mH. They also have a high resistance and so reduce eddy current losses.

6.5 Inductive reactance

(a) Effect on a.c. and d.c.

The opposition of an inductor to a.c. is called its *inductive reactance*, X_L, and increases if either the inductance L increases or the frequency f of the a.c. increases. It can be shown that it is given by:

$$X_L = 2\pi fL$$

X_L is in ohms if f is in hertz and L in henries. For example, if $f = 50$ Hz and $L = 10$ H, then $X_L = 2\pi \times 50 \times 10 = 1\,000\pi = 3.1\ \text{k}\Omega$.

An inductor also has resistance and its total opposition to a.c. is made up of its resistance R and its inductive reactance X_L.

We can now sum up the basic facts about resistors, capacitors and inductors. All are componments which, in their different ways oppose current flow and are used to control it in a circuit.

A resistor allows both a.c. and d.c. to flow through it.

A capacitor allows a.c. to flow 'through' it and its reactance decreases as the frequency of the a.c. increases. With d.c. the flow stops when the capacitor is charged.

An inductor allows a.c. and d.c. to pass but it opposes a.c. (and varying d.c.) more than steady d.c.; its reactance increases as the frequency of the a.c. increases.

(b) Phase shift

The current 'through' a capacitor in an a.c. circuit leads the voltage across it—see section 5.7. In an inductor the current I lags behind the voltage V, Fig. 6.09. The lag is due to the fact that when the current starts to flow, although it is small at first, it is increasing at its fastest rate, therefore so too is the build-up of the magnetic field. As a result the voltage induced in the inductor has its maximum value.

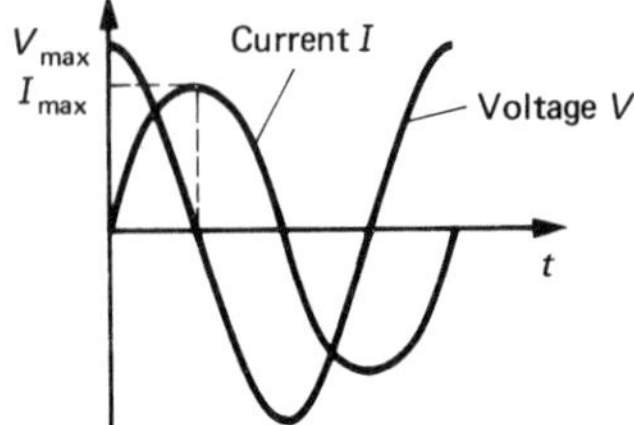

Fig. 6.09

The fact that currents and voltages are not in step in a.c. circuits has some important results which seem surprising. They will be considered in the next section.

6.6 Alternating current series circuit

(a) Impedance

If an a.c. supply of r.m.s. voltage V and frequency f is applied to a series circuit containing resistance R, capacitance C and inductance L, Fig. 6.10(*a*), each component offers some opposition to the current. The total opposition is called the *impedance* (Z) and is measured (like resistance and reactance) in ohms. It can be shown that:

$$Z = \sqrt{R^2 + (X_L - X_C)^2}$$

where $X_L = 2\pi fL$ is the inductive reactance and $X_C = 1/(2\pi fC)$ is the capacitive reactance.

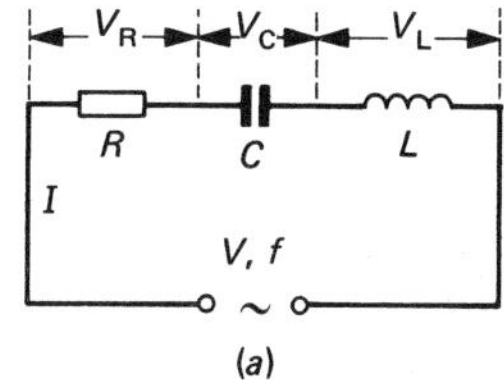

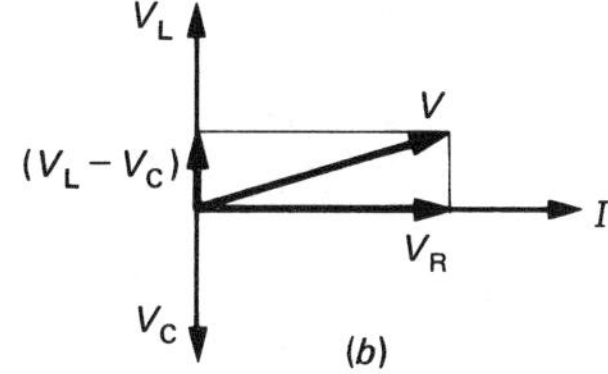

Fig. 6.10

The r.m.s. current I in the circuit is given by:

$$I = \frac{V}{Z}$$

This equation is the same as that connecting current and voltage in a d.c. circuit (i.e. $I = V/R$) with the impedance Z replacing the resistance R; it is used for a.c. circuit calculations.

(b) Worked example

In an a.c. series circuit $R = 100\ \Omega$, $L = 2.0$ H, $C = 10\ \mu$F and the supply has a r.m.s. voltage of 24 V and frequency 50 Hz. Calculate (a) the inductive reactance X_L, (b) the capacitive reactance X_C, (c) the impedance Z, and (d) the r.m.s. current I.

a) $f = 50$ Hz and $L = 2.0$ H.

$$\therefore \quad X_L = 2\pi fL = 2\pi \times 50 \times 2.0 = 200\pi$$

$$= 628\ \Omega$$

b) $f = 50$ Hz and $C = 10\ \mu\text{F} = 10 \times 10^{-6} = 10^{-5}$ F.

$$\therefore \quad X_C = \frac{1}{2\pi fC} = \frac{1}{2\pi \times 50 \times 10^{-5}} = \frac{10^5}{100\pi} = \frac{1\,000}{\pi}$$

$$= 318\ \Omega$$

c) $R = 100\ \Omega$ and $X_L - X_C = 628 - 318 = 310\ \Omega$.

$$\therefore \quad Z = \sqrt{R^2 + (X_L - X_C)^2} = \sqrt{(100^2 + 310^2)} = \sqrt{106\ 100}$$
$$= 330\ \Omega$$

d) $V = 24$ V and $Z = 330\ \Omega$.

$$\therefore \quad I = \frac{V}{Z} = \frac{24}{330} = 0.073 \text{ A} \quad (73 \text{ mA})$$

(c) Voltages

In a d.c. series circuit the voltages 'add up'. This is not so for a.c.

For example, consider the r.m.s. voltages in the a.c. series circuit of the worked example in section 6.6(b) above. The same r.m.s. current I passes through each component and the r.m.s. voltage across each one equals the current multiplied by the 'opposition' of the component. Therefore:

$$V_L = I \times X_L = 0.073 \times 630 = 46 \text{ V}$$
$$V_C = I \times X_C = 0.073 \times 320 = 23 \text{ V}$$
$$V_R = I \times R = 0.073 \times 100 = 7.3 \text{ V}$$

These are the voltages which would be recorded by an a.c. voltmeter connected across each component in turn. The surprising fact is that $V_L + V_C + V_R$ (= 76 V) is more than three times the r.m.s. voltage of 24 V applied to the whole circuit (and which an a.c. voltmeter connected across the whole circuit would read).

This apparent contradiction is due to the fact that whilst V_R and I are in phase, V_L leads I by 90° and V_C lags behind I by 90°. V_R, V_L and V_C are therefore acting as it were in different 'directions' with the phase shift between V_L and V_C being 180°, i.e. when one has its maximum positive value the other has its maximum negative value. The voltages cannot be added by ordinary addition but this can be done by applying the parallelogram law to the diagram (called a *phasor* diagram) in Fig. 6.10(*b*) in which V_L is assumed to be greater than V_C so that their resultant ($V_L - V_C$) is in the direction of V_L. The equation connecting them and the supply voltage V is:

$$V^2 = V_R^2 + (V_L - V_C)^2$$

(d) Power

In an a.c. circuit, power is only dissipated (i.e. electrical energy changed to heat) in parts having resistance. In a perfect capacitor (no leakage current) no power is used because the energy taken from the supply to charge the capacitor on the first quarter-cycle of current is returned to the supply on the second quarter-cycle when it discharges.

In a pure inductor (no resistance) the energy obtained during the first quarter-cycle of current supply and which is stored in the magnetic field round the inductor, is returned to the supply when the field collapses during the second quarter-cycle. In practice an inductor has some resistance and energy is 'lost' on this account.

6.7 About transformers

(a) Action

A transformer changes (transforms) an alternating voltage from one value to another. It consists of two coils, called the *primary* and *secondary* windings, which are not connected electrically. The windings are either one on top of the other or are side by side on an iron, iron-dust or air core. Transformer symbols are given in Fig. 6.11.

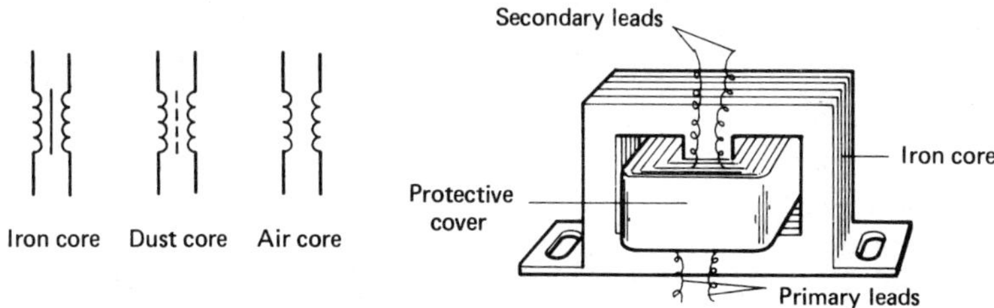

Fig. 6.11 **Fig. 6.12**

A transformer works by electromagnetic induction: a.c. is supplied to the primary and produces a *changing* magnetic field which *passes through* the secondary, thereby inducing a changing (alternating) voltage in the secondary. It is important that as much as possible of the magnetic field produced by the primary passes through the secondary. A practical arrangement designed to achieve this in an iron-cored transformer is shown in Fig. 6.12 in which the secondary is wound on top of the primary. You should also notice that the induced voltage in the secondary is always of opposite polarity to the primary voltage (Lenz's law), as shown in Fig. 6.13.

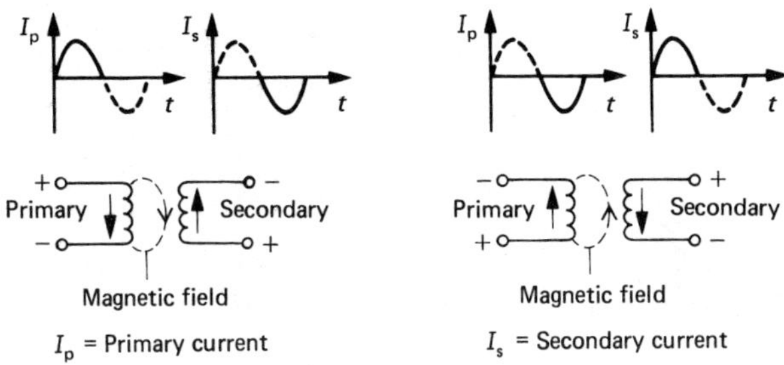

Fig. 6.13

Too large a current in a transformer causes magnetic saturation of the core, i.e. the magnetization of the core is a maximum and it is no longer able to follow changes of magnetizing current. Particular care is required when there is a d.c. component.

(b) Equations

It can be shown that if a transformer is 100% efficient at transferring electrical energy from primary to secondary (and many are nearly so), then:

$$\frac{\text{secondary a.c. voltage}}{\text{primary a.c. voltage}} = \frac{\text{turns on secondary}}{\text{turns on primary}}$$

In symbols

$$\frac{V_s}{V_p} = \frac{n_s}{n_p}$$

For example, if n_s is twice n_p, the transformer is a *step-up* one, and V_s will be twice V_p. In a *step-down* transformer there are fewer turns on the secondary than on the primary and V_s is less than V_p. The ratio n_s/n_p is called the *turns ratio*.

A transformer with two secondaries is represented in Fig. 6.14; it both steps-up and steps-down the primary voltage. The step-down secondary voltage is 6 V a.c. What is the step-up one?

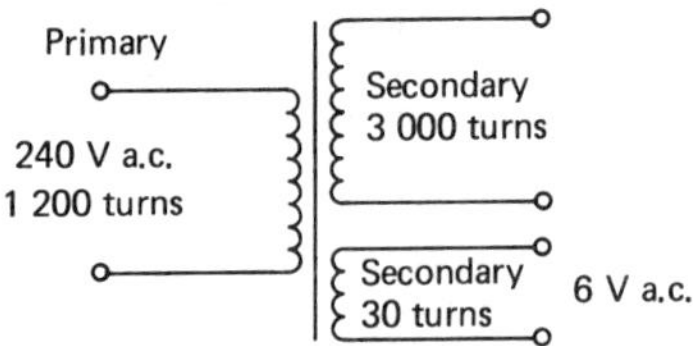

Fig. 6.14

If the voltage is stepped-up by a transformer the current is stepped-down in proportion and vice versa. This must be so if we assume that all the electrical energy given to the primary appears in the secondary. We can then say:

$$\text{power in primary} = \text{power in secondary}$$

or

$$V_p \times I_p = V_s \times I_s$$

where I_p and I_s are the primary and secondary currents respectively.

Therefore in a perfect transformer, if V_s is double V_p, I_s is half I_p. In practice I_s is less than half I_p because of small energy losses in the transformer arising from

(i) the *resistance of the windings* of copper wire, causing the current in them to produce heat,

(ii) *eddy currents* in iron and iron-dust cored transformers, (reduced by laminations), and

(iii) the *magnetic field* of the primary not passing entirely through the secondary.

6.8 Types of transformer

(a) Mains

Mains transformers—one is shown in Fig. 6.15(*a*)—are used at a.c. mains frequency (50 Hz in Britain), their primary coil being connected to the 230 V a.c. supply. Their secondary windings may be step-up or step-down or they may have one or more of each. They have laminated iron cores and are used in power supply units. Sometimes the secondary has a centre-tap—see section 21.2.

Step-down *toroidal* types, Fig. 6.15(*b*), are becoming popular. They have virtually no external magnetic field and a screen between primary and secondary windings gives safety and electrostatic screening. Their pin connections are brought out to a 0.1 inch grid which makes them ideal for printed circuit board (p.c.b.) mounting.

Isolating transformers have a one-to-one turns ratio (i.e. $n_s/n_p = 1/1$) and are safety devices for separating a piece of equipment from the mains supply. They do not change the voltage.

(*a*) (*b*) (*c*) (*d*)

Fig. 6.15

(b) Audio frequency

Audio frequency transformers, as illustrated in Fig. 6.15(*c*), also have laminated iron cores and are used as output matching transformers to ensure the maximum transfer of power from the a.f. output stage to the loudspeaker in, for example, a radio set or amplifier.

(c) Radio frequency

Radio frequency transformers, as in Fig. 6.15(*d*), usually have adjustable iron-dust cores and form part of the tuning circuits in a radio. They are enclosed in a small aluminium 'screening' can to stop them radiating energy to other parts of the circuit.

6.9 Revision questions

1. What does an inductor do?

2. Name two types of inductor and draw their symbols.

3. What is the unit of inductance? State two submultiples of it.

4. State two ways of increasing the inductance of a coil.

5. Supply the missing words in the following statements:
   a) A wire carrying a current produces a . . . . in the space around it.
   b) A coil carrying a steady current behaves as if it were a . . . .
   c) The inductance of a coil is caused by the current in it . . . . and producing a . . . . magnetic field which induces an opposing voltage in the coil itself.

6. What property do all inductors have as well as inductance?

7. Why do some inductors have a core which is
   a) made of soft iron,
   b) laminated?

8. At a certain frequency of a.c. a resistor, a capacitor and an inductor each offer the same 'opposition' to the a.c. How does the 'opposition' of each change (if it does) when the a.c. frequency increases?

9. a) Write down an equation for the impedance Z of a series circuit containing resistance R and reactances X_L and X_C.
   b) Give an equation for the r.m.s. current I in a circuit of impedance Z if the supply voltage has an r.m.s. value of V.
   c) Why do voltages not 'add up' by ordinary addition in a.c. series circuits?
   d) Why are pure inductors and perfect capacitors called 'wattless' components?

10. What does a transformer do?

11. Name and draw the symbols for three kinds of transformer.

12. In what two ways does the secondary voltage in a step-up transformer differ from the primary voltage?

13. If V_P and V_S are the primary and secondary voltages and n_P and n_S the number of turns on the primary and secondary windings respectively of a transformer, write down the equation connecting them.

14. a) State three ways in which electrical energy can be 'lost' in a transformer.
    b) What is done to minimize the loss in each case?

6.10 Problems

1. What is the inductive reactance of
 a) a 15 H smoothing choke at a frequency of 100 Hz,
 b) a 1 mH radio frequency choke at a frequency of 1 MHz?

2. Calculate the inductance of a coil having inductive reactance of 200 Ω at 1 kHz.

3. a) Calculate the impedance of a series circuit in which $R = 12\ \Omega$, $X_L = 96\ \Omega$ and $X_C = 80\ \Omega$.
 b) What is the r.m.s. current in the circuit if the r.m.s. value of the supply voltage is 10 V?
 c) Calculate the voltages V_R, V_L and V_C across each component.
 d) Check your answers to (c) by proving that:

$$\sqrt{V_R^2 + (V_L - V_C)^2} = 10\ \text{V}$$

4. A 12 V lamp is operated from a 240 V a.c. mains step-down transformer.
 a) What is the turns ratio of the windings?
 b) How many turns are on the primary if the secondary has 80 turns?
 c) What is the current in the primary coil if current through the lamp is 2 A?

5. An audio frequency output transformer in the output stage of a radio set has a step down turns ratio of 15/1. What current flows in the secondary when the primary current is 40 mA?

UNIT 7

Semiconductor diodes

7.1 About diodes

A diode allows current to flow easily in one direction but not in the other, i.e. its resistance is low in the conducting or 'forward' direction but very high in the opposing or 'reverse' direction. Most semiconductor diodes are made from silicon or germanium.

A diode has two leads, the *anode* and the *cathode*; its symbol is given in Fig. 7.01(*a*). The cathode is often marked by a band at one end, Fig. 7.01(*b*); it is the lead by which conventional current leaves the diode when forward biased—as the arrow on the symbol shows. In some cases the arrow is marked on the diode, Fig. 7.01(*c*), or the shape is different, Figs. 7.01(*d*), (*e*).

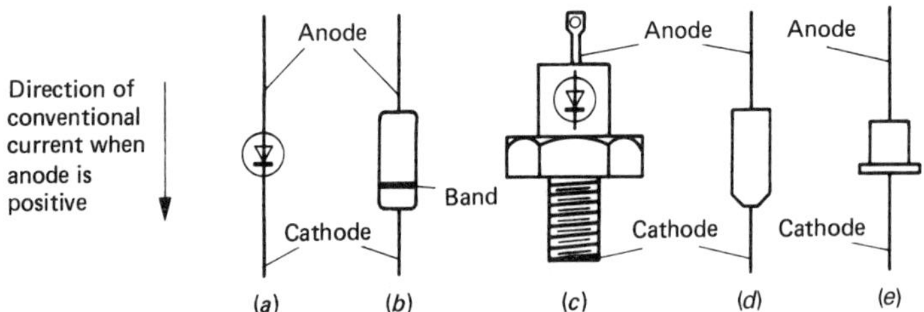

Fig. 7.01

There are several kinds of diode, each with features that suit it for a particular job. Three of the main types are:

(i) the junction diode,
(ii) the point-contact diode, and
(iii) the Zener diode.

Two identification codes are used for diodes. In the American system the code always starts with 1N and is followed by a serial number, e.g. 1N4001. In the Continental system the first letter gives the semiconductor material (A = germanium, B = silicon) and the second letter gives the use (A = signal diode, Y = rectifier diode, Z = Zener diode); for example, AA119 is a germanium signal diode. To complicate the situation some manufacturers have their own codes.

7.2 Intrinsic semiconductors

To understand how a diode works you need to know about semiconductors.

(a) Valence electrons

Many of the properties of materials can be explained if we assume that the electrons surrounding the nucleus of an atom are arranged in 'shells'. Those in the outermost shell are known as valence electrons because the chemical combining power or valency (as well as many physical properties) of the atom depends on them. The valence electrons form 'bonds' with the valence electrons of neighbouring atoms to produce in the case of most solids, a regularly repeating three dimensional pattern of atoms called a crystal lattice.

An atom of a semiconductor such as silicon or germanium has four valence electrons, Fig. 7.02(*a*), and in the lattice each one is *shared* with a nearby atom to form four *covalent bonds*, Fig. 7.02(*b*). Every atom thus has a 'half-share' in eight valence electrons and it so happens that this number of valence electrons gives a very stable arrangement (most of the inert gases have it). A strong crystal lattice results in which it is difficult for electrons to escape from their atoms. Pure silicon and germanium are therefore very good insulators, being perfect at near absolute zero (−273 °C).

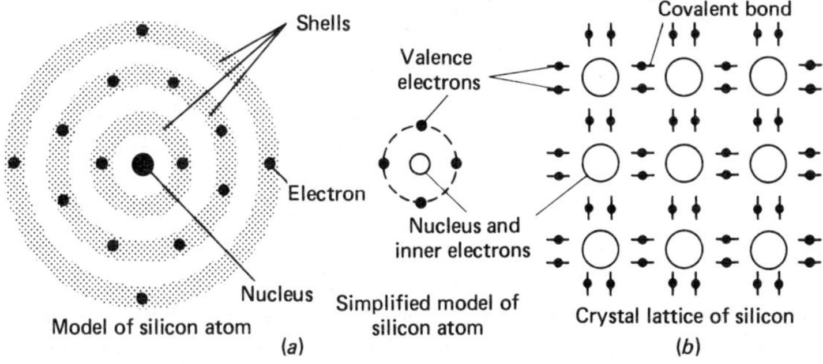

Fig. 7.02

(b) Electrons and holes

At ordinary temperatures the atoms are vibrating sufficiently in the lattice for a few bonds to break, setting free some valence electrons. When this happens, a deficit of negative charge, called a vacancy or *hole* is left behind by each free electron in the outermost shell of the atom from which it came, Fig. 7.03(*a*). The atom, now short of an electron, becomes a positive *ion*. We think of the hole as a positive charge equal in size to the negative charge on an electron.

Free electrons are attached to holes (because of their opposite charges) and if one should 'drop into' a hole and fill it, the ion concerned becomes a neutral atom again.

(c) Intrinsic conduction

When a battery is connected across a pure semiconductor (i.e. one with a perfect crystal lattice, it attracts free electrons to the positive terminal (which is short of these) and provides a supply of more free electrons at the negative terminal (where there is a surplus). The free electrons (from broken bonds) travel through the semiconductor by 'hopping' from one hole to another nearer the positive terminal, making it *seem* that positive holes are moving to the negative terminal. The current in a *pure* semiconductor is very small and can be *thought of* as streams of free electrons and holes going in opposite directions as in Fig. 7.03(*b*). It is called *intrinsic* conduction because the charge carriers come from inside the material.

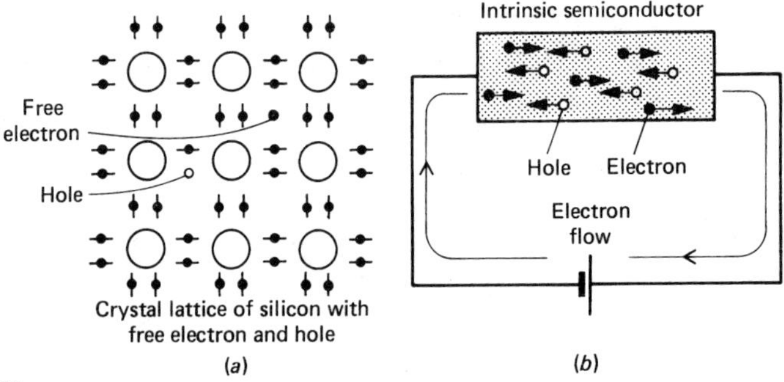

Fig. 7.03

The behaviour of electrons and holes in a semiconductor can be likened to a row of chairs, all occupied except one at the end. If everyone moves one place towards the vacant seat, the vacancy appears at the other end. The vacancy (hole) *seems* to have moved along the row but the motion has really been that of the occupants (electrons) of the chairs in the opposite direction. The 'hole' idea may be strange to you at first but it makes semiconductors easier to understand.

If the temperature of a semiconductor is raised, more bonds break and the intrinsic conductivity increases because more free electrons and holes are produced. The resistance of a semiconductor therefore decreases as the temperature rises; in a metallic conductor the opposite is true.

7.3 Extrinsic semiconductors

The use of semiconductors in devices such as diodes, transistors and integrated circuits, depends on increasing the conductivity of very pure (i.e. intrinsic) semiconducting materials (of impurity concentration less than 1 part in 10^{10}) by adding to them tiny but controlled amounts of certain 'impurities'. The process is called *doping* and the material obtained is known as an extrinsic semiconductor because the impurity supplies the charge carriers (which are present in addition to the intrinsic ones). The impurity atoms must have about the same size as the semiconductor atoms so that they fit into its crystal lattice without causing distortion. There are two kinds of extrinsic semiconductor.

(a) n-type

This type is made by doping silicon or germanium with, for example, phosphorus. A phosphorus atom has five valence electrons (i.e. is pentavalent) and Fig. 7.04(*a*) shows what occurs when one is introduced into the lattice of a silicon crystal. Four of its valence electrons form covalent bonds with four neighbouring silicon atoms but the fifth is spare and, being loosely held, it can take part in conduction. The impurity (phosphorus) atom is called a *donor* because it gives (donates) an electron for conduction.

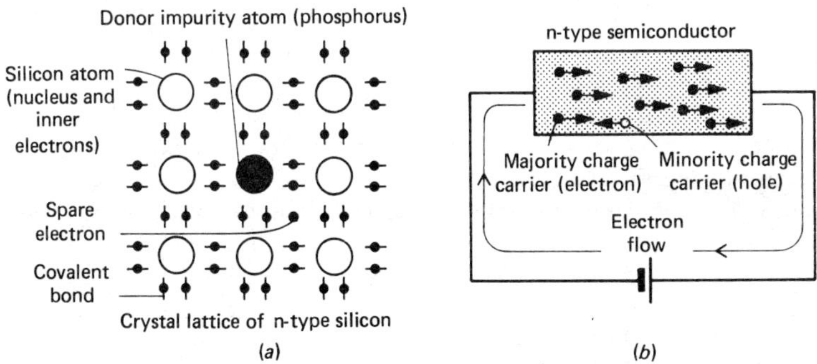

Fig. 7.04

The 'impure' silicon is an *n*-type semiconductor since the *majority* charge carriers are *n*egative electrons. (However, note that the overall charge in the crystal remains zero because every atom present is electrically neutral.) Impurity atoms are added in the amount that produces the required increase of conductivity.

A few positive holes are also present in n-type material (as intrinsic charge carriers formed by broken covalent bonds between silicon atoms); they are *minority* carriers. Fig. 7.04(*b*) shows conduction in an n-type semiconductor.

(b) p-type

To make this type, silicon or germanium is doped with, for example, boron. A boron atom has three valence electrons (i.e. is trivalent) and Fig. 7.05(*a*) shows what happens when one is introduced into the lattice of a silicon crystal. Its three valence electrons each share an electron with three of the four silicon atoms surrounding it. One bond is incomplete and the position of the missing electron, i.e. the hole, behaves like a positive charge since it can attract an electron from a nearby silicon atom, so forming another hole. For that reason the impurity (boron) atom is called an *acceptor*.

The 'impure' silicon is a *p*-type semiconductor since the *majority* charge carriers causing conduction are *p*ositive holes. (Note again that the semiconductor as a whole is electrically neutral.) In p-type material a few electrons are present as *minority* carriers; Fig. 7.05(*b*) shows conduction in such material.

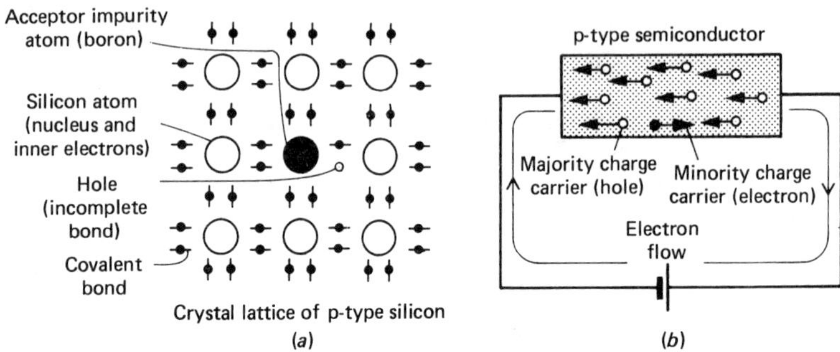

Fig. 7.05

In both n- and p-types of semiconductor a temperature rise increases the proportion of minority carriers (due to increased intrinsic conduction) and upsets the semiconductor device because its action depends on its extrinsic (majority) charge carriers.

7.4 The p-n junction

The operation of many semiconductor devices depends on effects which occur at the boundary (junction) between p- and n-type materials formed in the same continuous crystal lattice.

(a) Unbiased p-n junction

A p-n junction is represented in Fig. 7.06(*a*). As soon as the junction is produced, free electrons near the junction in the n-type material move (by diffusion) across the junction into the p-type material where they fill holes. (Diffusion occurs because the concentration of electrons in n-type material is large and in p-type material it is small. The electrons behave like scent molecules when a scent bottle is opened—they diffuse from the vapour in the bottle where their concentration is high into the air where it is initially zero.)

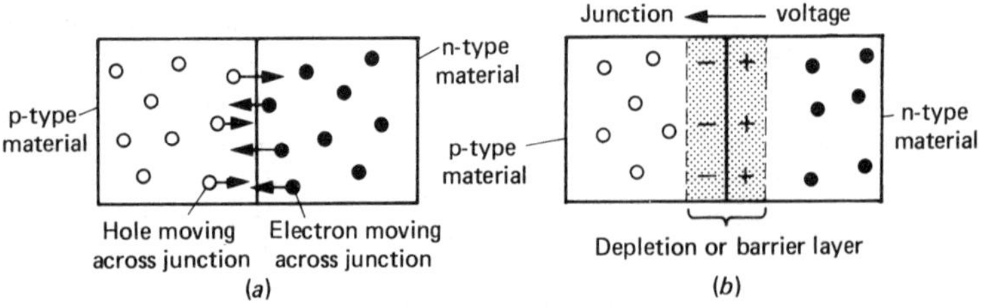

Fig. 7.06

As a result the n-type material near the junction becomes positively charged and the p-type material negatively charged (both previously being neutral). At the same time, and for a similar reason, holes diffuse from p-type to n-type, capturing electrons there. The exchange of charge soon stops because the negative charge on the p-type material opposes the further flow of electrons and the positive charge on the n-type opposes the further flow of holes. The region on either side of the junction becomes fairly free of majority charge carriers, as in Fig. 7.06(*b*), and is called the *depletion* (or *barrier*) *layer*; it is less than one-millionth of a metre wide and is, in effect, an insulator.

The situation is just as if there was a battery across the junction with a small voltage (about 0.1 V for germanium and 0.6 V for silicon) called the *junction voltage*, acting from n- to p-type.

(b) Reverse biased p-n junction

If a battery is connected across a p-n junction with its positive terminal joined to the n-type side and its negative terminal to the p-type side, it helps the junction voltage. Electrons and holes are repelled farther from the junction and the depletion layer widens, Fig. 7.07(*a*). Only a few minority carriers (produced by bonds breaking at ordinary temperatures in both p- and n-type materials) cross the junction and a tiny current, called the *leakage* or *reverse* current, flows. The resistance of the junction is very high.

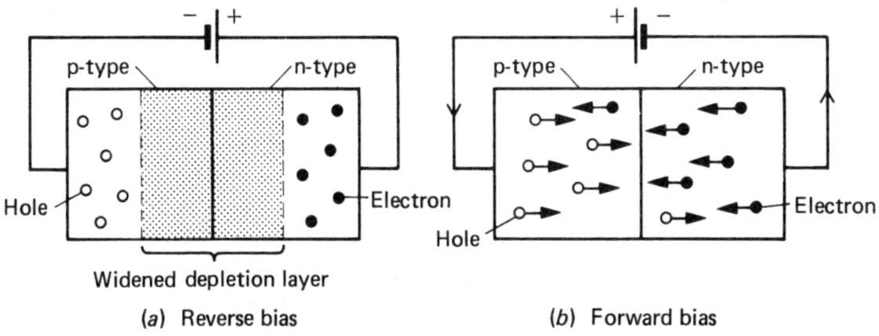

Fig. 7.07

(c) Forward biased p-n junction

If a battery is connected so as to oppose the junction voltage, the depletion layer narrows. When the battery voltage exceeds the junction voltage, appreciable current flows because majority carriers are able to cross the junction, electrons from the n- to the p-side and holes in the opposite direction, Fig. 7.07(*b*). The junction is then forward biased, i.e. the *p-type* side is connected to the *positive* terminal of the battery and the *n-type* side to the *negative* terminal. A small leakage current flows, because of minority carriers, and in this case adds to the main majority carrier flow. The resistance of this junction is very low.

7.5 Junction diode

(a) Construction

A junction diode consists of a p-n junction with one connection to the p-side (the anode A) and another to the n-side (the cathode K).

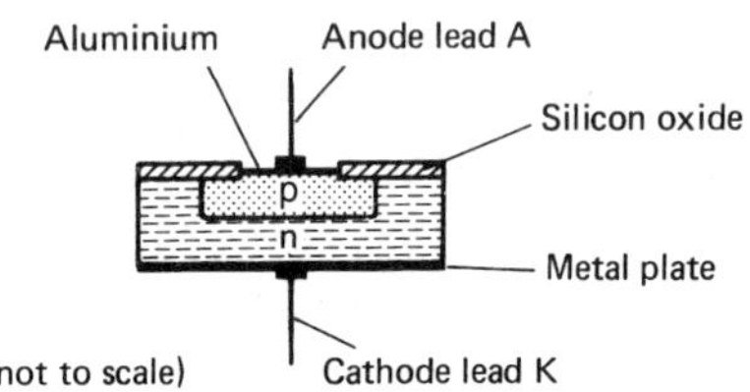

Fig. 7.08

A simplified section of a discrete-component silicon diode with 'planar-type' construction is shown in Fig. 7.08. (In practice the boundaries are neither straight nor well defined.) The manufacturing process involves first soldering a thin slice of n-type silicon to a metal base from which the cathode lead K is taken. A film of highly insulating silicon oxide is then formed on the surface of the slice by heating it in steam at about 1 100 °C. A 'window' is next etched chemically in the oxide film and vapour of the appropriate impurity allowed to diffuse through it so converting the top of the slice into p-type silicon. Aluminium is evaporated on to the p-type region to allow the anode lead A to be soldered. The diode is sealed in a case to exclude moisture and light.

Junction diodes are common in integrated circuits; their construction is similar to that of discrete components but on a much smaller scale—see section 13.1.

(b) Characteristics

Typical characteristic curves (see section 2.8) for silicon and germanium diodes at 25 °C are shown in Fig. 7.09. You can see from them that the foward current I_F is small until the forward voltage V_F is about 0.6 V for silicon and about 0.1 V for germanium; thereafter a very small change in V_F causes a large increase in I_F.

The reverse currents I_R are negligible (note the different scales on the negative axes of the graph) and remain so as the reverse voltage V_R is increased. However, if V_R is increased sufficiently, the insulation of the depletion layer breaks down and I_R increases suddenly and rapidly and permanent damage occurs to the diode. This breakdown voltage can have any value from a few volts up to 1 000 V for silicon and 100 V for germanium depending on the construction of the diode and the level of doping.

Two important electrical properties of a diode are the *average forward current* I_F(av) and the *maximum reverse voltage* V_{RRM} (previously called the 'peak inverse voltage'). For the 1N4001, I_F(av) = 1 A (giving $V_F \approx 1$ V) and V_{RRM} = 50 V. Under normal conditions these values should not be exceeded.

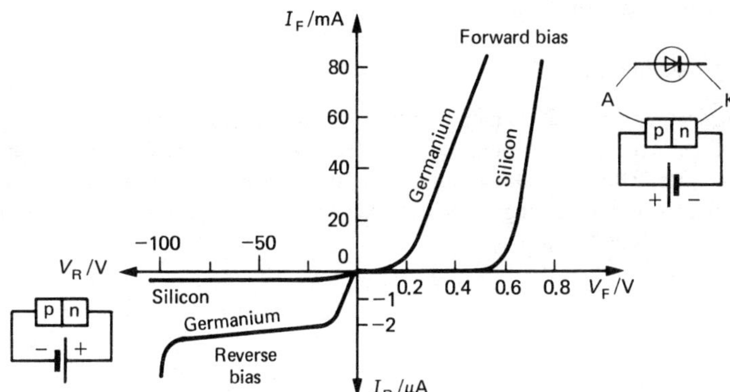

Fig. 7.09

(c) Uses

Junction diodes are used as *rectifiers* to change a.c. to d.c. in power supplies; the larger their junction areas the larger the forward current they pass. Silicon is preferred to germanium because its much lower reverse current makes it a more efficient rectifier (i.e. provides a more complete conversion of a.c. to d.c.). Silicon also has a higher breakdown voltage and can work at higher temperatures.

Junction diodes are also used to prevent damage to circuits by a reversed power supply. In Fig. 7.10(*a*) the battery is correctly connected and forward biases the diode. In Fig. 7.10(*b*) it is incorrectly connected but since it reverse biases the diode, no current passes and no damage occurs to the circuit.

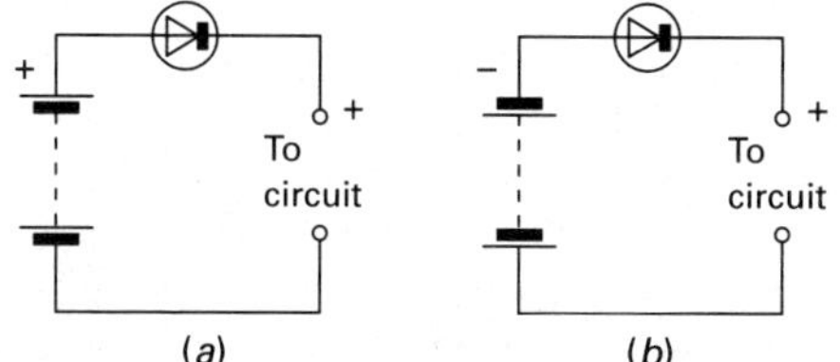

Fig. 7.10

A third use is as *clamp diodes* or *d.c. restorers* to prevent d.c. level shift problems in capacitor coupling circuits like that in Fig. 5.16(*a*) (p. 55) where a positive d.c. input, Fig. 5.16(*b*), becomes an a.c. output, Fig. 5.16(*c*) and (*d*). In the circuit of Fig. 7.11 diode D offers a fast discharge path for *C* and clamps the output to 0 V when the input is zero or negative, thereby stopping a d.c. level shift.

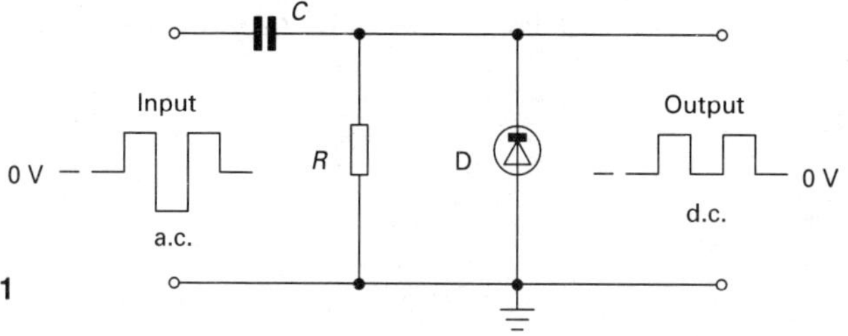

Fig. 7.11

7.6 Point-contact diode

(a) Construction

The construction of a germanium point-contact diode is shown in Fig. 7.12. The tip of a gold or tungsten wire (called a 'cat's whisker') presses on a pellet of n-type germanium. During manufacture a brief current is passed through the diode and produces a tiny p-type region in the pellet around the tip, so forming a p-n junction of very small area.

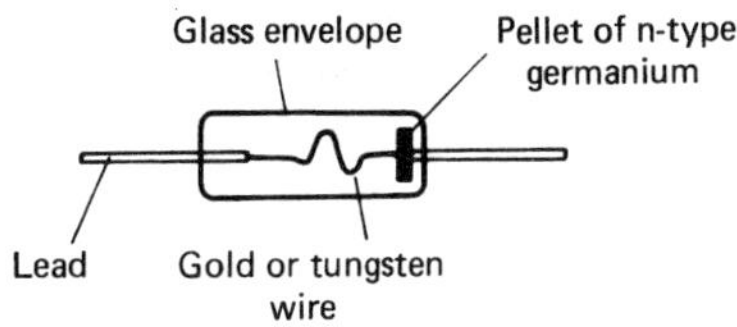

Fig. 7.12

(b) Use

Point-contact diodes are used as *signal diodes* to detect radio signals (a process similar to rectification in which radio frequency a.c. is converted to d.c.) because of their very low capacitance. When reverse biased, the depletion layer in a junction diode acts as an insulator sandwiched between two conducting 'plates' (the p- and n-regions). It therefore behaves as a capacitor and the larger its junction area and the thinner the depletion layer, the greater is its capacitance.

A capacitor 'passes' a.c. and the higher the frequency of the a.c. and the greater the capacitance, the less opposition it offers ($X_C = 1/(2\pi fC)$). At radio frequencies, therefore, a normal junction diode would not be a very efficient detector (rectifier) because of the comparatively large junction area: its opposition in the reverse direction would not be large enough. A point-contact diode on the other hand is more suited to high frequency signal detection because of its tiny junction area.

Germanium is used for signal diodes since it has a lower 'turn-on' voltage than silicon (about 0.1 V compared with 0.6 V) and so lower signal voltages start it conducting in the forward direction.

For the OA91 point-contact diode, for example, I_F(av) = 20 mA (giving $V_F \approx 1$ V) and V_{RRM} = 100 V.

7.7 Zener diode

In an ordinary junction diode if the reverse bias is increased until the depletion layer breaks down, the diode suffers permanent damage. A Zener diode is made to be used in the breakdown region so long as a resistor limits the current. It looks like a rectifier diode, the cathode often being marked by a band, Fig. 7.13(*a*); its symbol is given in Fig. 7.13(*b*).

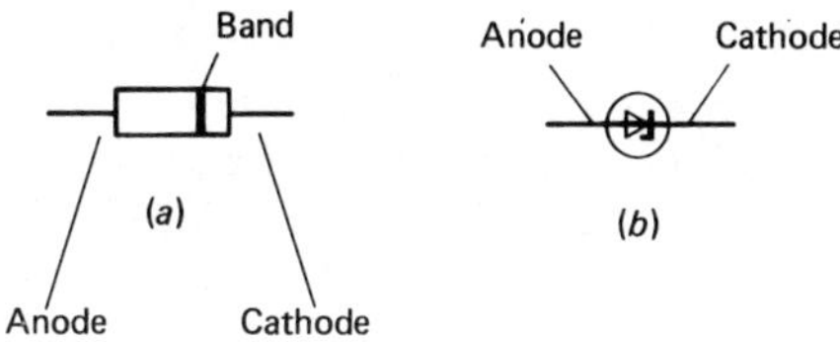

Fig. 7.13

A characteristic is shown in Fig. 7.14. You can see from it that as the reverse voltage V_F is increased, the reverse current is negligible until V_R reaches the breakdown voltage V_Z (here 5.1 V). I_R then increases suddenly. V_Z is called the *Zener* or *reference* voltage. If V_R falls below V_Z, I_R becomes negligible again.

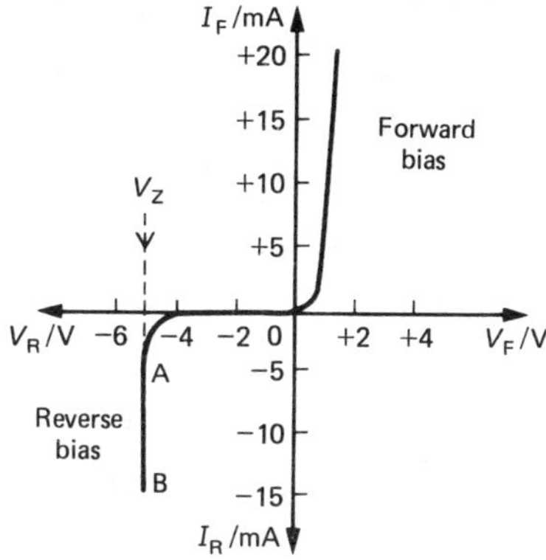

Fig. 7.14

The important thing is that the voltage across the diode remains almost constant at V_Z over a wide range of reverse currents, i.e. that part AB of the characteristic is nearly at right angles to the V_R axis. It is this property of a Zener diode which makes it useful in stabilized power supplies where it keeps the voltage output steady—as we shall see in section 21.4. To sum up, a Zener diode is used in *reverse bias* at its *breakdown voltage*.

To limit the reverse current at breakdown and prevent overheating, the power rating of the diode should not be exceeded. For example, a 500 mW (0.5 W) Zener diode with a breakdown voltage of 5.1 V can pass a maximum current I_{max} given by:

$$I_{max} = \frac{\text{power}}{\text{voltage}} \quad \text{(since power = voltage × current)}$$

$$= \frac{0.5\ \text{W}}{5.1\ \text{V}} = 0.1\ \text{A} = 100\ \text{mA}$$

If the diode is being used, for example, to supply a constant voltage of 5.1 V from a 9 V dry battery (whose voltage falls with use) as in Fig. 7.15, the maximum voltage V to be dropped across R is $(9 - 5.1) = 3.9$ V. The value of R is therefore given by:

$$R = \frac{V}{I_{max}} = \frac{3.9\ \text{V}}{0.1\ \text{A}} = 39\ \Omega$$

Zener diodes with specified Zener voltages are made in a range of values (e.g. 3.0 V, 3.9 V, 5.1 V, 6.2 V, 9.1 V, 10 V, 15 V, up to 200 V). Programmable Zener diodes are also available in which the reference voltage can be set to any desired value (e.g. between 3 and 30 V) by applying another voltage to a third lead on the diode.

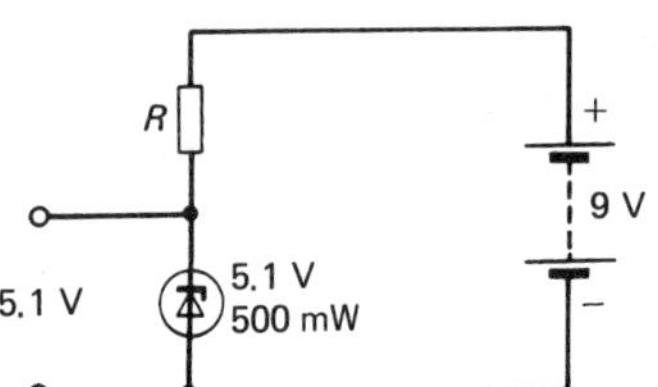

Fig. 7.15

7.8 Other diodes

There are several other special kinds of semiconductor diode, some of which we will consider later.

(i) Photodiode See section 9.5.

(ii) Light-emitting diode or LED See section 9.6.

(iii) Solar cell See section 9.4.

(iv) Varicap (varactor) diode When reverse biased, a junction diode behaves as a capacitor because its depletion layer acts as an insulator sandwiched between two conductors (the p- and n-type regions). The greater the area of the junction and the thinner the depletion layer the greater is the capacitance. Most diodes are made to have minimum capacitance but a varicap diode is designed to have a certain range of capacitance, e.g. 2–10 pF, the value being changed by varying the reverse voltage and therefore the width of the depletion layer.

Such diodes are used to tune TV and v.h.f. radio sets in special circuits which allow the set to lock automatically to the desired station. The process is called automatic frequency control (a.f.c.). A varicap diode is shown in Fig. 7.16 with its symbol.

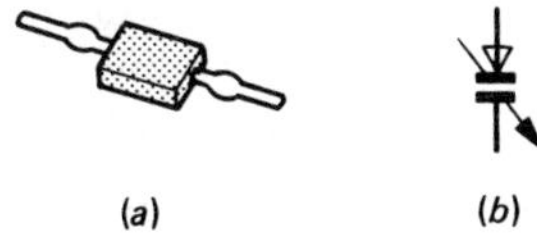

(a) (b)

Fig. 7.16

(v) Gunn diode This is made from n-type gallium arsenide sandwiched between metal electrodes and is used in microwave oscillators.

7.9 Revision questions

1. a) Name two semiconductor materials.
b) What is an intrinsic semiconductor?
c) What is an extrinsic semiconductor?
d) What are the majority charge carriers in
(i) n-type,
(ii) p-type, extrinsic semiconductors?
e) What are the minority charge carriers in
(i) n-type,
(ii) p-type, extrinsic semiconductors?
f) How is the number of
(i) majority,
(ii) minority, charge carriers affected by an increase in temperature of an extrinsic semiconductor?

2. a) Sketch an unbiased p-n junction and mark
(i) the depletion layer,
(ii) the majority charge carriers in each region, and
(iii) the direction of the junction voltage.
b) Sketch a reverse biased p-n junction and show
(i) the effect on the depletion layer,
(ii) the direction of the minority charge carriers, and
(iii) the direction of electron flow in the wires connecting the battery to the junction.
c) Sketch a forward biased p-n junction and show
(i) the effect on the depletion layer,
(ii) the directions of the majority and minority charge carriers, and
(iii) the direction of electron flow in the wires connecting the battery to the junction.

3. a) Draw on the same axes, choosing suitable scales, typical characteristics for silicon and germanium diodes.
   b) Why are the average forward current I_F(av) and the maximum reverse voltage V_{RRM} two important properties of a semiconductor diode?

4. a) State three reasons why silicon is used rather than germanium to make rectifier diodes for power supplies.
   b) Give two reasons why a point-contact germanium diode is more suitable than a silicon junction diode as a signal diode to detect radio signals.

5. a) State one use of a Zener diode.
   b) To what does a Zener diode owe its usefulness?
   c) At what voltage and bias conditions should a Zener diode be used?
   d) Why should a suitable resistor be connected in series with a Zener diode?

7.10 Problems

1. In the circuits of Figs. 7.17(*a*)–(*c*), say which of the 6 V 60 mA lamps are bright, dim or off.

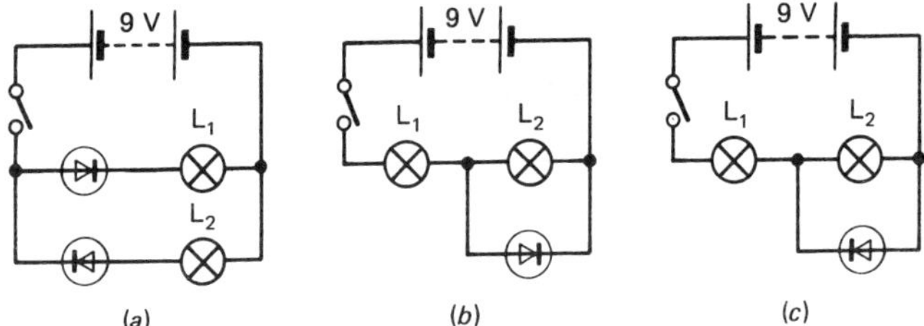

Fig. 7.17

2. The circuits in Fig. 7.18 show how a diode can be used to protect
   a) a circuit against wrong polarity connection of a d.c. supply,
   b) a delicate moving-coil meter, M, against a large voltage.

   How does each work?

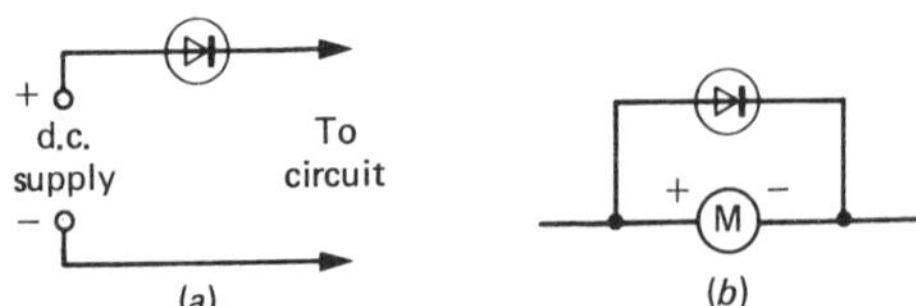

Fig. 7.18

3. From the data given below for a semiconductor diode, plot its characteristic.

Forward voltage (V)	0	0.2	0.4	0.6	0.8
Forward current (mA)	0	0	0.02	1.0	100

a) Is it a silicon or a germanium diode?
b) What value and power of resistor should be connected in series with it to limit the forward current to 100 mA on a 3.0 V supply?
c) If its maximum reverse voltage V_{RRM} = 50 V, what does this mean?

4. In the circuit of Fig. 7.19 a 6 V 1 W Zener diode D in parallel with a 6 V 60 mA lamp L is connected in reverse bias to a battery via a protective resistor *R*.

Explain how the lamp behaves as the battery voltage is increased gradually from 3 V to 9 V.

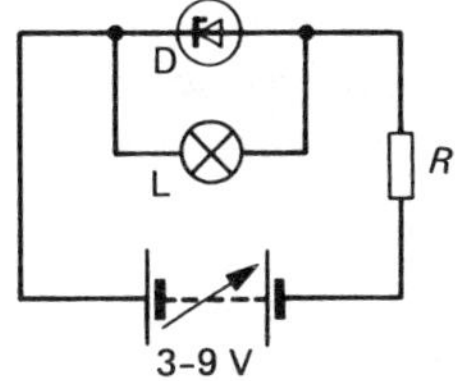

Fig. 7.19

5. Calculate the maximum current a Zener diode can pass without damage if its breakdown voltage is 10 V and its maximum power rating is 5 W.

6. If the Zener diode in question 5 is used to supply a constant voltage of 10 V from a 12 V battery, what is the value of the resistor which must be connected in series with it?

UNIT 8

Transistors

8.1 About transistors

Transistors are the most important devices in electronics today. As well as being made as discrete (separate) components, integrated circuits (ICs) may contain several thousands on a tiny slice of silicon.

They are three-terminal devices, used as *amplifiers* and as *switches*. Non-amplifying components such as resistors, capacitors, inductors and diodes are said to be 'passive'; transistors are 'active' components.

Hundreds of different transistors are available; some are shown in discrete form in Fig. 8.01. The same identification code is used as for diodes (see section 7.1) but in the American system, transistors always start with 2N followed by a number, e.g. 2N3053. In the Continental system the first letter gives the semiconductor material (A = germanium, B = silicon) and the second letter gives the use (C indicates an audio frequency amplifier, F a radio frequency amplifier and S a switching transistor); for example, BC108 is a silicon a.f. amplifier.

Fig. 8.01

The two basic types of transistor are:

(i) the *bipolar* or *junction transistor* (usually called *the* transistor); its operation depends on the flow of both majority and minority carriers;

(ii) the *unipolar* or *field effect transistor* (called the FET) in which the current is due to majority carriers only (either electrons or holes).

8.2 Junction transistor

(a) Construction

The bipolar or junction transistor consists of two p-n junctions in the same crystal. A very thin slice of lightly doped p- or n-type semiconductor (the *base* B) is sandwiched between two thicker, heavily doped materials of the opposite type (the *collector* C and *emitter* E).

The two possible arrangements are shown diagrammatically in Fig. 8.02, with their symbols. The arrow gives the direction in which conventional (positive) current flows; in the n-p-n type it points from B to E and in the p-n-p type from E to B.

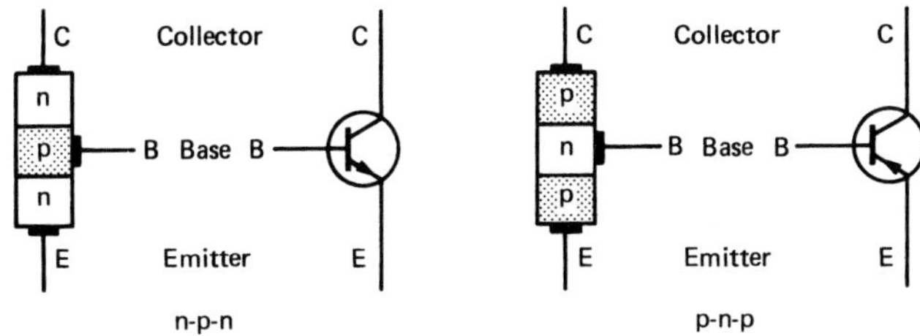

Fig. 8.02

As with diodes, silicon transistors are in general preferred to germanium ones because they withstand higher temperatures (up to about 175 °C compared with 75 °C) and higher voltages, have lower leakage currents and are better suited to high frequency circuits. Silicon n-p-n types are more easily mass-produced than p-n-p types; the opposite is true of germanium.

A simplified section is shown in Fig. 8.03(*a*) of an n-p-n discrete-component silicon transistor made by the *planar* process (outlined in section 7.5) in which the transistor is in effect created on one face (plane) of a piece of semiconducting material; Fig. 8.03(*b*) shows a transistor complete with case (called the 'encapsulation') and three wire leads.

Integrated transistors are constructed in a similar way by selective doping of silicon on a much smaller scale—see section 13.1.

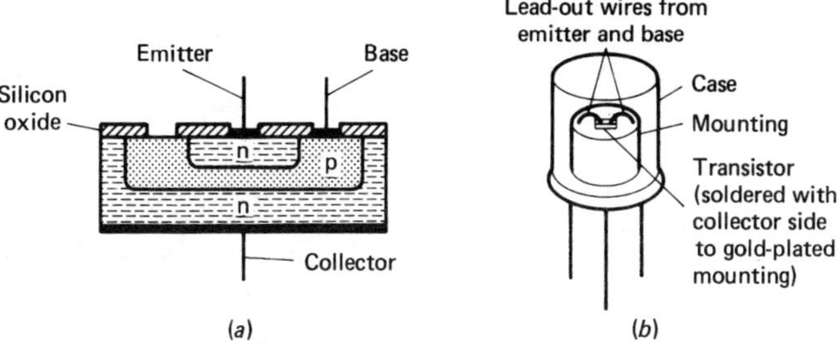

Fig. 8.03

(b) Action

In Fig. 8.04 an n-p-n silicon transistor is represented diagrammatically and is connected in a *common-emitter* circuit, i.e. the emitter is joined (via batteries B_1 and B_2) to both the base and the collector. For transistor action to occur the base-emitter junction must be *forward biased*, i.e. positive terminal of B_1 to p-type base, and the collector-base junction *reverse biased*, i.e. positive terminal of B_2 to n-type collector.

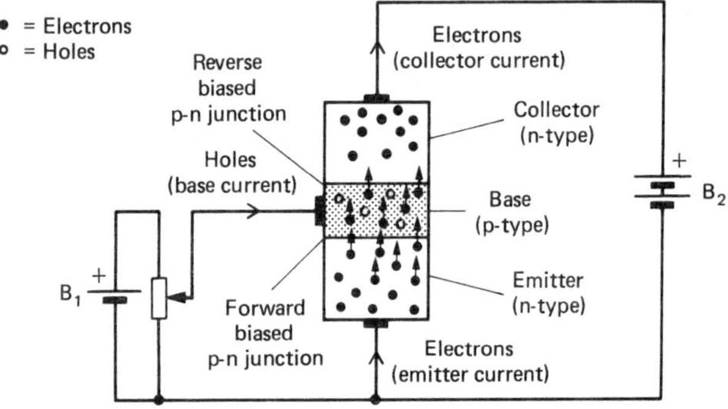

Fig. 8.04

When the base-emitter is about +0.6 V, electrons (the majority carriers in the heavily doped n-type emitter) cross the junction (as they would in any junction diode) into the base. Their loss is made good by electrons entering the emitter from the external circuit to form the *emitter current*. At the same time holes flow from the base to the emitter but, since the p-type base is lightly doped, this is small compared with the electron flow in the opposite direction, i.e. electrons are the majority carriers in an n-p-n transistor.

In the base, only a small proportion (about 1%) of the electrons from the emitter combine with holes in the base because the base is very thin (less than a millionth of a metre) and is lightly doped. Most of the electrons are swept through the base, because they are attracted by the positive voltage on the collector, and cross the base-collector junction to become the *collector current* in the external circuit.

The small amount of electron–hole recombination which occurs in the base gives it a momentary negative charge which is immediately compensated by battery B1 supplying it with (positive) holes. The flow of holes to the base from the external circuit creates a small *base current*. This keeps the base-emitter junction forward biased and so maintains the larger collector current.

Transistor action is the turning on (and controlling) of a large current through the high resistance (reverse biased) collector-base junction by a small current through the low resistance (forward biased) base-emitter junction. The term transistor refers to this effect and comes from the two words '*trans*fer-res*istor*'. Physically the collector is larger than the emitter and if one is used in place of the other the action is inefficient.

The behaviour of a p-n-p transistor is similar to that of the n-p-n type but it is holes that are the majority carriers which flow from the emitter to the collector and electrons are injected into the base to compensate for recombination. To obtain correct biasing the polarities of both batteries must be reversed. **Wrong battery connection can seriously damage transistors.**

8.3 Junction transistor as a current amplifier

(a) d.c. current gain

In the circuit of Fig. 8.05, when the base-emitter junction of the n-p-n silicon transistor is forward biased to about 0.6 V (0.1 V for a germanium transistor), a small base current I_B flows and, as we saw in the previous section, 'turns on' a larger collector current I_C. That is, I_C is zero until I_B flows (but see section 8.4). A junction transistor is therefore a current-operated device.

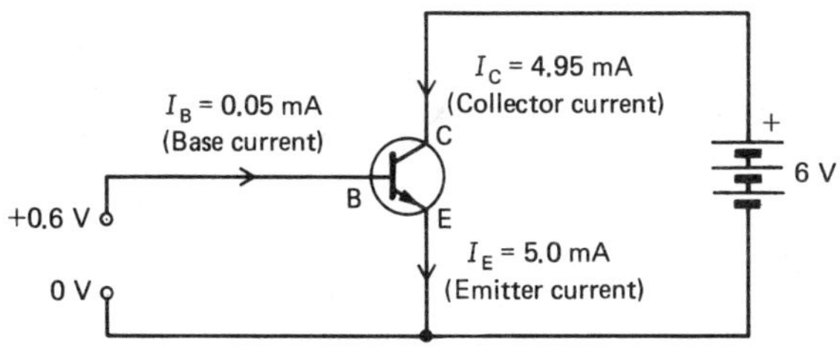

Fig. 8.05

Typically I_C may be 10 to 1 000 times greater than I_B depending on the type of transistor. If we look upon I_B as the input current to the transistor and I_C as the output current from it, then it is a *current amplifier*. The *d.c. current gain*, h_{FE}, is an important property of a transistor and is defined by:

$$h_{FE} = \frac{I_C}{I_B}$$

For the transistor in Fig. 8.05, where the arrows show the direction of conventional current flow, $I_C = 4.95$ mA and $I_B = 0.05$ mA (50 μA). Therefore:

$$h_{FE} = \frac{4.95}{0.05} = \frac{495}{5} = 99$$

More will be said about h_{FE} in section 8.5.

You should also note, since the current flowing out of a transistor must equal the current flowing into it, that:

$$I_E = I_B + I_C$$

In the above example $I_E = 0.05 + 4.95 = 5.0$ mA, i.e. $I_C = I_E$ approximately.

In the symbol h_{FE}, F indicates that we are considering forward currents and E that the transistor is connected in the common-emitter mode. Two other less usual methods of connection are 'common-base' and 'common-collector'.

(b) Demonstration

A simple circuit to show current amplification by a transistor is given in Fig. 8.06. The forward bias for the base-emitter junction and the reverse bias for the collector-base junction are both obtained from the same battery. The base current I_B flows through lamp L_1 and resistor R and 'turns on' the transistor. R **prevents excessive base currents which would destroy the transistor** and should always be present in any circuit.)

When the circuit is complete L_2 lights up but not L_1, showing that I_C is much greater than I_B. If L_1 is removed so that the base circuit is 'broken', I_B becomes zero as does I_C and L_2 goes out.

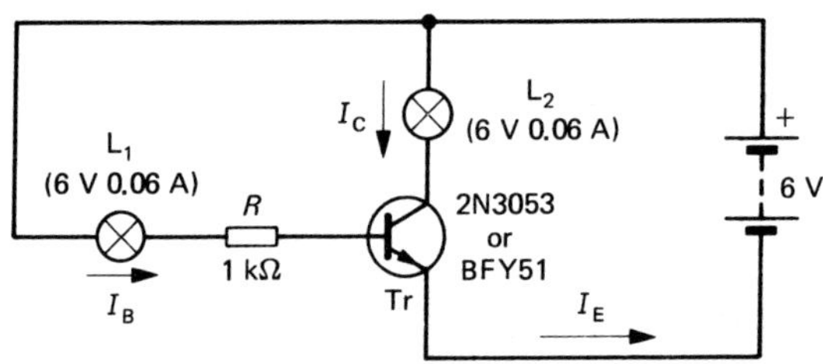

Fig. 8.06

8.4 Junction transistor—Characteristics

Characteristics are graphs, found by experiment, which show the relationships between various currents and voltages and enable us to see how best to use a transistor. A circuit for investigating an n-p-n transistor (e.g. BFY51 or 2N3053) in common-emitter connection is shown in Fig. 8.07. R_2 protects the transistor from excessive base currents. Three characteristics are important.

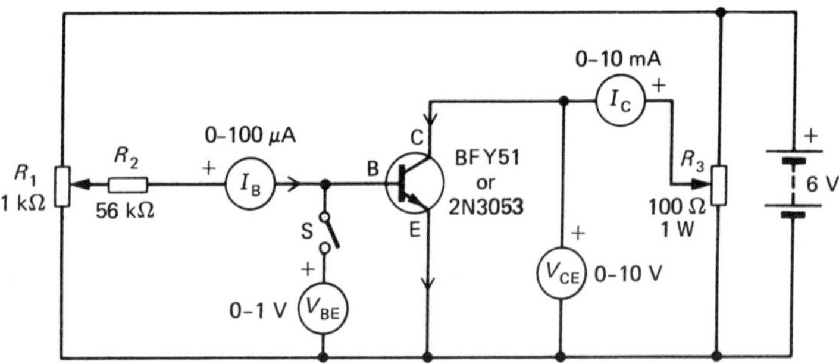

Fig. 8.07

(a) Input (base) characteristic (I_B–V_{BE})

The collector-emitter voltage V_{CE} is kept constant (e.g. at the battery voltage of 6 V) and the base-emitter voltage V_{BE} is measured for different values of the base current I_B, obtained by varying R_1.

A typical graph for a silicon transistor is given in Fig. 8.08(*a*). Note that I_B is negligibly small until V_{BE} exceeds about 0.6 V and thereafter small changes in V_{BE} cause large changes in I_B but V_{BE} is always near 0.6 V whatever the value of I_B. The *input resistance*, r_i, is defined as the ratio $\Delta V_{BE}/\Delta I_B$ where ΔI_B is the change in I_B due to a change of ΔV_{BE} in V_{BE}. ('Δ' means 'a small increase in' and is pronounced 'delta'—the Greek capital letter D.) Since the input characteristic is non-linear, r_i varies but is of the order of 1 to 5 kΩ.

(b) Output (collector) characteristic (I_C–V_{CE})

I_B is fixed at a low value, e.g. 10 μA, and I_C is measured as V_{CE} is increased in stages by varying R_3. This is repeated for different values of I_B to give a family of curves as in Fig. 8.08(*b*).

You can see that I_C depends almost entirely on I_B and hardly at all on V_{CE} (except when V_{CE} is less than about 0.5 V). As an *amplifier*, a transistor operates well to the right of the sharp bend or 'knee' of the characteristic, i.e. where I_C varies linearly with V_{CE} for a given I_B. The small slope of this part of the characteristic shows that the *output resistance*, r_o, of the transistor is fairly high, of the order 10 to 50 kΩ; it is given by $r_o = \Delta V_{CE}/\Delta I_C$ where ΔI_C is the change in I_C caused by a change of ΔV_{CE} in V_{CE}.

As a *switch*, a transistor operates in the shaded parts of Fig. 8.08(*b*) and changes over rapidly from the 'off' state in which $I_C = 0$ (cut-off) to the 'on' state in which I_C is a maximum (saturation).

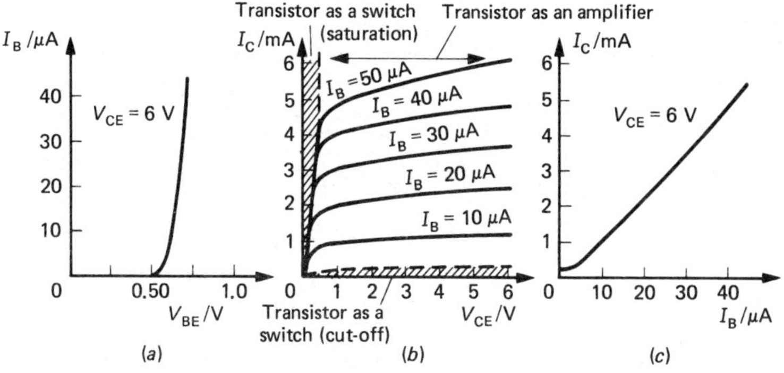

Fig. 8.08

(c) Transfer characteristic (I_C–I_B)

V_{CE} is kept fixed and I_C is measured for different values of I_B by varying R_1. The graph shown in Fig. 8.08(*c*) is almost a straight line, implying that I_C is directly proportional to I_B, i.e. the relation between I_C and I_B is linear.

The *a.c. current gain*, h_{fe}, is defined by:

$$h_{fe} = \frac{\Delta I_C}{\Delta I_B}$$

where ΔI_C is the change in I_C produced by a change of ΔI_B in I_B. For most purposes h_{fe} and h_{FE} (the d.c. current gain $= I_C/I_B$) can be considered as equal.

The characteristic also shows that when I_B is zero, I_C has a small value (about 0.01 μA for silicon and 2 μA for germanium at 15 °C). This is called the *leakage current,* I_{CEO}, and is due to minority carriers (holes for an n-p-n transistor and electrons for a p-n-p type) crossing the reverse biased collector-base junction from collector to base to emitter. Since minority carriers are produced by heat breaking bonds in the crystal lattice, the leakage current increases as the temperature rises (much more so for germanium than silicon) and upsets the working of the transistor.

Summing up, in a silicon junction transistor:

(i) I_C is zero until I_B flows;
(ii) I_B is zero until V_{BE} is about 0.6 V;
(iii) V_{BE} remains close to 0.6 V for a wide range of values of I_B.

8.5 Junction transistors—Data

While one type of transistor may replace another in many circuits, it is as well to study the data published by manufacturers when making a choice. Table 8.1 lists the main ratings—called *parameters*—for five popular n-p-n silicon transistors.

Table 8.1 Parameters of n-p-n silicon transistors

	BF194	BC108	ZTX300	2N3705	BFY51
I_C max/mA	30	100	500	800	1000
h_{FE} at I_C/mA	115 (typ) 1	110–900 2	50–300 10	50–130 50	40 (min) 150
P_{tot}/mW	220	360	300	360	800
V_{CEO} max/V	20	20	25	30	30
V_{EBO} max/V	5	5	5	5	6
f_T/MHz	260	250	150	100	50
Outline	MM10b	TO18	E-line	TO92a	TO5

All are general purpose types that can be used as audio frequency amplifiers or as switches, except the BF194 which is a radio frequency amplifier. The BC108 is a low current, high gain device; the ZTX300, 2N3705 and BFY51 are medium current types.

(i) Current, voltage and power ratings In most cases the symbols used are self-explanatory:

- $I_C max$ gives the maximum average value of collector current the transistor can pass without damage;
- $V_{CEO} max$ gives the maximum collector-emitter voltage that can be applied when the base is open-circuited;
- $V_{EBO} max$ gives the maximum emitter-base voltage that can be applied when the collector is open-circuited;
- P_{tot} is the maximum total power rating at 25 °C air temperature and equals $V_{CE} \times I_C$ approximately.

(ii) d.c. current gain (h_{FE}) Owing to the manufacturing spreads, h_{FE} is not the same for all transistors of the same type. Usually minimum and maximum values are quoted but sometimes only the minimum (e.g. BFY51) or the typical value (e.g. BF194) is given. Also, since h_{FE} decreases at high and low collector currents, the current at which it is measured is stated. It also increases with temperature.

When selecting a transistor type for a particular circuit, we have to ensure that its minimum h_{FE} will give the current gain required by that circuit: the fact that it may be greater does not usually matter. In many applications h_{FE} is the most important factor; this is true for a small-signal a.f. voltage amplifier because the current, voltage and power ratings will be within the capabilities of any transistor.

(iii) Transition frequency (f_T) This is the frequency at which h_{FE} falls to one; it is important in high frequency circuits—see section 12.5.

(iv) Outlines The case of a transistor is called the encapsulation and may be of metal or plastic. There are many shapes and sizes (i.e. outlines); those listed in Table 8.1 are shown in Fig. 8.09.

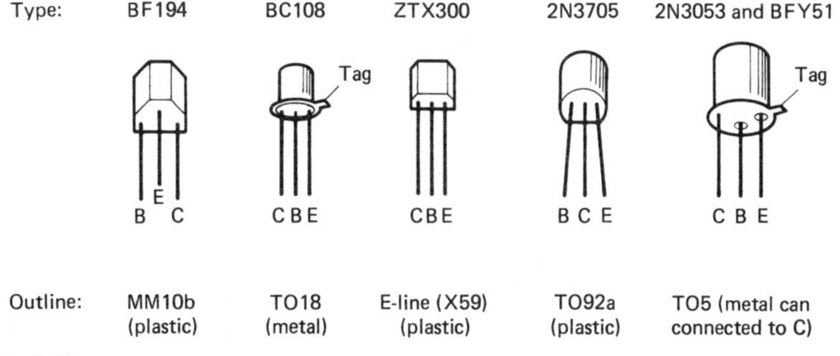

Fig. 8.09

8.6 Junction transistors—Testing

(a) Leakage current and d.c. current gain

The 'health' of a transistor can be checked from its collector-emitter leakage current I_{CEO} and d.c. current gain h_{FE}.

To measure I_{CEO}, the circuit of Fig. 8.10(*a*) is used for an n-p-n transistor. If the transistor is in good conditions, the microammeter should read almost zero current since there is no base current to turn it on, i.e. the resistance of the transistor should be very high.

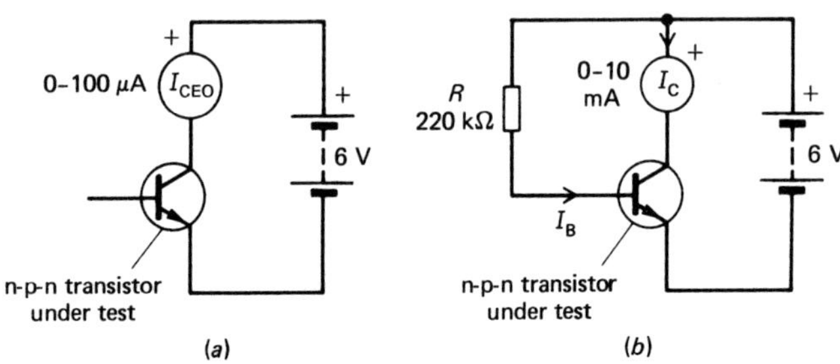

Fig. 8.10

To measure h_{FE}, the microammeter is replaced by a milliammeter and a resistor R is connected as in Fig. 8.10(*b*). I_C is read from the milliammeter and I_B calculated as follows. From the circuit we can say that:

voltage across R + voltage across base-emitter junction = battery voltage

$$\therefore \quad V_R + V_{BE} = 6 \text{ V}$$

For a silicon transistor $V_{BE} = 0.6$ V. Therefore:

$$V_R = 6 - 0.6 = 5.4 \text{ V}$$

$$V_R = \text{current through resistor} \times R$$

$$= \text{base current} \times R$$

$$= I_B \times R \text{ where } R = 220 \text{ k}\Omega$$

$$\therefore \quad 5.4 = I_B \times 220$$

$$\therefore \quad I_B = 5.4/220 = 0.025 \text{ mA} \quad (25\ \mu\text{A})$$

Then

$$h_{FE} = I_C/I_B = I_C/0.025$$

$$= 40 I_C$$

where I_C is in mA. You then know h_{FE} at a certain I_C and can compare its value with manufacturers' data.

To measure I_{CEO} and h_{FE} for a p-n-p transistor the battery and meter connections are reversed in both circuits.

(b) Identifying n-p-n and p-n-p types

An analogue multimeter which measures resistance (an ohmmeter) can be used for this test. When switched to the 'ohms × 1' range, the negative (−) terminal on most meters has positive polarity (due to a battery inside it); the positive terminal (+) has negative polarity. An n-p-n transistor gives a lower resistance reading when the − terminal is connected to the base and the + terminal to the collector or emitter than it does with the leads the other way round, Fig. 8.11(*a*). The reverse is true for a p-n-p transistor, Fig. 8.11(*b*). In each case the resistance is low when the p-n junction concerned is forward biased.

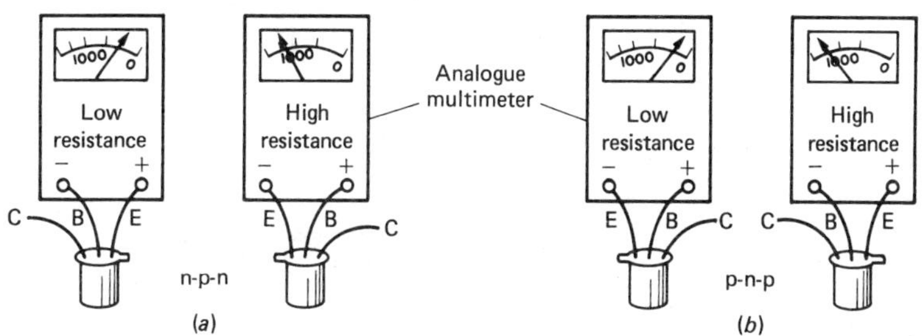

Fig. 8.11

A digital multimeter may also be used if the polarities of its terminals are found from the instruction booklet.

8.7 Field effect transistor—JUGFET

The junction (bipolar) transistor is a *current*-controlled amplifying device. In the field effect (unipolar) transistor it is the input *voltage* which controls the output current; the input current is usually negligible. This is a big advantage when the input cannot supply much current, e.g. a crystal microphone.

In a FET, a narrow channel of doped semiconductor connects two metal contacts called the *drain* D and the *source* S. The voltage (or more correctly the electric field it produces) applied to a third contact known as the *gate* G and located between S and D, determines the current which, as the terms suggest, flows from S to D.

There are two main types of FET, the JUGFET (standing for junction-gate FET) and the MOSFET (standing for metal-oxide-semiconductor FET). The JUGFET will now be considered.

(a) Action

An n-channel JUGFET is shown diagrammatically in Fig. 8.12(*a*) and its symbol in (*b*). Its action depends on the formation and control of the depletion layer at the p-n junction which is reverse biased. The p-type gate is more heavily doped than the n-type channel and so the depletion layer is mostly in the channel.

With respect to the source, the voltage of the drain V_{DS} is positive and that of the gate V_{GS} is negative. Making V_{GS} more negative widens the depletion layer (an insulator) and hence narrows the channel (a conductor). This reduces the electron flow, i.e. the drain current I_D, from S to D.

In an actual JUGFET there is usually a second p-type gate connected to and opposite the other gate so that the n-channel is sandwiched between two depletion layers.

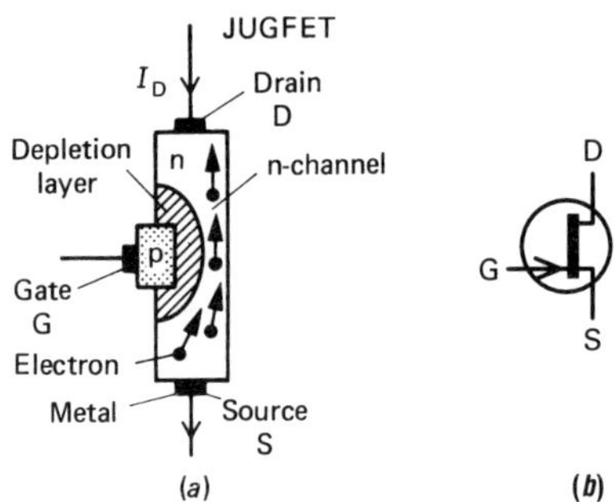

Fig. 8.12

(b) Characteristics

Normally only the *transfer* and *output* characteristics are plotted for FETs because the gate current is negligible. Those for an n-channel JUGFET such as the general purpose 2N3819 may be found using the circuit of Fig. 8.13. Since a FET is voltage-operated, a transfer characteristic (Fig. 8.14(*a*)) shows the relation between the gate *voltage* V_{GS} and I_D (for fixed V_{DS}); it is almost linear.

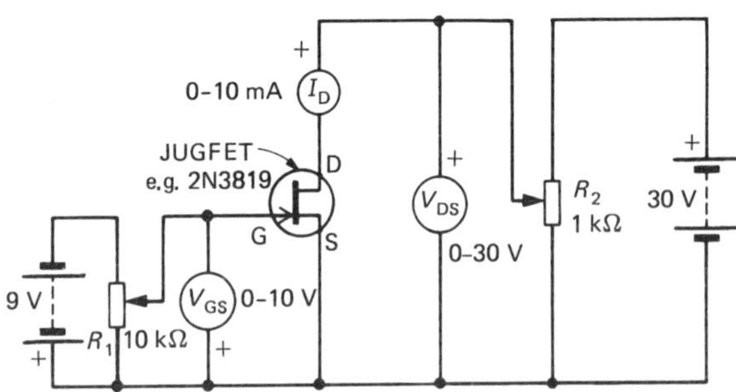

Fig. 8.13

The performance of a FET is measured by its *transconductance*, g_m, defined by:

$$g_m = \frac{\Delta I_D}{\Delta V_{GS}}$$

ΔI_D is the change in I_D caused by a change ΔV_{GS} in V_{GS}. The value of g_m can be found from a transfer characteristic and is roughly in the range 1 to 10 mA V^{-1}. It corresponds to h_{fe} of a junction transistor.

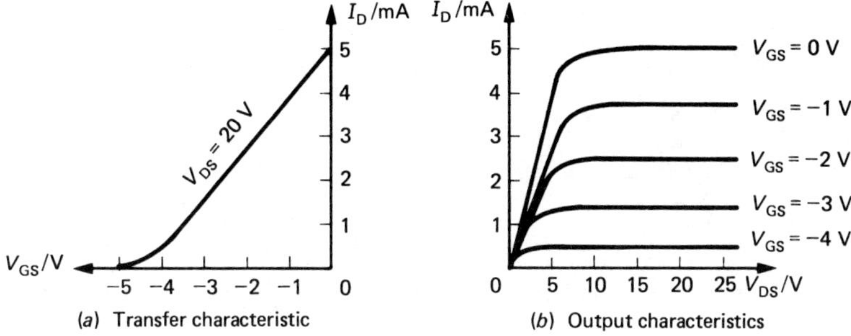

(*a*) Transfer characteristic (*b*) Output characteristics

Fig. 8.14

The output characteristics, Fig. 8.14(*b*), are similar to those of a junction transistor (Fig. 8.08(*b*)), but their slope to the right of the 'knee' is less, indicating a higher *output resistance* r_o (50 kΩ to 1 MΩ). The *input resistance* r_i is very high (greater than 10^9 Ω).

8.8 Field effect transistor—MOSFET

If the gate voltage of an n-channel JUGFET becomes positive, so forward biasing the gate-channel p-n junction, its input resistance falls to a few tens of ohms and a very large damaging current flows. The MOSFET does not have this disadvantage and both positive and negative voltages can be applied to the gate.

(a) Action

The basic construction (and the symbol) of a n-channel FET of this type is shown in Fig. 8.15(*a*). Its action depends on the existence of a very thin layer of highly insulating silicon oxide between the metal gate and the n-type channel connecting drain to source. The channel is formed in a chip of p-type silicon, called the *substrate* (or body), which is connected internally to the source.

The voltage V_{GS} between gate and source controls the charge carrier (electron) concentration in the channel. If D is positive with respect to S and if $V_{GS} = 0$, drain current I_D flows. When V_{GS} is negative, positive holes are attracted into the channel from the substrate, so in effect reducing (depleting) the concentration of free electrons in the channel and making it less conducting. I_D decreases and the MOSFET is said to be operating in the *depletion mode* (as does a JUGFET). When V_{GS} is positive, electrons are attracted into the n-channel, thereby increasing (enhancing) its conductivity. I_D increases; this is the *enhancement mode* of operation.

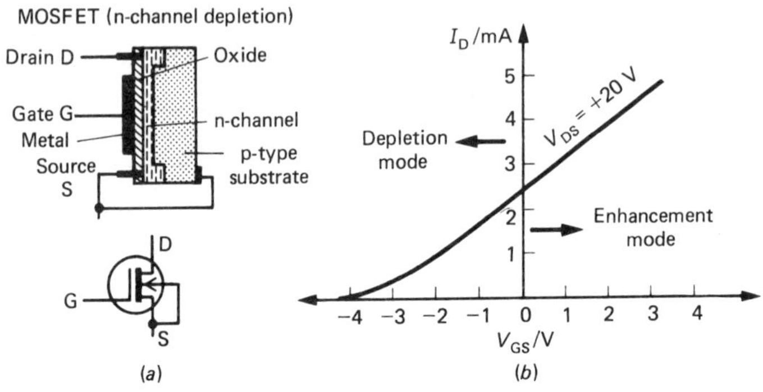

Fig. 8.15

(b) Characteristics

The transfer characteristic of an n-channel MOSFET (e.g. 2N3631) in Fig. 8.15(*b*) shows I_D for both positive and negative values of V_{GS}. It gives value for g_m similar to those for a JUGFET. The output characteristics are like junction transistor curves with output resistances r_o in the range 10 to 50 kΩ. Both types of characteristic can be found with the circuit of Fig. 8.13. The input resistance r_i is extremely high (greater than 10^{12} Ω).

(c) Further points

Power MOSFETs can drive electric motors and lamps which require large currents, taking their tiny input currents from, for example, an integrated circuit. Their high input resistance also makes them suitable for use in the output stages of audio power amplifiers (section 11.8). They can also act as high speed switches (section 14.2).

The MOSFET structure is very compact and lends itself to integrated circuit construction.

Great care is needed to protect MOSFET gates from static electric charges which can build up and break down the insulation of the oxide layer. They are therefore supplied with a metal clip short-circuiting their leads; it should not be removed until the device has been connected into the circuit.

Another kind of MOSFET which works only in the enhancement mode is shown diagrammatically with its symbol in Fig. 8.16(*a*). The conducting n-channel is not formed until V_{GS} is positive and repels holes in the p-type substrate away from the insulating layer, leaving behind the n-type channel between the drain and source. Fig. 8.16(*b*) shows its transfer characteristic.

In p-channel FETs, conduction is by positive holes and the polarity of the voltages applied to the electrodes is reversed, as are the arrows on symbols.

MOSFET (n-channel enhancement)

D
Oxide
G
S
n
n
p-type substrate

D
G
S

(*a*)

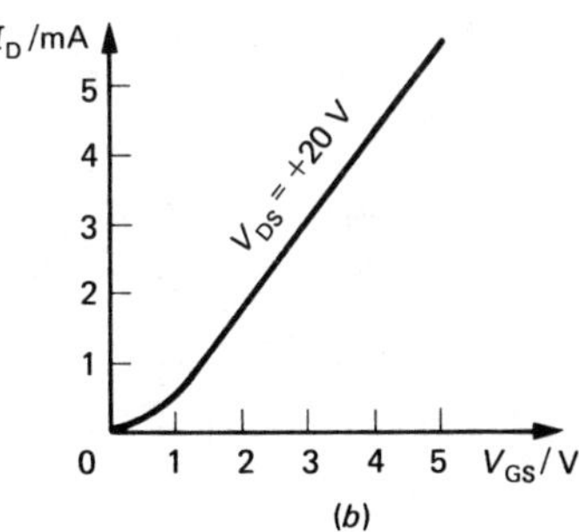

(*b*)

Fig. 8.16

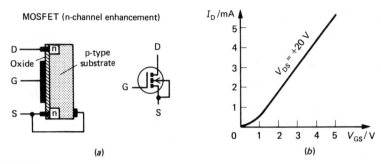

8.9 Comparison of transistors

Table 8.2 provides a summary of the main properties of the different transistor types.

Table 8.2 Properties of different types of transistor

Property	**Junction transistor (common emitter)**	**JUGFET**	**MOSFET**
h_{fe}	10 to 1 000	—	—
g_m	—	1 to 10 mA V^{-1}	1 to 10 mA V^{-1}
r_i	1 to 5 kΩ (medium)	>10^9 Ω (very high)	>10^{12} Ω (extremely high)
r_o	10 to 50 kΩ (medium)	50 kΩ to 1 MΩ (high)	10 to 50 kΩ (medium)

8.10 Revision questions

1. What types of transistor are represented by the symbols in Figs. 8.17(*a*) and (*b*)?

Fig. 8.17

2. Name the three parts of a bipolar junction transistor.
3. Why are most modern transistors made from silicon rather than germanium?
4. In a junction transistor
   a) how many p-n junctions are there,
   b) which is/are forward biased,
   c) which is/are reverse biased?

5. In which circuits of Fig. 8.18 will the lamp L light? (B_1 = 4.5 V, B_2 = 6 V, R = 1 kΩ, L = 6 V 60 mA.)

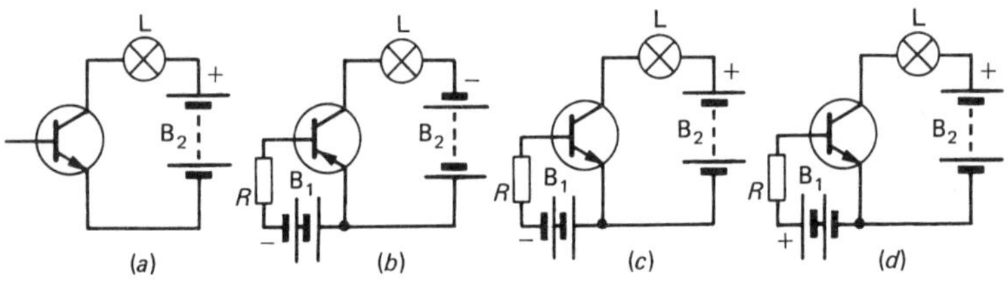

Fig. 8.18

6. Why should the base region of a junction transistor be very thin and lightly doped?

7. a) What is meant by transistor action?
b) Why is a junction transistor in common-emitter connection a current amplifier?

8. What is the value of the 'turn-on' voltage in a junction transistor made of
a) silicon,
b) germanium?

9. Define the d.c. current gain of a transistor. What is its symbol?

10. As applied to a transistor, what do the following terms mean:
a) input characteristic and input resistance,
b) output characteristic and output resistance,
c) transfer characteristic and a.c. current gain?

11. Name two parameters that indicate the condition of a junction transistor.

12. a) In a FET what corresponds to
(i) the collector,
(ii) the emitter and
(iii) the base of a junction transistor?
b) Why is a 'driving' circuit containing, for example, a crystal microphone, less likely to be upset if it is connected to a FET rather than to a bipolar transistor?

13. a) Draw the symbols for
(i) an n-channel JUGFET,
(ii) a p-channel MOSFET.
b) Compare the construction and method of operation of a JUGFET with a MOSFET.

14. What precautions must be taken when 'handling' MOSFETS?

8.11 Problems

1. For the transistor in Fig. 8.19, what is the value of
a) h_{FE} at 100 mA,
b) I_E?

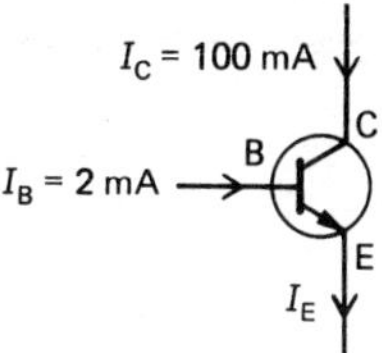

Fig. 8.19

2. A transistor has a d.c. current gain of 200 when the base current is 50 μA. What is the collector current in mA?

3. In the circuit of Fig. 8.20, does the transistor (silicon) turn on if:
a) R_1 = 10 kΩ and R_2 = 1 kΩ,
b) R_1 = 10 kΩ and R_2 = 100 kΩ?
Explain your answers.

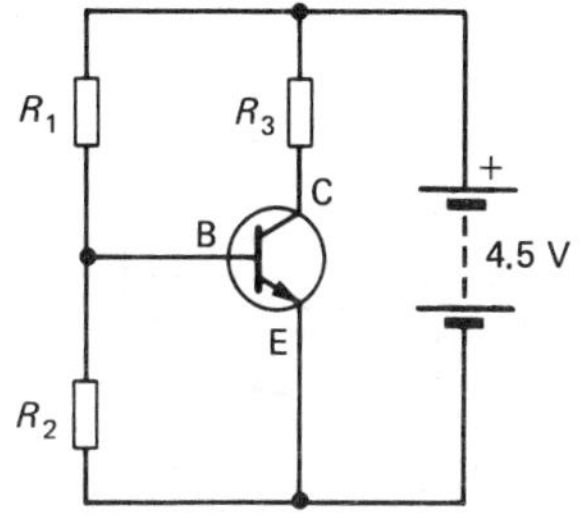

Fig. 8.20

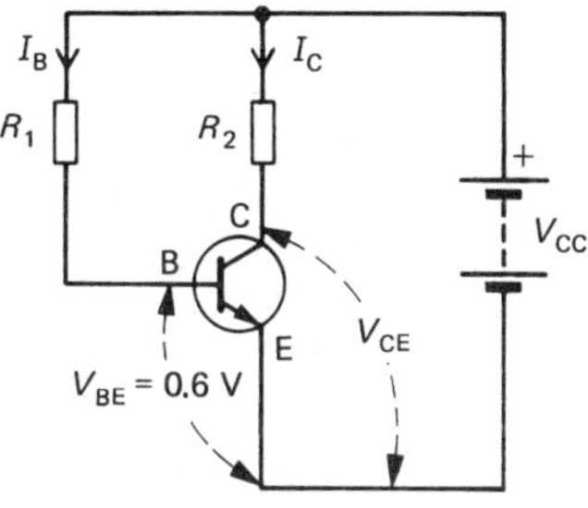

Fig. 8.21

4. What must be the value of R_1 in the circuit of Fig. 8.21 to give a base current I_B of 20 μA if the voltage across the base-emitter junction V_{BE} = 0.6 V and the battery voltage V_{CC} = 6 V? (Hint—see section 8.6.)

5. In the circuit of Fig. 8.21, R_1 = 3.9 kΩ, I_C = 25 mA, V_{CC} = 4.5 V and V_{BE} = 0.6 V. What is
a) the base current I_B, and
b) the d.c. current gain?

6. In the circuit of Fig. 8.21, R_1 = 100 kΩ, R_2 = 1 kΩ, V_{CC} = 6 V and V_{BE} = 0.6 V. Calculate
a) the voltage across R_1,
b) I_B,
c) I_C if h_{FE} = 60,
d) the voltage across R_2, and
e) the voltage across the collector-emitter.

UNIT 9

Transducers and switches

9.1 About transducers

Transducers change energy from one form into another. Those in which electrical energy is the input or the output, i.e. electrical transducers, are described in this Unit. They are the devices which enable electronic systems to 'communicate' with the outside world. For example, a microphone changes sound into electrical signals, a loudspeaker does the opposite.

Other transducers exist for dealing with physical quantities such as light intensity or temperature, as Fig. 9.01 shows.

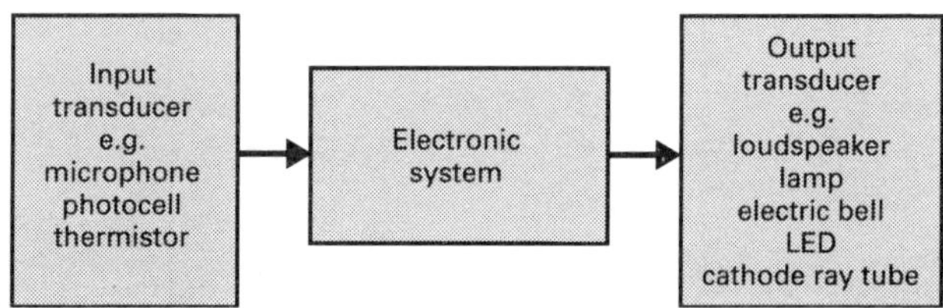

Fig. 9.01

A transducer must be 'matched' to the input or output of the system with which it is designed to work. Sometimes this means ensuring maximum voltage transfer between transducer and system (or vice versa), at other times we are concerned with transferring maximum power. In all cases the impedance of the transducer compared with the input (or output) impedance of the system is a critical factor. For example, maximum voltage transfer occurs between an input transducer (e.g. a microphone) and a system (e.g. an amplifier) when the input impedance of the system is *much larger* than that of the transducer—see section 11.5. Or again, the maximum transfer of power from a system (e.g. an amplifier) to an output transducer (e.g. a loudspeaker) occurs when the output impedance of the system *equals* the impedance of the transducer (as the maximum power theorem tells us—see section 2.12).

Impedance (rather than resistance, since in general we are dealing with a.c.) is an important property of a transducer.

9.2 Microphones

A good microphone should respond more or less equally to all sounds in the audio frequency range, i.e. from 20 Hz or so to about 20 kHz. Otherwise the electrical signals it passes on for amplification and conversion back to sound by the output transducer (e.g. a loudspeaker), will not be identical, or nearly so, to the original sound. The symbol for a microphone is shown in Fig. 9.02.

Fig. 9.02

(a) Moving-coil or dynamic microphone

This is the most popular type because of its good quality reproduction, robustness, omnidirectional properties (i.e. detection of sounds from all directions more or less equally) and reasonable cost. One is shown in Fig. 9.03(*a*). It consists of a small coil of many turns of very thin wire wound on a tube (the former) which is attached to a light disc (the diaphragm) as shown in Fig. 9.03(*b*). When sound strikes the diaphragm, it makes the coil and former move in and out of the circular gap between the poles of a strong permanent magnet. Electromagnetic induction occurs (see section 6.2) and the alternating voltage induced in the coil (typically 1 to 10 mV) has the same frequency as the sound. Its impedance is usually two or three hundred ohms.

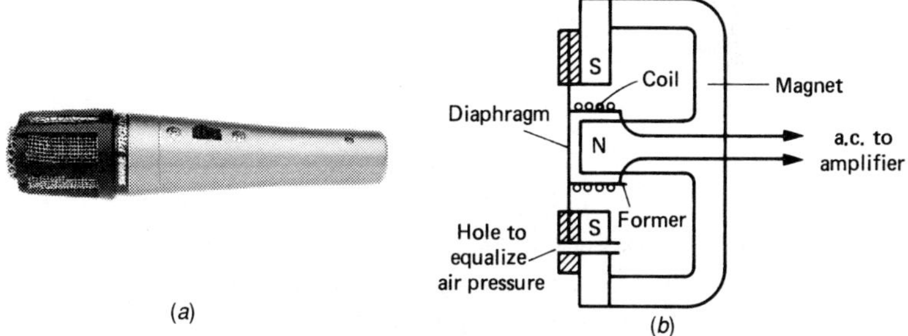

Fig. 9.03

(b) Electret capacitor microphone

This type is used in broadcasting studios, for public address systems and for concerts where the highest quality is necessary. Basically it consists of two capacitor plates A and B, Fig. 9.04(*a*). B is fixed while the metal foil disc A acts as the movable diaphragm. Sound waves make the diaphragm vibrate and as it alters position the capacitance changes. A small battery inside the microphone causes the resulting charging and discharging current to produce a small varying voltage across resistor *R*. This forms the input to a FET amplifier, also housed in the stem of the microphone. The small tie/lapel microphone in Fig. 9.04(*b*) is of this type.

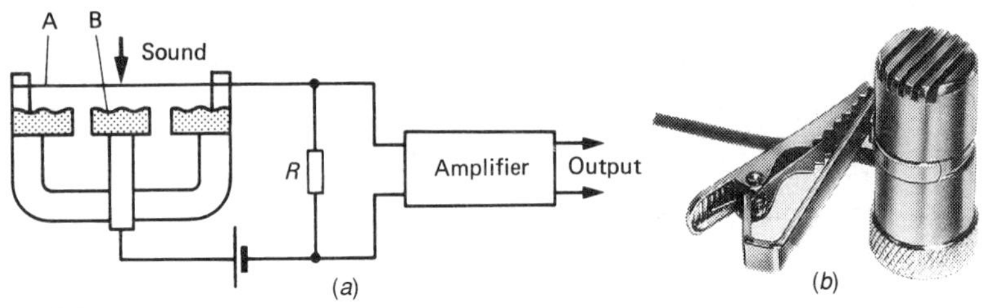

Fig. 9.04

(c) Other microphones

The *carbon microphone*, Fig. 9.05, is used in telephones. Sound makes the diaphragm vibrate and thus varies the pressure on the carbon granules between the front carbon block, which is attached to the diaphragm, and the fixed one at the back. An increase of pressure squeezes the granules closer together and their electrical resistance decreases. Reduced pressure increases the resistance. If there is a current passing through the microphone from a battery, this varies accordingly; it is a varying d.c. which can be thought of as a steady d.c. to which has been added a.c. having the frequency of the sound—see section 3.3.

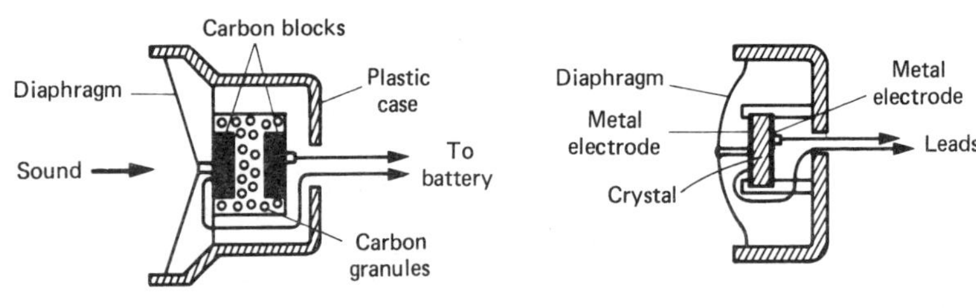

Fig. 9.05 **Fig. 9.06**

The *crystal microphone* shown in Fig. 9.06 has a high output (10 to 100 mV), is omnidirectional and inexpensive but it is easily damaged by moisture or heat. It is used in cassette recorders. The action depends on the *piezoelectric effect* in which certain crystals generate a charge and so also a voltage across opposite faces of a slice when they are bent, because of the displacement of ions in the crystal. The effect is given by natural crystals such as Rochelle salt (sodium potassium tartrate) and quartz and by synthetic ceramic crystals of lead zirconium titanate. (The latter is used in cigarette lighters where high pressures produce several thousand volts.) In a crystal microphone an alternating voltage is produced when sound makes the diaphragm vibrate. The impedance is very high, in the range 1–5 MΩ.

The *wireless microphone* incorporates a small radio transmitter which sends the sound falling on the microphone as a frequency modulated (FM) v.h.f. signal (see section 18.4) up to a distance of about 100 metres. Its short aerial transmits to any FM receiver and allows the performer/lecturer greater freedom of movement. It is battery operated.

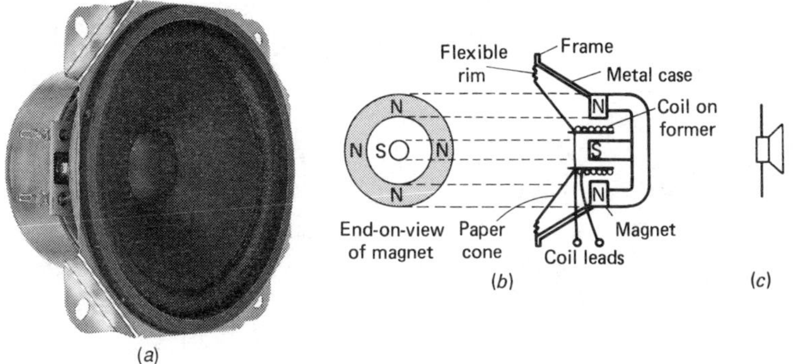

Fig. 9.07

9.3 Loudspeakers, headphones and earpieces

Loudspeakers, headphones and earpieces change electrical energy into sound.

(a) Moving-coil loudspeaker

Most loudspeakers used today are of the moving-coil type shown in Fig. 9.07(*a*). Their construction, Fig. 9.07(*b*), is essentially the same as that of the moving-coil microphone, in fact a loudspeaker can be used as a microphone. The symbol for a loudspeaker is given in Fig. 9.07(*c*).

When audio frequency a.c. (from an amplifier) passes through the coil of the speaker, it vibrates in and out between the poles of the magnet and produces sound of the same frequency. (In the moving-coil microphone, sound makes the coil *move* in the magnetic field and *current* is generated. Here current is fed into the coil in the magnetic field and motion results.)

To increase the mass of air set into motion and therefore the loudness of the sound, the coil is wound on a former (tube) fixed to a paper cone which also vibrates. When the cone moves forward, it produces a *compression* (i.e. greater air pressure) in front and a *rarefaction* (i.e. lower air pressure) behind. If these meet, they tend to cancel, thereby reducing the sound. Mounting the speaker on a baffle board, or, better still, enclosing it in a cabinet lined with sound absorbing material, helps to prevent this happening. However, cabinet design is complex if undesirable resonance effects are to be avoided and efficient coupling achieved between the speaker and the air in the room. In practice there are often two or three speakers in the same cabinet, each handling a certain range of audio frequencies.

A large speaker, called the *woofer*, deals with low (bass) frequencies, while a small speaker, the *tweeter*, handles high (treble) frequencies, as shown in Fig. 9.08(*a*). A mid-range speaker may also be used. A *crossover* network is needed between the amplifier and the speakers to feed the correct range of frequencies to each. In the simple arrangement of Fig. 9.08(*b*), L and C act as a voltage divider with high frequencies developing a voltage across L, for application to the tweeter, and low frequencies creating a voltage across C, for application to the woofer.

(Remember that $X_L = 2\pi fL$ and $X_C = 1/(2\pi fC)$.) A typical *crossover frequency* is about 3 kHz, i.e. frequencies above this go mostly to the tweeter, those below mostly to the woofer.

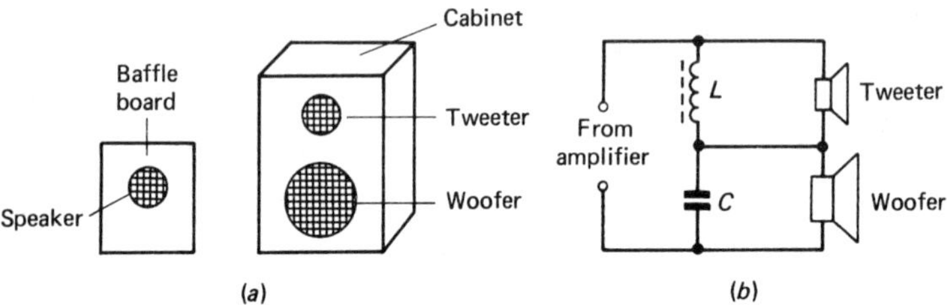

Fig. 9.08

The speaker coil has both inductive reactance X_L and resistance R and its impedance Z is given by $Z = \sqrt{(R^2 + X_L^2)}$ (see section 6.6). Z varies with the frequency of the alternating current but it is usually quoted for 1 kHz; common values at this frequency are 4 Ω, 8 Ω, 15 Ω, the maximum being about 80 Ω. At 1 kHz, X_L is small and an approximate value for Z can be obtained by measuring the resistance R of the coil using a multimeter. For example, if $R = 12\ \Omega$, then we can assume that $Z = 15\ \Omega$.

(b) Headphones

High quality headphones as used for stereo listening are *moving-coil* (dynamic) types. Often each earphone is fitted with its own slider volume control as in Fig. 9.09(*a*). The symbol for headphones is given in Fig. 9.09(*b*).

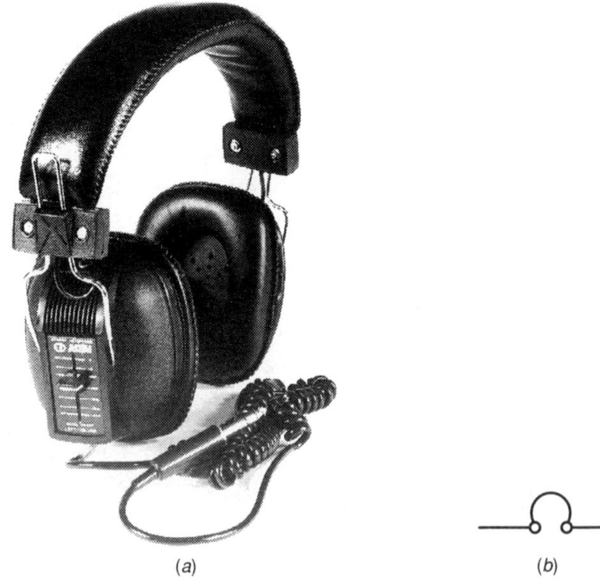

(*a*) (*b*)

Fig. 9.09

Headphones for general use, e.g. mono listening and telephone headsets, work on a different principle; they have high impedance (typically 1 kΩ per earphone compared with 8 Ω for the moving-coil type) and their frequency response is lower (e.g. 30 Hz to 15 kHz compared with 20 Hz to 20 kHz). The action of this *magnetic* type, Fig. 9.10(*a*), can be explained from Fig. 9.10(*b*). Current passes through the coils of an electromagnet and attracts the iron diaphragm more, or less, depending on the value of the current. As a result the diaphragm vibrates and produces sound.

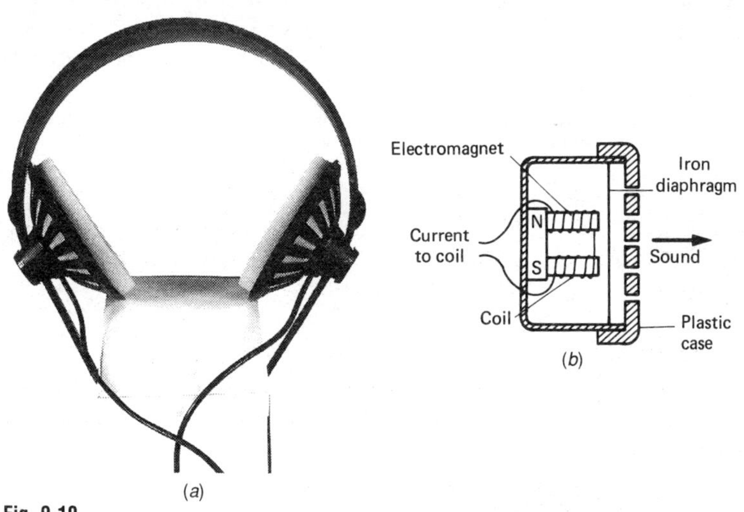

Fig. 9.10

(c) Earpieces

These are used in deaf-aids and with transistor radios. The *magnetic* type (impedance 8 Ω shown in Fig. 9.11(*a*) works on the moving-coil principle. The *crystal* type shown in Fig. 9.11(*b*) (impedance several megohms) depends on the reverse piezoelectric effect—see section 9.2. That is, an alternating voltage applied across opposite faces of the crystal makes it vibrate and produce sound of the same frequency as the a.c. The symbol for an earpiece is given in Fig. 9.11(*c*).

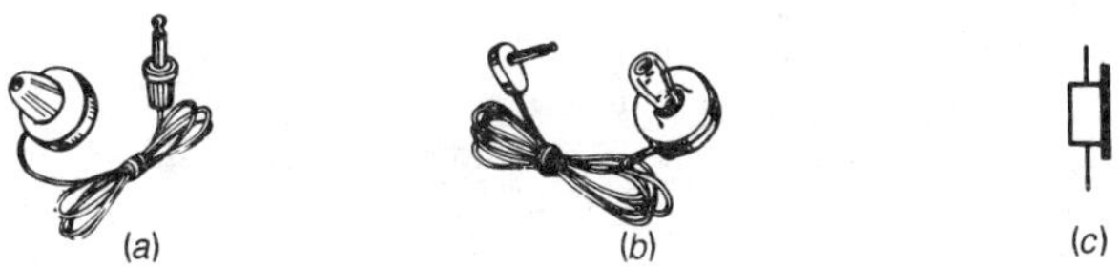

Fig. 9.11

9.4 Photocells

Photocells change light (also infrared and ultraviolet radiation) into electrical signals and are useful in burglar and fire alarms as well as in counting and automatic control systems.

(a) Photoconductive cell or light dependent resistor (LDR)

The resistance of certain semiconductors such as cadmium sulphide decreases as the intensity of the light falling on them increases. The effect is due to the energy of the light setting free electrons from donor atoms in the semiconductor, so increasing its conductivity, i.e. reducing its resistance.

A popular LDR is the ORP12 shown in Fig. 9.12(*a*); there is a 'window' over the grid-like metal structure to allow light to fall on a thin layer of cadmium sulphide, Fig. 9.12(*b*). Its resistance varies from 10 MΩ in the dark to 1 kΩ or so in daylight. The symbol for an LDR is given in Fig. 9.12(*c*).

Fig. 9.12

(b) Photovoltaic cells

When illuminated, a photovoltaic cell produces a voltage, i.e. it is a true cell. It consists of a p-n semiconductor junction with the p-layer being thin enough to allow the incident light to reach the junction. There it creates electron–hole pairs by breaking bonds between atoms in the depletion layer which exists at the junction. The junction voltage then sweeps the positive holes to the p-side and the negative electrons to the n-side, as shown in Fig. 9.13 together with the symbol for a photovoltaic cell. If there is an external circuit, current flows through it, with the p-side acting as the positive terminal of the cell. The source of energy is the incident light.

The voltage available depends on the junction materials, the intensity of the light and the current taken. For a silicon cell in full sunlight the voltage on open circuit is 0.45 V approximately with a maximum current of about 35 mA for each square centimetre of cell and an efficiency of about 10%, i.e. only 10% of the light is changed to electrical energy. For gallium arsenide the efficiency is nearer 20%.

Solar cells, Fig. 9.14, are photovoltaic types; they are used in panels on artificial satellites to power the electronic equipment, in calculators, and for small-scale electricity generation in places of high sunshine.

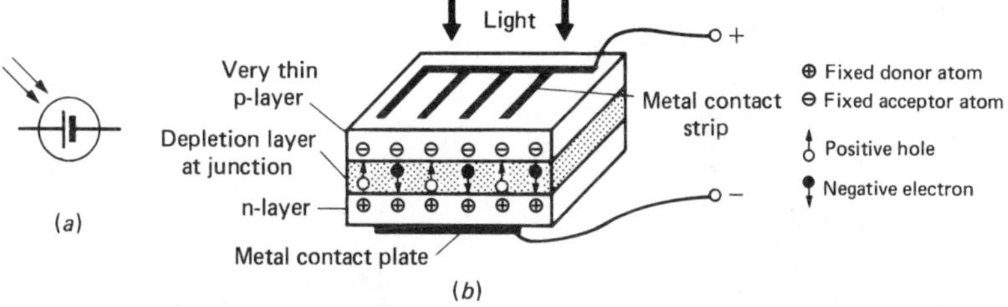

Fig. 9.13

Fig. 9.14

9.5 Photodiode and phototransistor

(a) Photodiode

This consists of a normal p-n junction in a case with a transparent 'window' through which light can enter. One is shown in Fig. 9.15(*a*) and its symbol in Fig. 9.15(*b*). A photodiode is operated in reverse bias and the leakage (minority carrier) current increases in proportion to the amount of light falling on the junction. This effect is due to the light energy breaking bonds in the crystal lattice of the semiconductor and producing electrons and holes (as in the photovoltaic cell).

Photodiodes are used as fast 'counters' which generate a pulse of current every time a beam of light is interrupted. They are also used in light-meters to measure light intensity.

(b) Phototransistor

You can think of this as a photodiode giving current amplification due to transistor action. Some are moulded in transparent plastic cases with the top convex acting as a lens focusing light on the transistor. As a result, extra minority carriers are liberated at the reverse biased collector-base junction. The leakage current so produced is then amplified. When used in this way, no connection to the base is needed and many phototransistors do not have a base lead.

A phototransistor is about 100 times more sensitive than a photodiode. Its symbol is shown in Fig. 9.15(*c*).

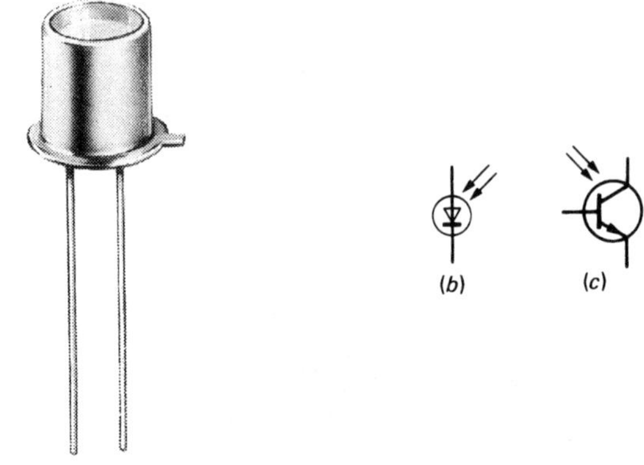

Fig. 9.15

9.6 Light-emitting diode (LED)

The transducers in sections 9.4 and 9.5 produce electrical signals from light; the opposite is done by a light-emitting diode, shown in Fig. 9.16 with its symbol. The cathode lead is nearer the 'flat' at the base of the LED and some, but by no means all, manufacturers make it shorter than the anode lead.

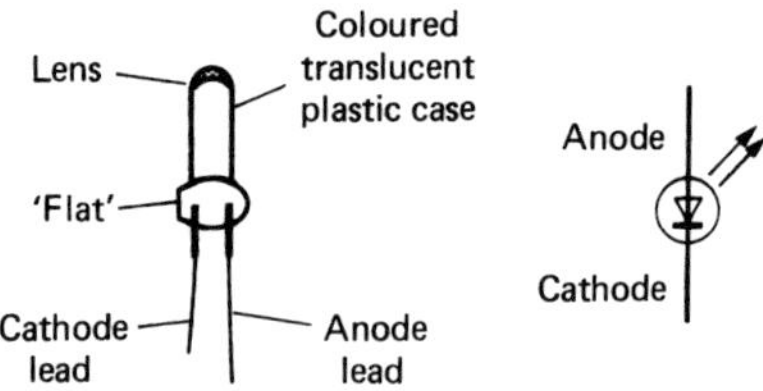

Fig. 9.16

(a) Action

A LED consists of a junction diode made from the semiconducting compound gallium arsenide phosphide. It emits light when *forward biased*, the colour depending on the composition and impurity content of the compound: red, yellow, green and blue LEDs are available. When a p-n junction diode is forward biased, electrons move across the junction from the n-type side to the p-side where they recombine with holes near the junction. The same occurs with holes going across the junction from the p-type side. Every recombination results in the release of a certain amount of energy, causing, in most semiconductors, a temperature rise. In gallium arsendide phosphide some of the energy is emitted as light which gets out of the LED because the junction is formed very close to the surface of the material. A LED does not emit light when reverse biased and if the bias is 5 V or more it may be damaged.

(b) External resistor

Unless a LED is of the 'constant-current type' (incorporating an integrated circuit regulator—see section 21.4—for use on a 2 to 18 V d.c. or a.c. supply), it must have an external resistor R connected in series to limit the forward current which, typically, may be 10 mA (0.01 A). Taking the voltage drop (V_F) across a conducting LED to be about 1.7 V, R can be calculated approximately from:

$$R = \frac{(\text{supply voltage} - 1.7)\ \text{V}}{0.01\ \text{A}}$$

For example, on a 5 V supply, $R = 3.3/0.01 = 330\ \Omega$.

(c) Uses

LEDs are used as indicator lamps, particularly in digital electronic circuits to show whether the output is 'high' or 'low'. One way of using a LED to test for a 'high' output (9 V in this case) is shown in Fig. 9.17(*a*) and for a 'low' output (0 V) in Fig. 9.17(*b*). In the first case the output acts as the 'source' of the LED current and in the second it has to be able to accept or 'sink' the current. If the output is unable to supply the current required by the LED, the circuit of Fig. 9.17(*c*) can be employed. Here the output supplies the small base current to the transistor which then drives the LED.

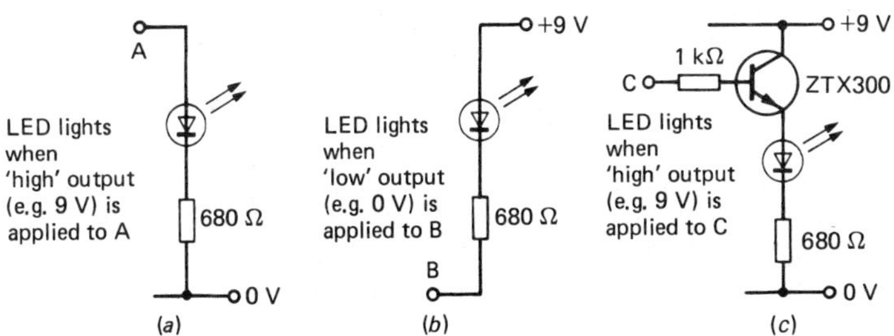

Fig. 9.17

(d) Decimal display

Many electronic calculators, clocks, cash registers and measuring instruments have seven-segment red or green LED displays as numerical indicators, Fig. 9.18(*a*). Each segment is a LED and depending on which segments are energized, the display lights up the numbers 0 to 9 as in Fig. 9.18(*b*). Such displays are usually designed to work on a 5 V supply. Each segment needs a separate current-limiting resistor and all the cathodes (or anodes) are joined together to form a common connection.

The advantages of LEDs are small size, reliability, long life, small current requirement and high operating speed.

(e) Opto-switch (opto-isolator)

This consists of a LED combined with a phototransistor in the same package and is shown in Figs. 9.19(*a*) and (*b*).

It allows the transfer of signals (on-off digital or continuoulsy varying analogue) from one circuit to another that cannot be connected electrically to the first, because, for example, it works at a different voltage. Light (or infrared) from the LED falls on the phototransistor which is shielded from outside light. The insulation between the two is typically 2 kV. Slotted opto-switches like the one shown are used for the detection of liquid levels and as event counters to indicate, for instance, when the end of a tape has passed through the slot.

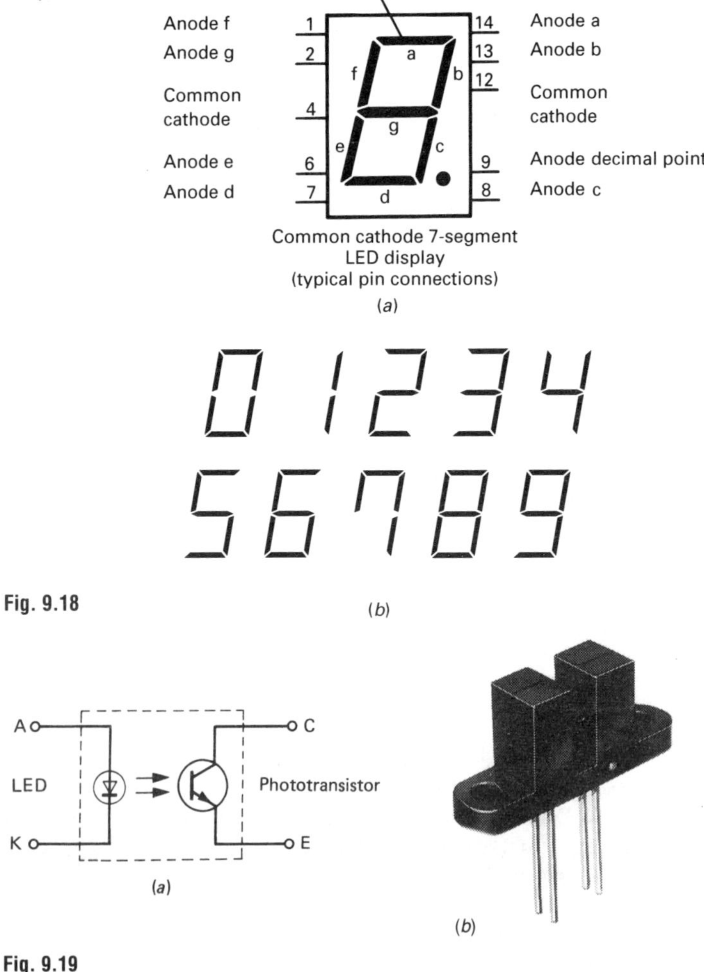

Fig. 9.18

Fig. 9.19

9.7 Liquid crystal and other numerical displays

(a) Liquid crystal display (LCD)

LCDs also are used as numerical indicators, especially in digital watches where their much smaller current needs than LED displays (microamperes compared with milliamperes) prolong battery life. Liquid crystals are organic (carbon) compounds which exhibit both solid and liquid properties. A 'cell' with transparent metallic conductors, called electrodes, on opposite faces, containing a liquid crystal, Fig. 9.20(*a*), and on which light falls, goes 'dark' when a voltage is applied across the electrodes. The effect is due to molecular rearrangement within the liquid crystal.

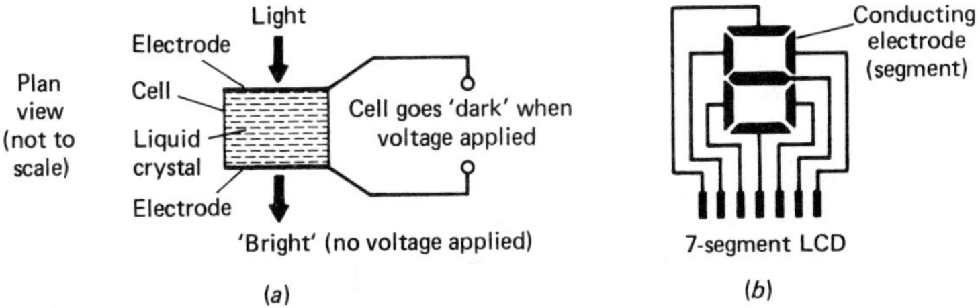

Fig. 9.20

The pattern of the conducting electrodes on a seven-segment LCD decimal display for producing the numbers 0 to 9 is shown in Fig. 9.20(*b*). Only the liquid crystal under those electrodes to which the voltage is applied goes 'dark'. The display has a silvered background which reflects back incident light and it is against this continuously visible background (except in darkness when it has to be illuminated) that the numbers show up as dark segments. LCDs require an a.c. supply to drive them using special circuitry. They are more expensive than LED displays.

Thousands of tiny LCDs are used to form the picture elements (pixels) of the flat screens in small TV receivers and portable personal computers (section 20.3).

(b) Gas discharge display (GDD)

Each segment of a GDD consists of a glass tube containing mainly neon gas at low pressure which produces a bright orange glow when a current is passed through it. Although about 170 V is required to start the display, the current taken by a segment may only be about 200 μA.

(c) Fluorescent vacuum display

Blue-green light is produced when electrons, emitted from an electrically heated filament (see section 9.8) and controlled by the voltage on a wire mesh (the grid), strike a fluorescent screen in an evacuated glass tube. This type of display is used for some calculators.

LED, gas discharge and fluorescent displays generate their own light, i.e. are 'active' displays and are most easily seen if the surrounding (ambient) light level is low. LCDs are 'passive' displays, viewed best in bright light since they are seen by reflected ambient light.

9.8 Cathode ray tube (CRT)

A cathode ray tube changes electrical energy into light and is used in oscilloscopes (see section 21.8), television receivers (see section 18.5), computer monitors (see section 20.3), electrocardiograph (ECG) machines, radar and other electronic systems. It consists of an electron gun for producing a narrow beam of high-speed electrons (a cathode ray) in an evacuated glass tube, a beam deflection system (electrostatic or magnetic) and a screen at the end of the tube which emits light where the beam falls on it. A typical CRT with electrostatic deflection is shown in Fig. 9.21(*a*); a voltage divider provides appropriate voltages from a high voltage power supply (e.h.t. denotes extra high tension).

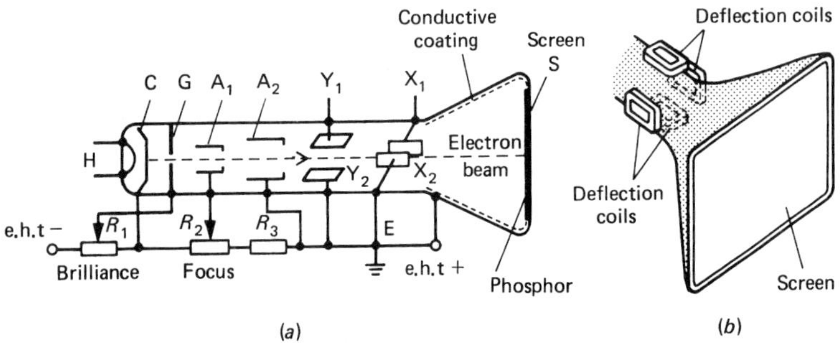

Fig. 9.21

(a) Electron gun

The *heater*, H, is a tungsten wire inside, but electrically insulated from, the hollow cylindrical nickel *cathode*, C. When current passes through H, it raises the temperature of C to dull red heat. This makes the mixture of barium and strontium oxides on the tip of C emit electrons copiously—a process called *thermionic emission*.

The negatively charged electrons are accelerated towards the *anodes* A_1 and A_2 (metal discs or cylinders), which have positive voltages with respect to C, A_2 more so than A_1. The *modulator* or *grid*, G, another hollow nickel cylinder, has, relative to C, a negative voltage which can be varied by R_1. The more negative it is, the fewer is the number of electrons emerging from its central hole and the less bright is the spot produced on the *screen*, S, by the beam after it has shot through the central holes in A_1 and A_2. R_1 is the *brilliance* control. Focusing of the beam to give a small luminous spot on S is achieved by changing the voltage between A_1 and A_2, i.e. by altering the *focus* control R_2. Typical voltages for a small tube are 1 kV for A_2, 200 to 300 V for A_1 and −50 to 0V for G.

There must be a return path for the electrons, from S to C, otherwise unwanted negative charge would build up on S. This does not happen because, when struck by electrons, S emits a more or less equal number of *secondary* electrons. These are attracted to and collected by a conductive coating (e.g. of graphite) on the inside of the tube near S and which is connected to A_2. From there the circuit is completed through the power supply to C. Therefore S and A_2 have about the same voltage relative to C and so the electron beam travels with constant speed between A_2 and S.

It is common practice to earth A_2 (and the conductive coating), thus preventing earthed objects (e.g. people) near S upsetting the electron beam. The other electrodes in the gun are then negative with respect to A_2 but the electrons are still accelerated through the same voltage between C and A_2.

(b) Deflection system

In electrostatic deflection (used in oscilloscopes) the beam from A_2 passes between two pairs of metal plates, Y_1Y_2 and X_1X_2, at right angles to each other. If one plate of each pair is at a positive voltage with respect to the other, the beam moves towards it. The Y-plates, which are farther from the screen, are horizontal and deflect the beam vertically i.e. along the *y*-axis. The X-plates are vertical and produce horizontal deflections, i.e. along the *x*-axis.

To minimize the defocusing which the voltages on the deflecting plates might have on the beam, one plate of each pair is at the same voltage (earth) as A_2. In practice X_2, Y_2 and A_2 are connected internally and brought out to one terminal, sometimes marked E (for earth). Deflecting voltages are then applied to Y_1 (often marked Y or input) and E or to X_1 (marked X) and E.

In magnetic deflection (used in television receiver tubes whose very large deflecting angle of 110° would require extremely high electrostatic deflecting voltages) two pairs of current-carrying coils are placed outside tube at right angles to each other, Fig. 9.21(*b*). The magnetic field produced by one pair causes deflections in the *y*-direction and by the other pair in the *x*-direction.

(c) Screen

The inside of the wide end of the tube is coated with a *phosphor* which emits light where it is hit by electrons. The colour of the light emitted and the 'afterglow', or 'persistance', i.e. the time for which the emission is visible after the electron bombardment has stopped, depends on the phosphor. Phosphors emitting bluish-green light are popular for oscilloscopes; for radar tubes long persistence types producing orange light are common; for black-and-white TV tubes white phosphors of short persistence are used, while for colour TV red, green and blue phosphors are employed—see section 18.6.

(d) Mono and PDA tubes

Two desirable properties of a CRT are high light output from the phosphor and good deflection sensitivity, i.e. large deflections from small voltages on the X- and Y-plates in the electrostatic system and by small currents in the coils in the magnetic system. The first property requires the electrons to be accelerated through a large voltage so that they hit the screen at *high* speed (and so have lots of energy). The second is achieved if the electrons travel at *low* speed between the deflecting plates or coils.

The CRT described above is a mono-accelerator tube (*mono* for short) in which the electrons are accelerated to their final speed before deflection; this does not give high deflection sensitivity. In a post-deflection acceleration tube (PDA for short), by having comparatively low voltages on A_1 and A_2, the electrons enter the deflection region at fairly low speed. The main acceleration occurs after the deflection by using the conductive coating inside the tube as the final 'anode'. The deflection sensitivity is then affected less adversely than in a mono tube with the same overall accelerating voltage.

9.9 Thermistors

A thermistor or *therm*al res*istor* is a semiconductor transducer whose resistance changes markedly when its temperature changes. Rod-, disc- and bead-shaped varieties are shown in Fig. 9.22(*a*) and the symbol for a thermistor in Fig. 9.22(*b*).

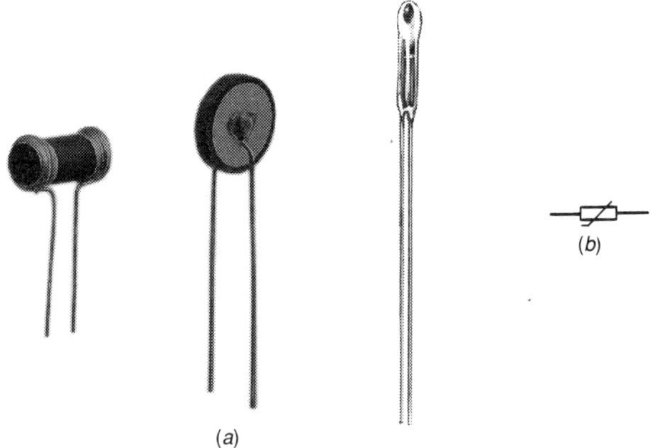

Fig. 9.22

The resistance of most thermistors decreases as their temperature increases. These are called n.t.c. types (standing for *n*egative *t*emperature *c*oefficient) and are made from oxides of nickel, manganese, copper, cobalt and other materials. They are used for temperature control and measurement; they are heated either externally from the surroundings or internally by the current flowing through them. Resistance changes cause current or voltage changes which supply the input to the electronic system where they are 'processed'.

With p.t.c. thermistors (standing for *p*ositive *t*emperature *c*oefficient) the resistance increases as the temperature increases. They are based on barium titanate and are used mainly to prevent damage in circuits which might experience a large temperature rise. This could happen, for example, to a much-overloaded electric motor.

9.10 Stepper motors

Electric motors convert electrical energy into mechanical energy. Stepper motors, driven by a series of electrical pulses, rotate by a small exact amount for each pulse. They are used in robots and in magnetic memory disc drives for computers (see section 20.4) where very precise movement is required.

The *rotor* is a permanent cylindrical magnet with a large number of poles round its perimeter. It rotates inside two sets of *stator* (field) coils which each have a row of metal teeth, Fig. 9.23(*a*). The teeth are magnetized with alternate north and south poles when current is sent through a coil. The rotor poles line up with a pair of opposite poles on the stator teeth, Fig. 9.23(*b*). When the current in one stator coil is reversed, say the upper one, the polarity of each tooth on the upper stator changes. As a result the rotor turns by one tooth as its poles realign themselves with the new arrangement.

A small stepper motor typically rotates through angular 'steps' of 7.5°, i.e. it makes 48 'steps' per revolution.

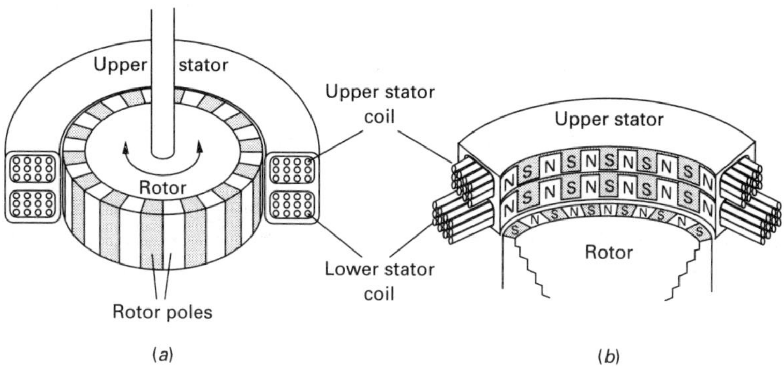

Fig. 9.23

9.11 Relays and reed switches

(a) Relays

A relay is a switch operated by an electromagnet. It is useful if we want a small current in one circuit to control another circuit containing a device such as a lamp or electric motor which requires a large current, or if we wish several different switch contacts to be operated simultaneously.

The structure of a relay is shown in Fig. 9.24(*a*) and its symbol in (*b*). When the controlling current flows through the coil, the soft iron core is magnetized and attracts the L-shaped soft iron armature. This rocks on its pivot and opens, closes or changes over, the electrical contacts in the circuit being controlled—in Fig. 9.24(*a*) it closes the contacts.

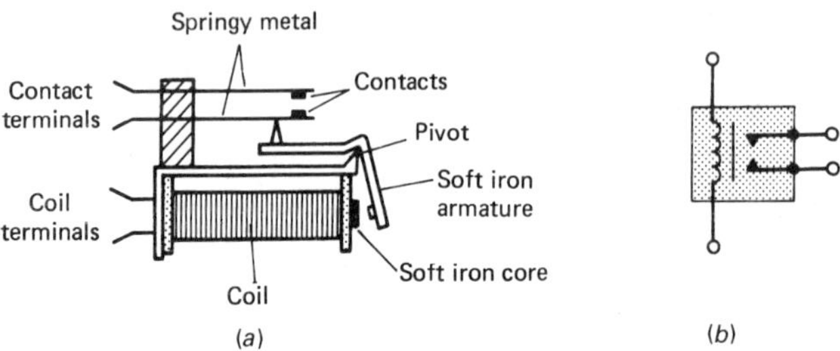

Fig. 9.24

The current needed to operate a relay is called the *pull-in* current, and the *drop-out* current is the current in the coil when the relay just stops working. If the coil resistance R of a relay is 185 Ω and its operating voltage V is 12 V, the pull-in current I is given by:

$$I = \frac{V}{R} = \frac{12}{185} = 0.065\ \text{A} = 65\ \text{mA}$$

(b) Reed switches

Relays operate comparatively slowly, so for fast switching of a single circuit reed switches are used. A reed switch with normally open contacts is shown in Fig. 9.25(*a*). The reeds are thin strips of easily magnetizable and demagnetizable material. They are sealed in a glass tube containing an inert gas such as nitrogen to reduce corrosion of the contacts.

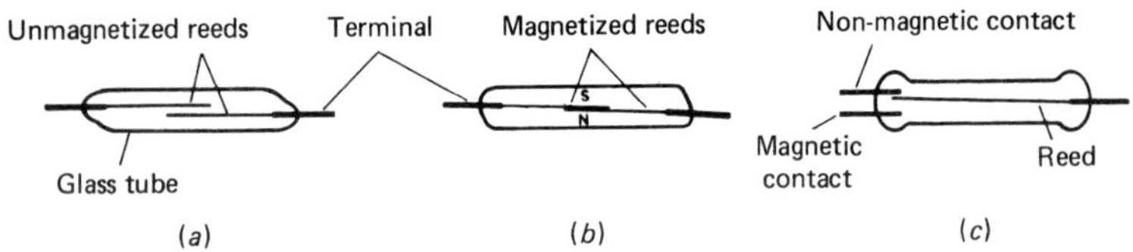

Fig. 9.25

The switch is operated either by bringing a magnet near or by passing a current through a coil surrounding it. In both cases the reeds become magnetized, attract each other and on touching they complete the circuit connected to the terminals, as shown in Fig. 9.25(*b*). They separate when the magnet is removed or the current stops flowing in the coil.

When the change-over reed switch of Fig. 9.25(*c*) operates, the reed is attracted from the non-magnetic contact to the magnetic one.

(c) Protection of transistor-controlled relays and reed switches

When the current in the coil of a relay or a reed switch falls to zero, a large voltage is induced in the coil due to its inductance—see section 6.3. This voltage could damage any transistor used to control the current in the coil. However, if a diode is connected in reverse bias relative to the supply voltage, Fig. 9.26, it offers a low resistance to the induced voltage and stops it building up to a high value.

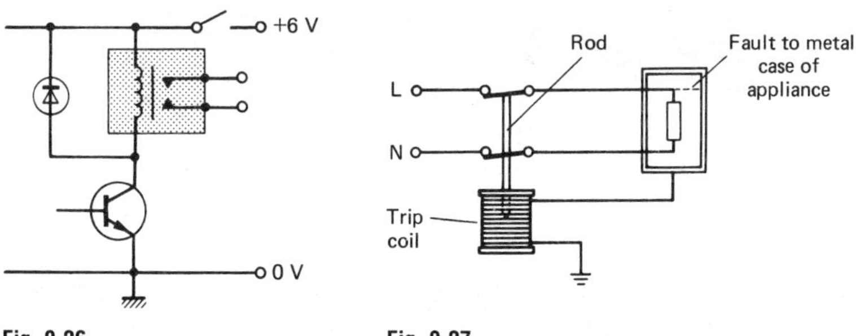

Fig. 9.26 **Fig. 9.27**

(d) Earth-leakage (or residual current) circuit breaker

This is sometimes present as a safety device in mains electrical circuits. In one variety, current passes to earth through a relay-type 'trip coil' when, for example, the metal case of the appliance becomes 'live' due to a fault, as in Fig. 9.27. As a result the rod in the coil opens the switch, which can be set to break the circuit before the case rises above say 25 V.

9.12 Mechanical switches

In a mechanical switch a force has to be applied to bring together or separate electrically conducting metal contacts. Various types are used in electronic circuits.

(i) Push-button The switch in Fig. 9.28(*a*) is a 'push-on, release-off' type; its symbol is given in Fig. 9.28(*b*) and that for the 'push-off, release-on' variety in Fig. 9.28(*c*).

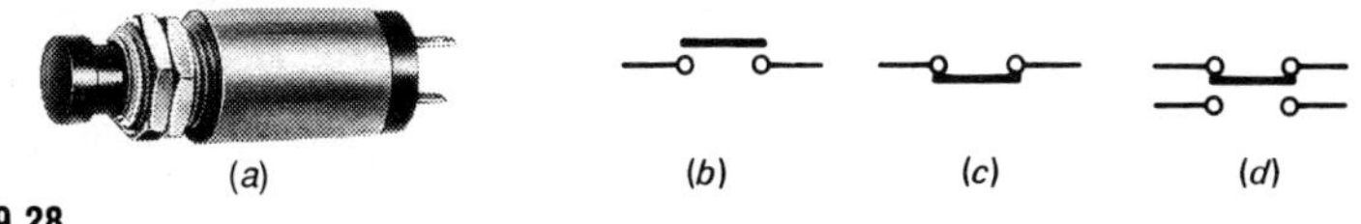

Fig. 9.28

Another common type is the 'push-on, push-off' switch. 'Push-to-change-over' switches are also made; their symbol is shown in Fig. 9.28(*d*).

(ii) Slide The one shown in Fig. 9.29(*a*) is a 'change-over' or 'single-pole-double-throw' (SPDT) switch. The 'poles' are the number of separate circuits the switch makes or breaks at the same time. The 'throws' are the number of positions to which each pole can be switched. For example, in the symbol for an SPDT switch, Fig. 9.29(*b*), there are two positions for the switch (B or C) and only one circuit (that joined to A) is switched.

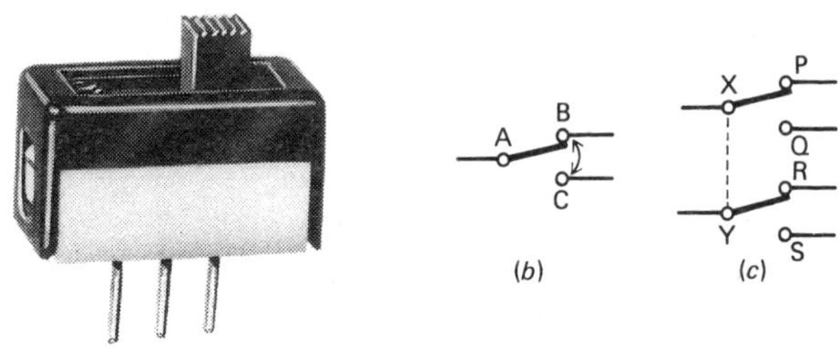

Fig. 9.29

A 'double-pole-double-throw' (DPDT) switch operates two circuits simultaneously. In Fig. 9.29(*c*) the circuits are those connected to X and Y and each one can go to either of two positions—to P or Q for X and to R or S for Y.

(iii) Toggle This is often used on equipment as a power supply 'on-off' switch, either in the 'single-pole-single-throw' (SPST) form (shown with its symbol in Figs. 9.30(*a*) and (*b*)) or as an SPDT or DPDT type.

Fig. 9.30

The rating for a particular switch depends on whether it is to be used in a.c. or d.c. circuits. For example, a certain toggle switch is rated at 250 V a.c. 1.5 A or 20 V d.c. 3 A—if these values are exceeded the life of the switch is shortened. When a circuit is switched off, sparking occurs at the switch contacts and vaporizes the metal. In general this lasts longer with d.c. than a.c. because the latter falls to zero twice per cycle.

(iv) Rotary wafer One or more discs (wafers) of paxolin (an insulator) are mounted on a twelve-position spindle as shown in Fig. 9.31(*a*). The wafers have metal contact strips on one or both sides and rotate between a similar number of fixed wafers with springy contact strips.

The contacts on the wafers can be arranged to give 1-pole 12-way, 2-pole 6-way, 3-pole 4-way, 4-pole 3-way (as in Figs. 9.31(*b*) and (*c*)) or 6-pole 2-way switching.

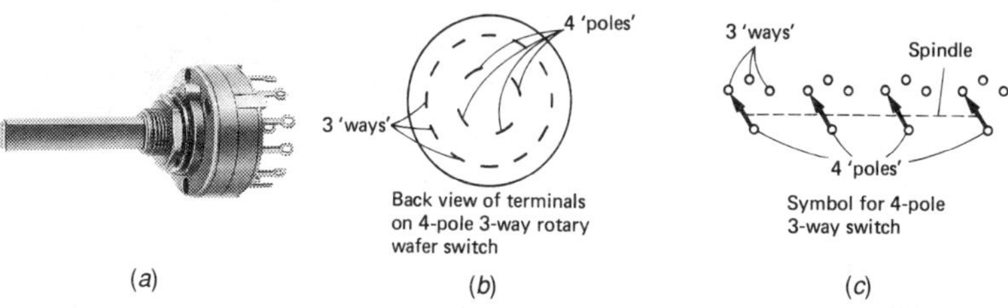

Fig. 9.31

(v) Keyboard switch The one shown in Fig. 9.32 is an SPST push-to-make momentary type which can be mounted on a printed circuit board (p.c.b.).

Fig. 9.32

9.13 Revision questions

1. What does a transducer do?

2. Name four kinds of microphone. State one use for each.

3. a) Draw a diagram of a moving-coil loudspeaker and explain how it works.
b) What is meant by saying that a loudspeaker has an impedance of 8 Ω?
c) Why is a loudspeaker often mounted in a cabinet?

4. A *crossover network* supplies a *woofer* and a *tweeter* and has a *crossover frequency* of 2.8 kHz. Explain the terms in italics.

5. What is the difference between a photovoltaic and a photoconductive cell?

6. a) What do the letters LED stand for?
b) What are typical values for
(i) the voltage drop across a forward biased LED, and
(ii) the forward current?
c) Draw two circuits to show how a *lit* LED could be used to indicate that the ouput from a circuit is
(i) high, (ii) low.

7. Give one advantage and one disadvantage of
a) LED displays,
b) liquid crystal displays.

8. a) Name the three main parts of a CRT.
b) Draw a diagram of a CRT labelling
(i) the heater, (iv) the anodes,
(ii) the cathode, (v) the Y-plates,
(iii) the modulator, (vi) the X-plates.
c) State two functions of the anodes.

9. Explain the following terms as they apply to a CRT: thermionic emission, secondary electrons, PDA, afterglow.

10. What is a thermistor? State two uses.

11. a) Describe the action of a stepper motor briefly.
b) State two uses for stepper motors.

12. What does each symbol represent in Fig. 9.33?

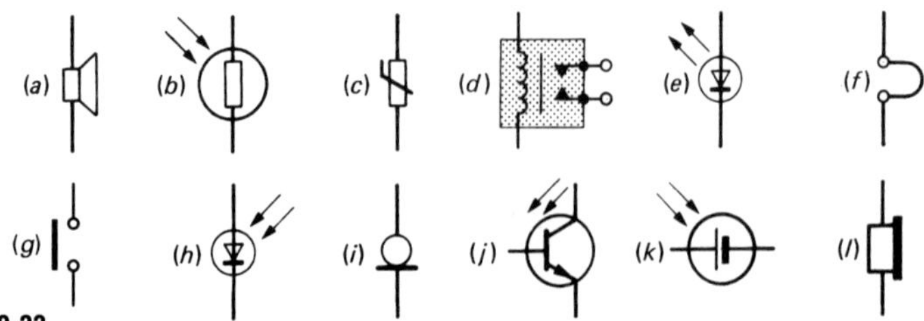

Fig. 9.33

13. a) What does a relay do?
b) Why are the armature and core of a relay made of 'soft' iron and not steel?

14. Name five types of mechanical switch.

15. a) What is meant by the terms
(i) SPST,
(ii) SPDT,
(iii) DPDT,
(iv) 3-pole 4-way, as applied to a switch?
b) Draw a symbol for each type of switch in a).

9.14 Problems

1. In Fig. 9.34 the LED is bright when a current of 10 mA flows and the forward voltage across it (V_F) is 2 V. Calculate the value (preferred) of R if a 9 V supply is used.

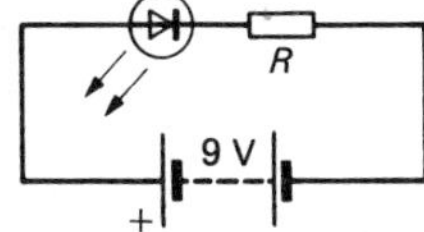

Fig. 9.34

2. a) What voltage is required to operate a relay having a coil of resistance 1 kΩ and a pull-in current of 10 mA?
b) Calculate the pull-in current of a reed switch designed to operate with a coil of resistance 700 Ω on a 6 V battery.

3. In Fig. 9.35 there are two circuits in which the operation of a relay is delayed intentionally for a certain time by using a large capacitor. Such circuits have their uses in electronics.
a) In Fig. 9.35(*a*) the normally open contacts of a relay close and a lamp *comes on a few seconds after* S is switched on (and stays on so long as S is on). Explain why.
b) In Fig. 9.35(*b*) the lamp comes on as soon as S is switched on and *goes off a few seconds later* (with S still on). Why is this?

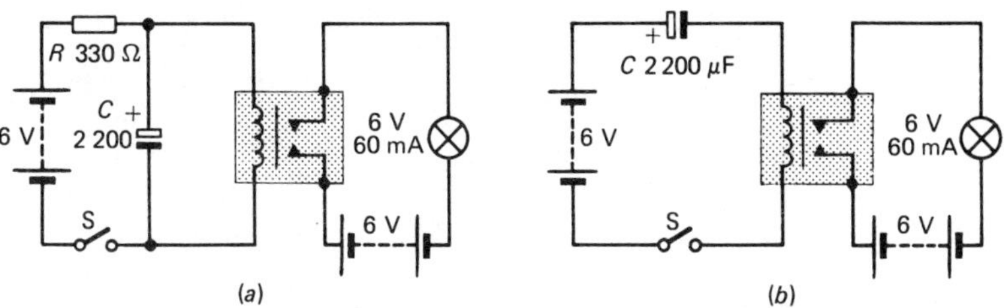

Fig. 9.35

PART THREE

Linear circuits

UNIT 10

Basic audio frequency amplifiers

10.1 Introduction

Linear (or analogue) circuits are *amplifier-type* circuits. They handle signals that are electrical representations (analogues) of physical quantities which vary continuously with time and take on all values between a certain maximum and minimum. The information carried by such signals (e.g. the loudness and pitch of a sound) is in the amplitude and shape of their waveforms. The circuit design results in a linear relationship between the input and the output i.e. doubling the input doubles the output and so on.

In general the job of an amplifier is to produce an output which is an enlarged copy of the input. The symbol for an amplifier is shown in Fig. 10.01. Amplifiers can be classified according to their function and the frequency range they cover. In a *voltage* (or small-signal) amplifier the output voltage is greater than the input voltage. A *power* (or large-signal) amplifier is designed to deliver power to an output transducer (although it may also have a voltage gain).

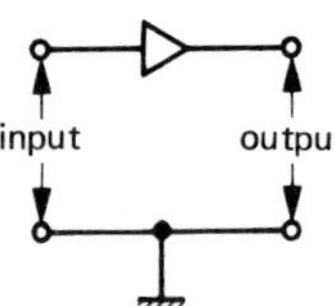

Fig. 10.01

An *audio frequency* (a.f.) amplifier amplifies a.c. signals in the audio frequency range, i.e. 20 Hz to 20 kHz. *Radio frequency* (r.f.) amplifiers operate above 20 kHz, at radio and television signal frequencies.

A complete audio amplifier for, say, a CD player consists of several amplifier stages coupled together so that the output of one stage becomes the input to the next stage. The early stages are voltage amplifiers and the last one a power amplifier. Each stage is built round one or more transistors.

This Unit gives an introduction to a.f. voltage amplifiers. Unit 11 deals with more advanced aspects, including a.f. power amplifiers, while in Unit 12 radio frequency amplifiers are considered.

10.2 Voltage amplifier using a junction transistor

Although a junction transistor in the common-emitter mode is basically a current amplifier (see section 8.3), it can act as a voltage amplifier if a suitable resistor, called the *load*, is connected in the collector circuit. The small alternating voltage to be amplified, i.e. the *input voltage* v_i, is applied to the base-emitter circuit and causes small changes of base current which produce large changes in the collector current flowing through the load. The load converts these current changes into voltage changes which form the alternating *output voltage* v_o, v_o being much greater than v_i. (Note the use of small italic letters to represent instantaneous values of varying quantities.)

The simplest circuit for a transistor amplifier is shown in Fig. 10.02. To see in more detail just why voltage amplification occurs, consider first the situation when there is no input (i.e. $v_i = 0$)—called the quiescent state (quiescent means 'quiet').

(a) Quiescent state

For transistor action to take place, the base-emitter junction must be forward biased (and has to remain so even when v_i is applied and goes negative). As we saw in section 8.3, a simple way of doing this is to connect a resistor R_B called the *base bias resistor*, as shown. A steady (d.c.) base current I_B flows from + V_{CC}, through R_B into the base and back to 0 V via the emitter. The value of R_B can be calculated (see section 10.3) once the value of I_B for the best amplifier performance has been decided.

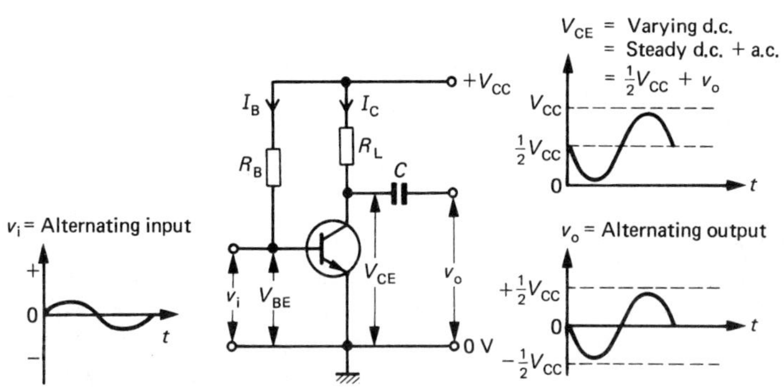

Fig. 10.02

If V_{CC} is the power supply voltage and V_{BE} is the base-emitter junction voltage (always about +0.6 V for an n-p-n silicon transistor), then for the *base-emitter* circuit, since d.c. voltages add up, we can write:

$$V_{CC} = I_B \times R_B + V_{BE} \tag{1}$$

I_B causes a much larger collector current I_C which produces a voltage drop $I_C \times R_L$ across the load R_L. The voltage at the end of R_L joined to $+V_{CC}$ is fixed and so the voltage drop must be at the end connected to the collector. If V_{CE} is the collector-emitter voltage, then for the *collector-emitter* circuit we can write:

$$V_{CC} = I_C \times R_L + V_{CE} \qquad (2)$$

Component values are chosen so that the steady base bias current I_B makes the quiescent collector-emitter voltage V_{CE} about half the power supply voltage V_{CC}. This allows v_o to have its maximum swing capability, in theory from 0 V to V_{CC}.

(b) Applied input signal

When v_i is applied and goes positive, it increases the base-emitter voltage slightly (e.g. to +0.61 V). When v_i swings negative, the base-emitter voltage decreases slightly (e.g. to +0.59 V). As a result a small alternating current is superimposed on the quiescent base current I_B which in effect becomes a varying d.c.

When the base-current *increases*, large proportionate increases occur in the collector current. From equation (2) it follows that there is a corresponding large *decrease* in the collector-emitter voltage (since V_{CC} is fixed). A decrease of base current causes a large increase of collector-emitter voltage. In practice, positive and negative swings of a few millivolts in v_i can result in a fall or rise of several volts in the voltage across R_L and therefore in the collector-emitter voltage as well.

The collector-emitter voltage is a varying direct voltage and may be regarded as an alternating voltage superimposed on a steady direct voltage, i.e. on the quiescent value of V_{CE}—see section 3.3. Only the alternating part is wanted and capacitor C blocks the direct part but allows the alternating part, i.e. the output v_o, to pass on to the next stage.

Summing up, a transistor will act as a voltage amplifier if (i) it has a suitable collector load, and (ii) it is biased so that the quiescent value of $V_{CE} \approx \frac{1}{2} V_{CC}$, known as the *class A* biasing condition.

(c) Further points

The transistor and load together bring about voltage amplification.

The output is 180° out of phase with the input, i.e. when the input has its maximum positive value, the output has its maximum negative value, as the graphs in Fig. 10.02 show, i.e. the amplifier is an *inverter*.

The emitter is common to the input, output and power supply circuits and is usually taken as the reference point for all voltages, i.e. 0 V. It is called 'common' or 'ground', or 'earth' if connected to earth.

10.3 Worked example

A silicon transistor in the simple voltage amplifier circuit of Fig. 10.02 operates satisfactorily on a quiescent (no input) collector current (I_C) of 3 mA. If the power supply (V_{CC}) is 6 V, what must be the value of (a) the load resistor (R_L) and (b) the base bias resistor (R_B), for the quiescent collector-emitter voltage (V_{CE}) to be half the power supply voltage? The transistor d.c. current gain (h_{FE}) is 100.

a) The collector-emitter circuit equation is:

$$V_{CC} = I_C \times R_L + V_{CE}$$

Rearranging we get:

$$I_C \times R_L = V_{CC} - V_{CE}$$

That is:

$$R_L = (V_{CC} - V_{CE})/I_C$$

Substituting $V_{CC} = 6$ V, $V_{CE} = \frac{1}{2}V_{CC} = 3$ V and $I_C = 3$ mA gives:

$$R_L = (6 - 3)\ V/3\ \text{mA} = 1\ \text{k}\Omega$$

b) The d.c. current gain is given by:

$$h_{FE} = \frac{I_C}{I_B}$$

I_B is the quiescent base current to produce the quiescent collector current I_C. Rearranging:

$$I_B = \frac{I_C}{h_{FE}}$$

Substituting $I_C = 3$ mA and $h_{FE} = 100$, we get:

$$I_B = 3\ \text{mA}/100 = 0.03\ \text{mA}\ (30\ \mu\text{A})$$

The base-emitter circuit equation is:

$$V_{CC} = I_B \times R_B + V_{BE}$$

V_{BE} is the base-emitter voltage.
Rearranging gives:

$$R_B = (V_{CC} - V_{BE})/I_B$$

Substituting $V_{CC} = 6$ V, $V_{BE} = 0.6$ V (for a silicon transistor) and $I_B = 0.03$ mA gives:

$$R_B = (6 - 0.6)\ \text{V}/0.03\ \text{mA} = 5.4\ \text{V}/0.03\ \text{mA}$$
$$= 540/3\ \text{k}\Omega = 180\ \text{k}\Omega$$

10.4 Load lines, operating point and voltage gain

When designing a voltage amplifier, the aim is usually to obtain:

(i) a certain voltage gain;
(ii) minimum distortion of the output so that it is a good copy of the input;
(iii) operation within the current, voltage and power limits for the transistor.

The choice of the quiescent or d.c. operating point (i.e. the values of I_C and V_{CE}) determines whether these requirements will be met. This choice is made by constructing a *load line*, as we will now see.

If you look back to Fig. 8.08(*b*) (p. 95), you will see the output characteristics of a transistor showing the relation between V_{CE} and I_C with *no load* in the collector circuit. With a load R_L, the equation connecting them is (see section 10.2):

$$V_{CC} = I_C \times R_L + V_{CE}$$

where V_{CC} is the power supply voltage.

Rearranging we get:

$$V_{CE} = V_{CC} - I_C \times R_L \qquad (3)$$

Knowing V_{CC} and R_L this equation enables us to calculate V_{CE} for different values of I_C. If a graph of I_C (on the *y*-axis) is plotted against V_{CE} (on the *x*-axis) we get a straight line, called a load line. The line can be drawn if we know just two points. The easiest to find are the end points where the line cuts the V_{CE}- and I_C-axes. We shall call these points A and B respectively.

For A we put $I_C = 0$ in (3) and get $V_{CE} = V_{CC} = 6$ V (say).

For B we put $V_{CE} = 0$ in (3) and get $I_C = V_{CC}/R_L$. If $R_L = 1$ kΩ say, then $I_C = 6\text{V}/1\text{ k}\Omega = 6$ mA.

In Fig. 10.03, AB is the load line for $V_{CC} = 6$ V and $R_L = 1$ kΩ. It is shown superimposed on the output characteristics of the transistor used in the circuit of Fig. 10.02. We can regard a load line as the output characteristic of the *transistor and load* for particular values of V_{CC} and R_L. Different values of either give a different load line; for example, a smaller value of R_L gives a steeper line.

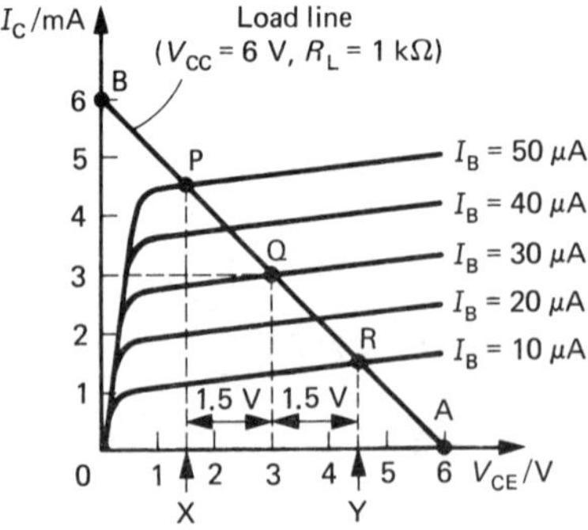

Fig. 10.03

The choice of load line and d.c. operating point affects the shape and size of the output waveform. Choosing a line which cuts the characteristics where they are not linear (straight) or where they are not equally spaced can cause distortion. Selecting an operating point too near either the I_C- or the V_{CE}-axis can have the same effect, shown in Figs. 10.04(*a*) and (*b*) respectively. The best position for the d.c. operating point is near the *middle* of the chosen load line, e.g. at Q in Fig. 10.03. The 'swing' capability of the output is then a maximum (from near V_{CC} to near 0 V) and distortion a minimum as shown in Fig. 10.04(*c*). But also note that too large an input can cause distortion even if the operating point has been correctly chosen, as in Fig. 10.04(*d*). In severe cases a sine wave input would be 'clipped' so much as to give a square wave output.

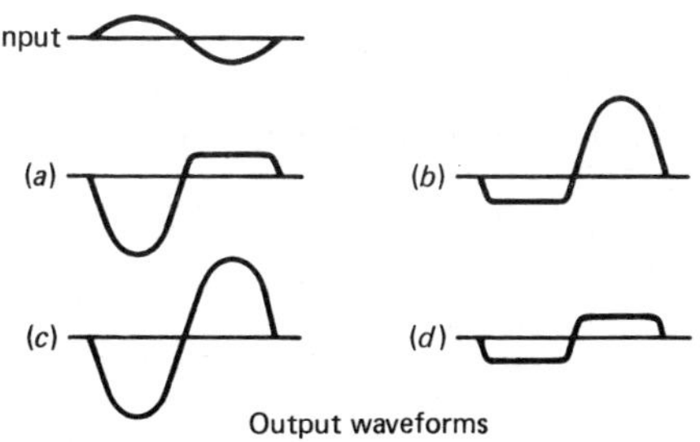

Fig. 10.04

Having chosen Q, the quiescent values of V_{CE} and I_C can be read off. In Fig. 10.03 they are $V_{CE} = 3$ V (i.e. half V_{CC}) and $I_C = 3$ mA. The value of I_B which gives these values is obtained from the transistor output characteristics passing through Q (since I_C and V_{CE} have to satisfy *at the same time* both the characteristic and the load line). Here it is the 30 μA characteristic. R_B can then be calculated as in section 10.3.

The *voltage gain A* is given by:

$$A = \frac{\text{output voltage}}{\text{input voltage}} = \frac{\text{change in } V_{CE}}{\text{change in } V_{BE}} = \frac{\Delta V_{CE}}{\Delta V_{BE}}$$

It can be obtained from the load line by noting that when the input causes I_B to vary from 10 to 50 μA (from R to P), V_{CE} varies from 4.5 to 1.5 V (from Y to X). From the input characteristic of the transistor in Fig. 8.08(*a*) we can find the change in V_{BE} to cause this change of 40 μA in I_B. From the graph, it is approximately 40 mV (0.04 V). We have:

$$A = \frac{4.5 - 1.5}{0.04} = \frac{3.0}{0.04} = 75$$

10.5 Stability and bias

(a) Thermal runaway

If the temperature of a transistor rises, there is greater vibration of the semiconductor atoms, resulting in the production of more free electrons and holes. The collector current increases causing further heating of the transistor until eventually the transistor is damaged or destroyed. Initial temperature rise may be due to the heating effect of the collector current or to a rise in the surrounding temperature.

To stop this 'thermal runaway' effect and stabilize the d.c. operating point, special bias circuits have been designed which automatically compensate for variations of collector current. The simplest of these will now be considered; a more complex one is discussed in section 11.1.

(b) Collector-to-base bias

The basic circuit of Fig. 10.02 for a voltage amplifier can be adequately stabilized for many applications by halving the value of R_B and connecting it between the collector and base as in Fig. 10.05, rather than between $+V_{CC}$ and base.

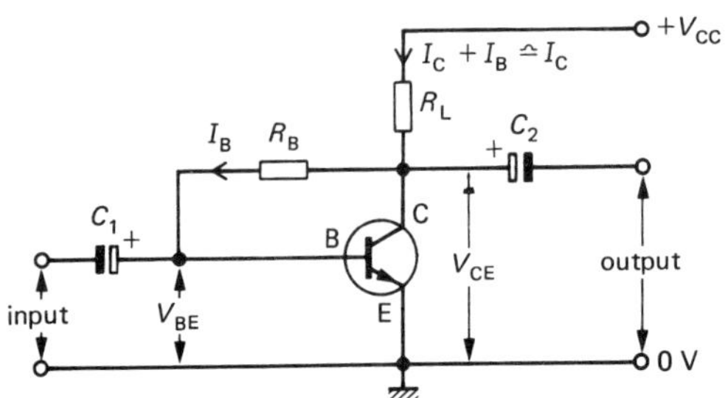

Fig. 10.05

For the circuit in Fig. 10.05 we can write (since I_C is much greater than I_B):

$$V_{CC} = I_C \times R_L + V_{CE} \quad (4)$$

where:

$$V_{CE} = I_B \times R_B + V_{BE} \quad (5)$$

From (4) you can see that if I_C increases for any reason, V_{CE} decreases since V_{CC} is fixed. From (5) it therefore follows that since V_{BE} is constant (0.6 V or so), I_B must also decrease and in so doing tends to bring back I_C to its original value.

Taking the quiescent conditions to be (as in the worked example in section 10.3): $V_{CE} = \frac{1}{2}V_{CC} = 3$ V, $I_C = 3$ mA and $I_B = 0.03$ mA, the value of R_B in Fig. 10.05 is found by rearranging equation (5) to give:

$$R_B = \frac{V_{CE} - V_{BE}}{I_B} = \frac{(3 - 0.6)\text{ V}}{0.03\text{ mA}}$$

$$= 2.4\text{ V}/0.03\text{ mA} = 82\text{ k}\Omega \quad \text{(preferred value)}$$

This is about half the value of 180 kΩ for R_B in the unstabilized circuit of Fig. 10.02. The collector-to-base bias circuit is a useful general purpose voltage amplifier circuit.

10.6 Simple two-stage voltage amplifier

When greater gain is required, two (or more) amplifier stages are coupled. The circuit in Fig. 10.06 is for a two-stage capacitor-coupled a.f. voltage amplifier using collector-to-base bias.

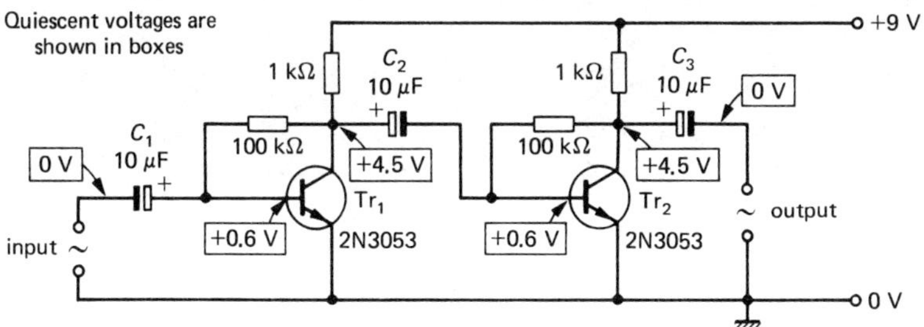

Fig. 10.06

(a) Capacitor coupling

The output from Tr_1 is applied to the input of Tr_2 via C_2 which connects the collector of Tr_1 (it must have a quiescent d.c. voltage of +4.5 V for correct operation) and the base of Tr_2 (it is +0.6 V above the emitter at 0 V, being a forward biased junction). A direct connection, without C_2, would have the disastrous effect of fixing the collector of Tr_1 at only +0.6 V above ground and would also send a base current of several mA into Tr_2 (through the 1 kΩ Tr_1 collector load) that would saturate Tr_2 permanently. At most audio frequencies the reactances of C_2 ($X_C = 1/(2\pi f C_2)$) and also of the input and output coupling capacitors C_1 and C_3, are small and so the alternating part of the output voltage from Tr_1 is transferred with little loss to the base of Tr_2.

C_1, C_2 and C_3 are electrolytics and must be connected with the correct polarities. For example, as the collector voltage of Tr_1 is more positive than the base voltage of Tr_2, the + terminal of C_2 therefore goes to the collector of Tr_1.

In an amplifier with an even number of stages, as in Fig. 10.06, the output and input are in phase, i.e. it is a *non-inverting* amplifier. An inverting amplifier has an odd number of stages.

(b) Frequency response

The voltage gain A of a capacitor-coupled amplifier is fairly constant over most of the a.f. range but it falls off at the lower and upper limits. At low frequencies, the reactances of the coupling capacitors increase and less of the low frequency part of the input is passed on. At high frequencies various stray capacitances in the active devices and between the connections can cause the fall.

A typical voltage gain–frequency curve is shown in Fig. 10.07. The *bandwidth* is the range of frequencies within which the voltage gain does not fall below $1/\sqrt{2}$ (i.e. 0.7) of its maximum value A_{max}. (To fit in the large frequency range, frequencies are not plotted on the usual linear scale but on a logarithmic scale in which equal divisions represent equal changes in the 'log of the frequency f'.) The two points P and Q at which this happens are called the '3 dB points'. The decibel (dB) scale compares signal levels and is explained in section 11.9.

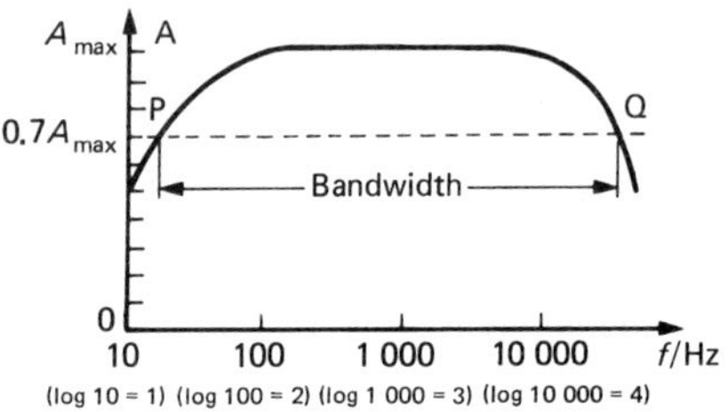

Fig. 10.07

10.7 Voltage amplifier using a FET

In a FET voltage amplifier, changes in the gate (input) voltage causes changes in the drain current which are converted into larger voltage changes by a load resistor in the drain (output) circuit. The load line and d.c. operating point are selected as for a junction transistor voltage amplifier—see section 10.4. The chosen operating point is realized in practice by applying the correct quiescent bias voltage to the gate.

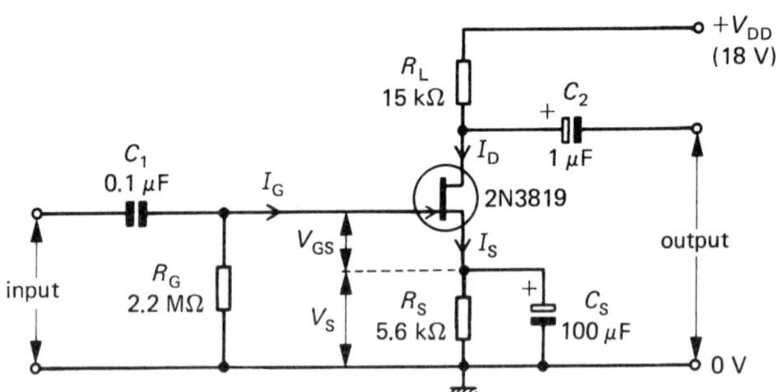

Fig. 10.08

The circuit for an n-channel JUGFET voltage amplifier in common-source connection is shown in Fig. 10.08. The values of the load resistor R_L and the supply voltage V_{DD} (note the symbol: V_{CC} is used for the supply voltage to a junction transistor) are both higher than for a bipolar transistor to obtain a reasonable gain. Negative (quiescent) bias is required for the gate since the FET works in the depletion mode—see section 8.7. It is provided as follows.

(i) Source and gate resistors (R_S and R_G) In the quiescent state, the source current I_S ($= I_D + I_G \approx I_D$ since $I_G \approx 0$) is steady and causes a voltage drop across resistor R_S. The source end of R_S is therefore positive with respect to the other end connected by the high resistor R_G to the gate. R_G ensures that the gate has the same voltage as the lower end of R_S on the diagram. This is so because there is negligible current through R_G and hence practically no voltage across it. Both ends of R_G have the same voltage, namely that of the lower end of R_S, i.e. 'ground' or 0 V. The source is therefore at a higher voltage than the gate, i.e. the gate is negative with respect to the source. (If you find this difficult to understand, refer back to section 2.4.)

The circuit automatically compensates for any change of I_S and helps to stabilize the d.c. operating point because any increase in I_S increases the voltage V_S ($= I_S \times R_S$) across R_S. The voltage of the source end of R_S (and so of the source) rises and since the lower end of R_S and the gate are tied to 0 V, the gate-source voltage V_{GS} must go more negative, tending to reduce I_S to its previous value.

(ii) Decoupling capacitor (C_S) The large decoupling capacitor C_S provides a bypass (i.e. a low impedance) round R_S for the a.c. part of the source current (which becomes a varying d.c.) when an alternating input is applied. Otherwise the varying voltage developed across R_S would cause unwanted changes in the value (quiescent) of V_{GS} required to give the chosen operating point.

(iii) Blocking capacitor (C_1) This blocks any d.c. voltage from the input which would affect the operating point. With R_G it forms a voltage divider across the input. The alternating voltage developed across R_G is applied to the gate for amplification and for this voltage to be as large as possible, the value of R_G should be large compared with the reactance of C_1. Since the resistance of R_G is usually in the range of 2.2 to 10 MΩ, the capacitance C_1 can be small (e.g. 0.1 μF).

FET voltage amplifiers give lower gains than junction transistor types (typically ten compared with up to a thousand). It can be shown that the voltage gain A is given approximately by $A = g_m \times R_L$ where g_m is the transconductance of the FET—see section 8.7. However, their much greater input impedances make them better for certain applications, e.g. as impedance-matching devices (see section 11.6) and as r.f. amplifiers (see section 12.4).

10.8 Revision questions

1. Explain the following terms:
   a) voltage amplifier,
   b) power amplifier,
   c) a.f. amplifier,
   d) r.f. amplifier.

2. Is a junction transistor in common-emitter connection basically a current or a voltage amplifier?

3. Answer the following questions about the voltage amplifier circuit in Fig. 10.09.
   a) Which are the input terminals?
   b) Which are the output terminals?
   c) Where is the power supply connected (give polarities)?
   d) What is the purpose of R_L?
   e) What is the purpose of R_B?
   f) What function does C_1 serve?
   g) What does C_2 do?
   h) What is the phase relationship between the input and output voltages?

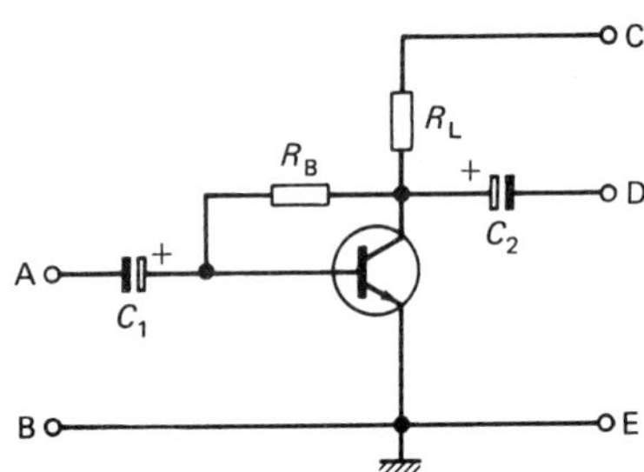

Fig. 10.09

4. Explain the following terms as applied to an amplifier:
   a) quiescent state,
   b) d.c. operating point,
   c) voltage gain.

5. When designing a voltage amplifier what are the three main aims?

6. a) What is the purpose of drawing a load line?
   b) Name the factors which should be considered when choosing (i) a load line, (ii) the d.c. operating point.
   c) In Fig. 10.10, comment on the choice of (i) AB, (ii) AC, as load lines.
   d) In Fig. 10.10, comment on the choice of (i) P, (ii) Q, (iii) R, (iv) S, as d.c. operating points.

7. What is meant by thermal runaway? How is it prevented?

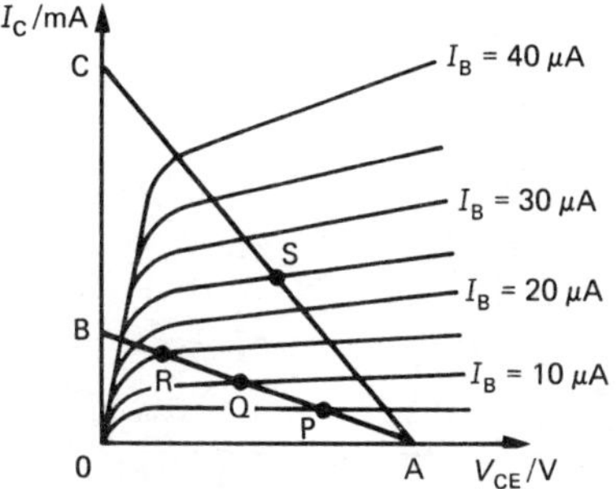

Fig. 10.10

8. Answer the following questions about the FET voltage amplifier in Fig. 10.11.

a) What is the purpose of R_S?

b) What does R_G ensure?

c) What does C_S do?

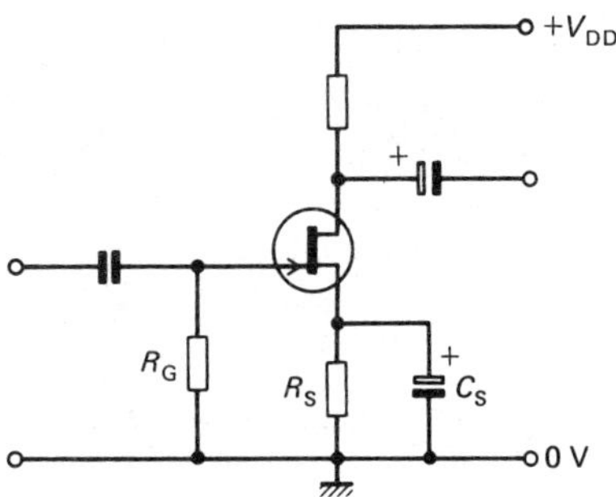

Fig. 10.11

10.9 Problems

1. In the circuit of Fig. 10.12, if V_{CC} = 9 V, R_L = 1 kΩ, I_C = 3 mA and h_{FE} = 200 for the transistor, calculate

a) V_{CE},

b) I_B, and

c) R_B, if V_{BE} = 0.6 V.

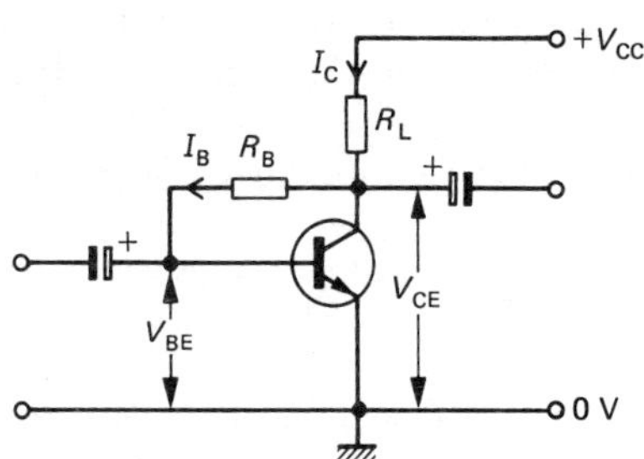

Fig. 10.12

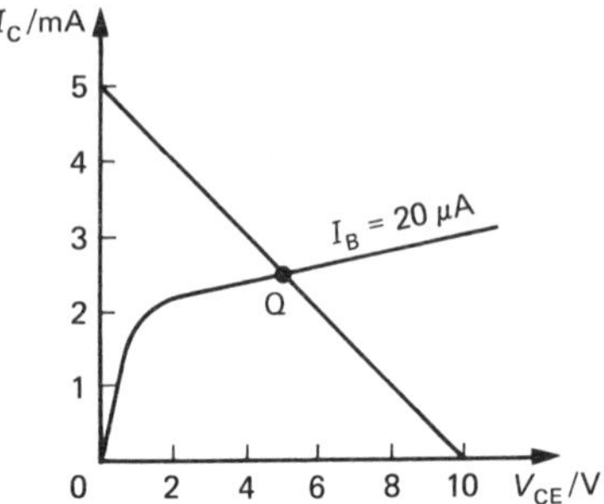

Fig. 10.13

2. The load line for the amplifier in Fig. 10.12 is shown in Fig. 10.13.
   a) Use it to find the values of V_{CC} and R_L.
   b) If Q is the d.c. operating point, what are the quiescent values of V_{CE}, I_C and I_B?
   c) Calculate R_B if the transistor is a silicon type.

3. The output characteristics of a junction transistor in common-emitter connection are shown in Fig. 10.14(*a*). The transistor is used in an amplifier with a 9 V supply and a load resistor of 1.8 kΩ.
   a) Copy the graph and draw the load line.
   b) Choose a suitable d.c. operating point and read off the quiescent values of I_C, I_B and V_{CE}.

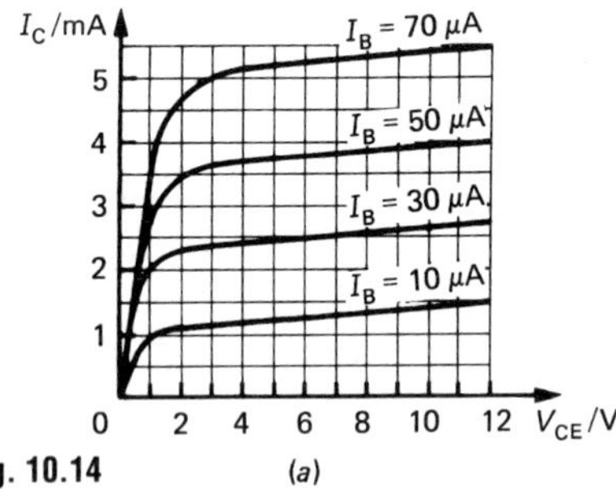

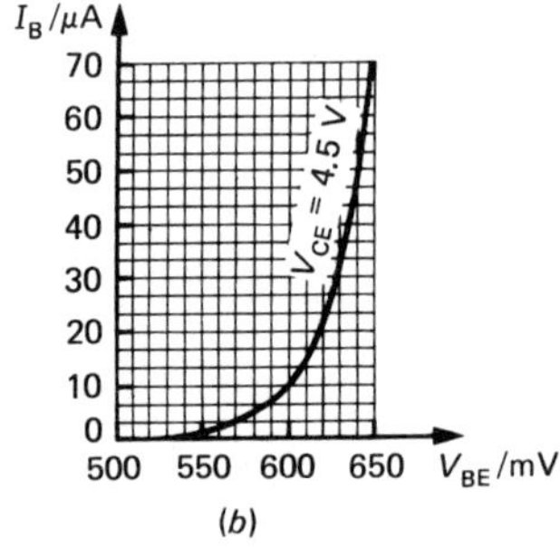

Fig. 10.14 (*a*) (*b*)

   c) What is the quiescent power consumption of the amplifier?
   d) If an alternating input voltage varies the base current by ±20 μA about its quiescent value, what is
      (i) the variation in the collector-emitter voltage, and
      (ii) the peak output voltage?
   e) An input characteristic of the transistor is given in Fig. 10.14(*b*). Use it to find the base-emitter voltage variation which causes a change of ±20 μA in the base current.
   f) Using your answers from d) and e) find the voltage gain of the amplifier.
   g) If the amplifier uses the collector-to-base bias circuit of Fig. 10.12, calculate the value of R_B to give the quiescent value of I_B. (Assume V_{BE} = 0.6 V.)

UNIT 11

More audio frequency amplifiers

This Unit deals with more advanced aspects of audio amplifier circuit design.

11.1 Fully stabilized voltage amplifier

When complete stabilization of the d.c. operating point is required, voltage divider and emitter resistor bias is used. The circuit for a junction transistor is shown in Fig. 11.01. It has three main features.

(i) Voltage divider (R_1, R_2) The junction of R_1 and R_2 fixes the base *voltage* at a value sufficient to forward bias the base-emitter junction. (In collector-to-base bias the base *current* was fixed.) The current through R_1 and R_2 is usually made about ten times greater than the quiescent base current so that when the latter varies with the input, the base voltage is hardly affected—see section 4.7.

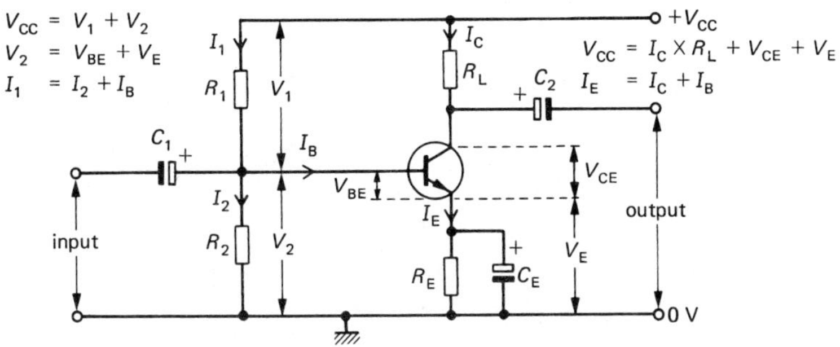

Fig. 11.01

(ii) Emitter resistor (R_E) If the collector current I_C increases due to say, a temperature rise, the emitter current I_E also increases (since $I_E = I_C + I_B$), as does V_E which equals $I_E \times R_E$. The voltage across R_2 is given by $V_2 = V_{BE} + V_E$. Therefore if V_2 is fixed (by the voltage divider), V_{BE} must decrease when V_E increases. As a result I_B decreases, the fall being sufficient to restore I_C to its original value.

(iii) Decoupling capacitor (C_E) When an alternating input is applied, the collector and emitter currents become varying d.c., i.e. steady d.c. + a.c. The large capacitor C_E across R_E offers an easy path for the a.c. The combined impedance of C_E and R_E in parallel is negligible to the a.c., and the a.c. voltage across them is also negligible. Without C_E an a.c. voltage would be developed across R_E and reduce the gain considerably (as we will see in section 11.4).

(iv) Further point The output voltage in this circuit can swing only between V_{CC} and V_E instead of between V_{CC} and ground. The value of R_E is therefore chosen so that V_E is about one-fifth to one-tenth of the supply voltage V_{CC}, e.g. 1 V for a 6 V supply.

11.2 Worked example

The voltage amplifier of Fig. 11.01 has quiescent currents $I_B = 30\ \mu\text{A}$, $I_C = 3$ mA and quiescent voltages $V_{CE} = 2$ V, $V_{BE} = 0.6$ V when the supply voltage $V_{CC} = 6$ V. If $V_E = 1$ V and $I_2 = 10I_B$, calculate (a) R_E, (b) R_L, (c) R_2 and (d) R_1.

a)

$$V_E = I_E \times R_E = I_C \times R_E \text{ (approx.)}$$

$$\therefore\quad R_E = V_E/I_C = 1\text{ V}/3\text{ mA}$$

$$= 330\ \Omega$$

b)

$$\text{Voltage across } R_L = I_C \times R_L$$

$$= V_{CC} - V_{CE} - V_E$$

$$= 6 - 2 - 1 = 3\text{ V}$$

$$\therefore\quad I_C \times R_L = 3\text{V}$$

$$\therefore\quad R_L = 3\text{ V}/3\text{ mA}$$

$$= 1\text{ k}\Omega$$

c)

$$I_2 = 10 \times I_B = 10 \times 30 = 300\ \mu\text{A} = 0.3\text{ mA}$$

$$V_2 = V_{BE} + V_E = 0.6 + 1 = 1.6\text{ V}$$

$$= I_2 \times R_2$$

$$\therefore\quad R_2 = V_2/I_2 = 1.6\text{ V}/0.3\text{ mA}$$

$$= 5.3\text{ k}\Omega\ (5.6\text{ k}\Omega\text{ preferred value})$$

d)

$$I_1 = I_2 + I_B = 0.3 + 0.03 = 0.33\text{ mA}$$

$$V_1 = V_{CC} - V_2 = 6 - 1.6 = 4.4\text{ V}$$

$$= I_1 \times R_1$$

$$\therefore\quad R_1 = V_1/I_1 = 4.4\text{ V}/0.33\text{ mA}$$

$$= 13.4\text{ k}\Omega\quad (15\text{ k}\Omega\text{ preferred value})$$

11.3 Feedback equation

The performance of an amplifier can be changed by feeding part of the output back to the input. The feedback is *positive* if it is in phase with the input, i.e. adds to it, as shown in Fig. 11.02(*a*). This is the situation in a non-inverting amplifier. The gain *increases*, often in an uncontrollable way, leading to instability and oscillation. The feedback is *negative* if it is 180° out of phase (in antiphase) with the input, i.e. subtracts from it, as shown in Fig. 11.02(*b*). This is the situation in an inverting amplifier, and in this case the gain *decreases* as we will now see.

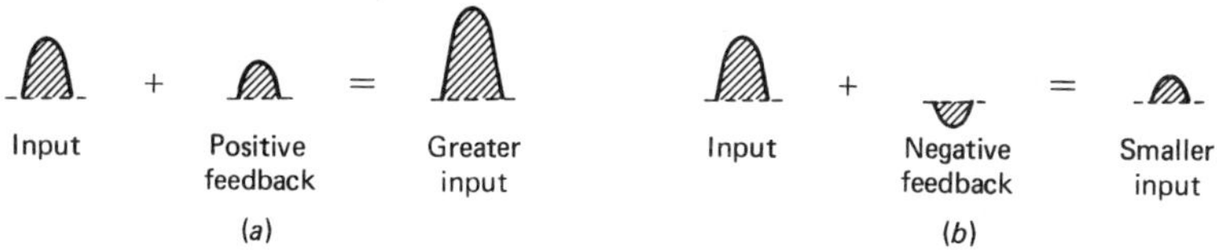

Input + Positive feedback = Greater input

(*a*)

Input + Negative feedback = Smaller input

(*b*)

Fig. 11.02

The block diagram in Fig. 11.03(*a*) is of an amplifier of voltage gain A_o with no feedback, which produces an output voltage V_o' when the input voltage is V_i, i.e. $V_o' = A_oV_i$. In Fig. 11.03(*b*), negative feedback has been introduced (by a suitable

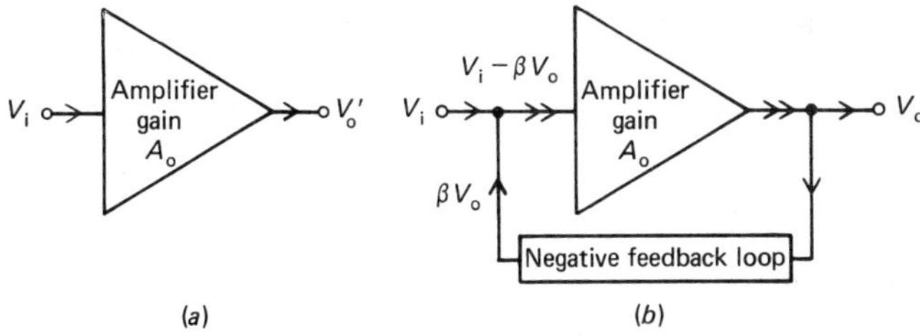

Fig. 11.03

circuit) so that the effective input is now $V_i - \beta V_o$ where β (beta) is the *feedback factor*, i.e. the proportion of the new output V_o which is fed back. Hence, since the gain of the amplifier itself is still A_o, we have:

$$V_o = A_o(V_i - \beta V_o)$$

Rearranging:

$$V_o + \beta A_o V_o = A_o V_i$$

That is:

$$V_o(1 + \beta A_o) = A_o V_i$$

$$\therefore \quad V_o = \frac{A_o}{1 + \beta A_o} \times V_i$$

The voltage gain, A, of the amplifier with negative feedback is given by:

$$A = \frac{V_o}{V_i} = \frac{A_o}{1 + \beta A_o}$$

This is the *negative feedback equation*. It shows that A, called the *closed-loop voltage gain*, is less than A_o, the *open-loop voltage gain*, and decreases as β increases. For example, if $A_o = 100$ and $\beta = 1/100$, then $\beta A_o = 1$ and $A = 100/2 = 50$.

In cases where βA_o is much greater than one, we can say:

$$A \approx \frac{A_o}{\beta A_o} = \frac{1}{\beta}$$

11.4 Negative feedback

(a) Advantages

Although negative feedback (n.f.b.) reduces the gain of an amplifier and may necessitate more stages of amplification, it has several highly desirable effects. These can be predicted from theoretical considerations. For our purposes it is enough to know that they include:

(i) increased *stability* of the overall gain, i.e. the gain is less affected by changes of transistor parameters (e.g. h_{FE}) and by variations of temperature and supply voltage;
(ii) less *distortion* of the output;
(iii) greater *bandwidth*, i.e. a wider range of frequencies is amplified by the same amount;
(iv) alteration of the *input and output impedances* of the amplifier to almost any desired values.

There are various n.f.b. circuits—only two common ones will be considered.

(b) Voltage feedback

The collector-to-base bias circuit of the junction transistor voltage amplifier of Fig. 10.05 is shown again in Fig. 11.04 with the alternating components of currents marked. Resistor R_B provides not only the quiescent (d.c.) bias to the base-emitter junction but it also applies voltage negative feedback (of the a.c. input) as follows.

When an alternating input is applied, an alternating current I_f is superimposed on the d.c. base current and is fed back through R_B to the base. I_f is in antiphase with the alternating input current since the output voltage in a common-emitter amplifier is 180° out of phase with the input—see section 10.2.

If I_o (r.m.s. value) is the a.c. part of the collector (output) current, then the alternating part of the collector-emitter voltage, i.e. the output voltage, is $I_o \times R_L$ where R_L is the value of the load resistor. This is the alternating *voltage* which drives I_f through R_B (hence *voltage* feedback). Assuming the resistance R_B is much larger than the input resistance of the amplifier, then:

$$I_f = I_o \times R_L / R_B$$

$$\therefore \quad I_f / I_o = R_L / R_B$$

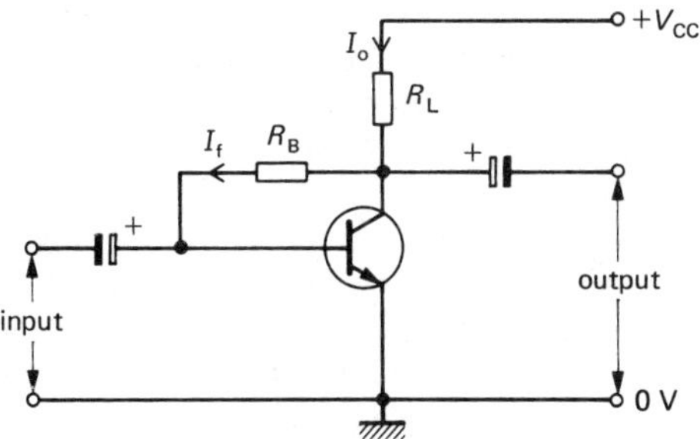

Fig. 11.04

The ratio I_f/I_o is the *feedback factor* β, hence:

$$\beta = \frac{R_L}{R_B}$$

If the value of R_B to provide the correct d.c. bias is not the same as that needed to give the required amount of feedback, the bias is obtained by other means (e.g. by the 'voltage divider and emitter resistor' method) and a capacitor is joined in series with R_B to stop d.c. reaching the base.

(c) Current feedback

This type is used in the FET voltage amplifier of Fig. 11.05; the circuit is the same as Fig. 10.08 but with the source decoupling capacitor C_S omitted.

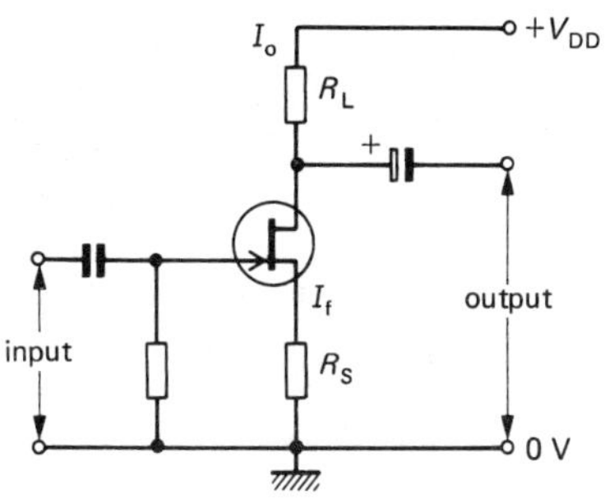

Fig. 11.05

When an alternating input voltage is applied, alternating components I_o and I_f (r.m.s. values) are superimposed on the d.c. drain and source currents respectively. I_o creates the output voltage V_o across the load R_L. The alternating voltage V_f produced by I_f across the source resistor R_S is fed back to the gate in antiphase with the input voltage as follows.

A positive half cycle of the input makes the gate less negative, the source current increases as does the voltage across R_S. The source therefore becomes more positive with respect to the gate (as explained in section 10.7), i.e. the gate tends to go more negative and thereby reduces the effective input to the amplifier. The opposite happens on a negative half cycle of input.

The feedback voltage $V_f = I_f \times R_S = I_o \times R_S$ since source and drain currents are almost equal, i.e. V_f is proportional to the output *current* I_o (hence *current* feedback). The output voltage $V_o = I_o \times R_L$, therefore:

$$\frac{V_f}{V_o} = \frac{I_o \times R_S}{I_o \times R_L} = \frac{R_S}{R_L}$$

But V_f/V_o is the *feedback factor* β, hence:

$$\beta = \frac{R_S}{R_L}$$

Current feedback is also used in junction transistor amplifiers.

(d) Equalization and tone control by frequency-selective n.f.b.

In sound reproduction the electrical signal applied to an a.f. amplifier may not be a faithful representation of the original sound because the input transducer does not respond equally to all frequencies. To compensate for this the amplifier has to boost some frequencies more than others. The process is called *equalization* and is achieved by using a n.f.b. circuit whose impedance is not purely resistive, as in the two previous cases, but depends on the frequency.

For example, if the n.f.b. circuit consists of a capacitor C in parallel with a resistor R (both of the correct value), low (bass) frequencies can be amplified more than high (treble) frequencies. At low frequencies the reactance X_C of C ($= 1/(2\pi fC)$) is large (since f is small), giving a small amount of n.f.b. (i.e. β is small) and so the reduction in gain A, due to the n.f.b., is small—see section 11.3. At high frequencies, X_C is small, n.f.b. is greater and there is a larger fall in A. In effect this means bass notes are boosted.

A circuit for tone control is designed on the same principles and is often incorporated with that for frequency-selective n.f.b. in the preamplifier which precedes the power amplifier. For example, the LM381 IC preamplifier circuit of Fig. 13.19 contains a CR equalizing network for audio signals.

11.5 Input and output impedances

The input and output impedances are two other important properties of an amplifier.

The *input impedance* Z_{in} equals V_{in}/I_{in}, where I_{in} is the a.c. input current when V_{in} is the a.c. input voltage. It depends not only on the input resistance r_i of the transistor but also on the presence of capacitors, resistors, etc., in the circuit.

The *output impedance* Z_{out} is the a.c. equivalent of the internal or source resistance of a battery—see section 2.12. It causes a 'loss' of voltage at the output terminals when another circuit or device is connected to them. That is, the output side of the amplifier behaves like a voltage generator and its terminal voltage on closed circuit is less (the greater Z_{out} is) than on open circuit—as in the d.c. case. Both Z_{out} and Z_{in} can be measured (in ohms) experimentally; this is considered in section 11.11.

In many electronic circuits, signals are in the form of voltages which have to be passed on from one part of the circuit to the next with minimum loss. To achieve *maximum voltage transfer*, the value of Z_{in} of the receiving circuit or device should be large compared with Z_{out} of the supplying circuit or device. As a rough rule, we aim to have Z_{in} at least *ten* times greater than Z_{out}. You can understand this from the following example.

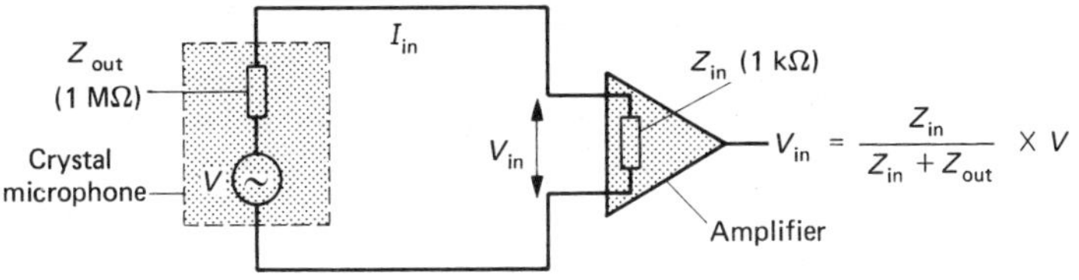

Fig. 11.06

In Fig. 11.06 suppose Z_{out} of the crystal microphone is 1 MΩ and that the amplifier is a single-stage junction transistor type with Z_{in} = 1 kΩ. Of the alternating voltage V generated by the microphone only one-thousandth of it (i.e. 1 kΩ/1 MΩ) will be available at the amplifier input terminals. The rest is 'lost' across Z_{out} of the microphone. Some means of increasing the value of Z_{in} is required if the *impedance matching* is to be improved.

In the two-stage junction transistor, common-emitter amplifier of Fig. 10.06 about half of the voltage signal is lost in the coupling between the two stages because the values of Z_{in} and Z_{out} of each stage are of the same order (generally a few thousand ohms). So far as impedance matching is concerned it would be better to use two FETs since they have very high Z_{in} and moderate Z_{out}.

The impedances of output transducers are often quite low, for example, 4 Ω is typical for a loudspeaker. Impedance-matching problems have to be solved here too, as we will see in section 11.7.

11.6 Impedance-matching circuits

There are various ways of changing the input and output impedances of an amplifier to secure maximum voltage transfer.

(a) Darlington pair

The input impedance can be increased by connecting two transistors as in Fig. 11.07, called a Darlington pair. The emitter current of Tr_1 is almost the same as its collector current and equals h_{FE1} times the base (input) current. This emitter currents forms the base current of Tr_2. The collector (output) current of Tr_2 equals h_{FE2} times its base current. The overall *current gain* is $h_{FE} = h_{FE1} \times h_{FE2}$ (typically 10^4).

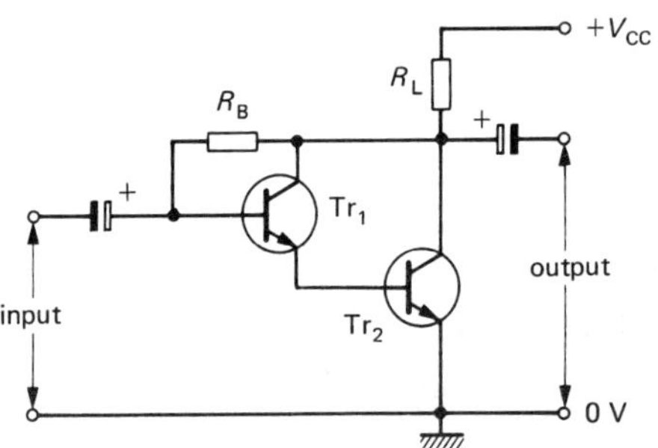

Fig. 11.07

The high gain means that only a tiny input current is needed to obtain a certain output current. In effect this makes the input impedance very high.

(b) Emitter-follower

As mentioned in section 11.4, negative feedback alters the input and output impedances of an amplifier. The emitter-follower (or common-collector amplifier), shown in Fig. 11.08, uses 100% n.f.b. Its special features are:

(i) a high input impedance (e.g. 500 kΩ);
(ii) a low output impedance (e.g. 30 Ω);
(iii) a voltage gain less than, but nearly, one; and
(iv) an output signal in phase with the input.

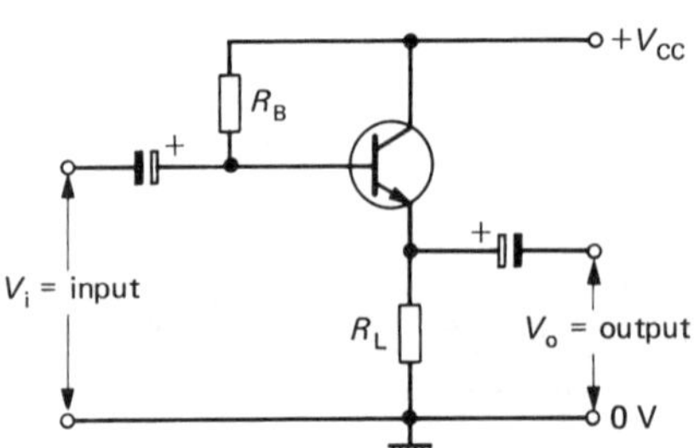

Fig. 11.08

The input V_i is applied as usual between base and ground but the load resistor R_L, across which the output V_o is taken, is the emitter circuit. Since R_L is not decoupled, all of V_o is fed back in opposition to V_i (see section 11.1). This reduces the input current for the same V_i, thereby, in effect, increasing the input impedance. R_B fixes the d.c. operating point.

The low output impedance is due to the fact that the input and output circuits are connected by the base-emitter junction, which, being forward biased, has a low resistance. (By contrast, in a common-emitter amplifier the input and output circuits are separated by the high resistance of the reverse biased collector-base junction, giving it a much higher output impedance.)

The voltage gain is always just less than one because the input is applied across the small resistance r_e of the base-emitter junction in series with the much larger emitter load resistor R_L and the output is taken from across R_L, i.e. r_e and R_L act as a voltage divider to the input.

Also note that, as with common-emitter voltage amplifiers, the output voltage will have maximum swing capability (i.e. from $+V_{CC}$ to 0 V) only if the quiescent (no a.c. input) emitter voltage is $\frac{1}{2}V_{CC}$, achieved by choosing the value of R_B to give the correct quiescent base bias current.

The circuit is called an *emitter-follower* since the emitter voltage 'follows' the base (input) voltage. The term common-collector amplifier arises because, from the point of view of a.c., the collector is common to both input and output circuits via the negligible resistance of the power supply.

The corresponding FET circuit is a *source-follower* (or common-drain amplifier). The emitter-follower has a lower output impedance but the source-follower is better for acting as a *buffer amplifier* to match a very high impedance input transducer such as a capacitor microphone to a conventional amplifier stage.

11.7 Power amplifiers—Single-ended

The amplifiers considered so far have been voltage or small-signal amplifiers designed to give large undistorted output voltages from small input voltages. While these do produce small amounts of power in their collector loads, this would be insufficient to operate, for example, a loudspeaker. A power or large-signal amplifier is then required which generates large swings of output current and voltage (i.e. large a.c. power) and needs a large alternating input voltage.

(a) The matching problem

A power amplifier, in effect, converts d.c. power from the supply into a.c. power in the load. It can be regarded as a 'robust' voltage amplifier which uses a transistor (a power transistor) capable of supplying high output currents. However, its matching requirement is for maximum power transfer rather than for maximum voltage transfer, the latter usually being the case where one circuit or device is connected to another.

The maximum power theorem (see section 2.12) states that a generator delivers maximum power to an external load when the output impedance of the generator equals the input impedance of the load. Thus a transistor supplies maximum power when its output impedance equals the impedance of its load.

The impedance of a loudspeaker is typically 4 Ω, at 1 kHz. The output impedance of a transistor is appreciably greater than this. Consequently direct connection of the speaker as the collector load would give poor results.

(b) Transformer matching

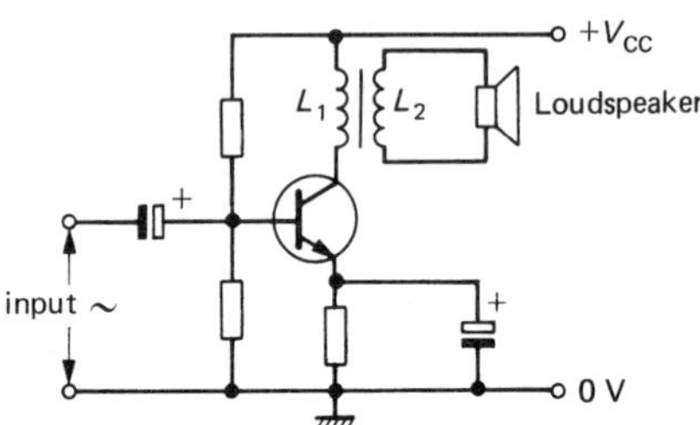

Fig. 11.09

One solution is to connect a step-down transformer with its primary coil L_1 as the collector load (Fig. 11.09). If the impedance Z_1 of the primary is about the same as the output impedance of the transistor, the latter delivers maximum power to L_1. Also if the impedance Z_2 of the secondary L_2 more or less equals the impedance of the speaker, L_2 delivers maximum power to the speaker. The theory of the transformer (see section 6.7) shows that, if n_1 and n_2 are the number of turns on L_1 and L_2 respectively and if V_1 and V_2 are the primary and secondary voltages, we have:

$$\frac{n_1}{n_2} = \frac{V_1}{V_2}$$

From section 6.6 we know $V = I \times Z$ so we may write:

$$\frac{n_1}{n_2} = \frac{V_1}{V_2} = \frac{I_1 \times Z_1}{I_2 \times Z_2} \qquad (1)$$

Also, if there is no power loss, $I_1 \times V_1 = I_2 \times V_2$ and we have:

$$I_1 = \frac{V_2}{V_1} \times I_2 = \frac{n_2}{n_1} \times I_2$$

Substituting for I_1 in (1) gives:

$$\frac{n_1}{n_2} = \left(\frac{n_2}{n_1} \times I_2\right)\left(\frac{Z_1}{I_2 \times Z_2}\right) = \frac{n_2}{n_1} \times \frac{Z_1}{Z_2}$$

$$\frac{n_1^2}{n_2^2} = \frac{Z_1}{Z_2} \text{ or } \frac{n_1}{n_2} = \sqrt{\left(\frac{Z_1}{Z_2}\right)}$$

Therefore, if $Z_1 = 2$ kΩ and $Z_2 = 4$ Ω, $n_1/n_2 = \sqrt{(2000/4)} = \sqrt{(500/1)} \approx 22/1$. A *matching* transformer with a step-down ratio of about 22 to 1 would be suitable in this case.

(c) Efficiency

A power amplifier using one output transistor is said to be *single-ended*. In general it is only used where low powers (up to about 1 W) are required because its efficiency η, (pronounced eta) defined by:

$$\eta = \frac{\text{a.c. power output to load}}{\text{d.c. power taken from supply}} \times 100\%$$

cannot exceed 50% even with correct matching. This is the result of having to bias the transistor to work near the middle of its load line under *class A* conditions (like a voltage amplifier—see section 10.2), to prevent distortion of the output. The quiescent (no input) collector current is therefore large and represents wasted d.c. power which produces unwanted heat, possibly requiring the transistor to be mounted on cooling fins, called a *heat sink*, Fig. 11.10.

Fig. 11.10

Further disadvantages are the cost, size and weight of the transformer and the distortion which it itself introduces.

11.8 Power amplifiers—Push-pull

Most a.f. power amplifiers, including those in radio receivers, employ the push-pull principle. Two transistors are used and are called a *complementary pair* because one is an n-p-n type and the other a p-n-p type. The basic circuit is given in Fig. 11.11.

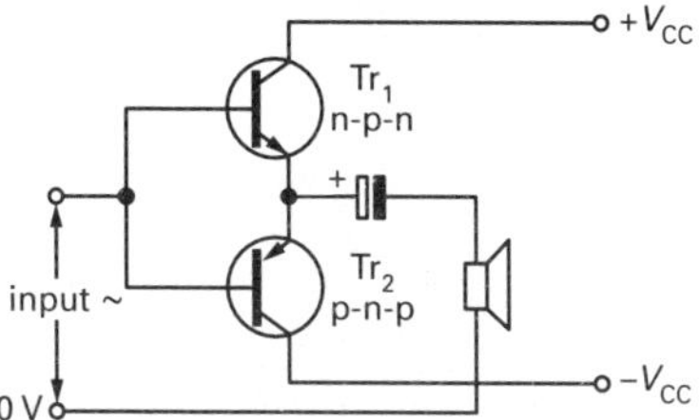

Fig. 11.11

(a) Emitter-follower matching

Tr_1 is the n-p-n transistor, its collector being positive with respect to the emitter; the opposite is the case for Tr_2, the p-n-p type. The bases of both transistors are connected to the input terminal. The load, shown as a loudspeaker, is joined to both emitters via a d.c. blocking capacitor. The circuit thus consists of two emitter-followers, which, as we said in section 11.6, have a low output impedance. In fact, their impedance is comparable with that of a loudspeaker, so solving the impedance-matching problem.

(b) Push-pull action

When an input is applied, the positive half cycles forward bias the base-emitter junction of Tr_1, which conducts and 'pushes' a series of positive half cycles of output current through the emitter load, i.e. the loudspeaker. During this time Tr_2 is cut off because its base-emitter junction is reverse biased when the base is positive with respect to the emitter. Negative half cycles of the input reverse bias Tr_1 (i.e. base negative with respect to emitter) and cut it off but forward bias Tr_2. Tr_2 conducts and pulls a series of negative half cycles of output current through the speaker. Both halves of the input cycles are thus 'processed'.

When a transistor only conducts for one half of each complete input cycle and is cut off during the other half, it is said to be working under *class B* conditions (in class A it conducts for the whole input cycle). The quiescent (no input) base and collector currents are zero since the base-emitter junctions of neither Tr_1 nor Tr_2 are forward biased. Therefore, no d.c. power is wasted when there is no input. This explains why the maximum efficiency of a class B power amplifier is high, 78% (as compared with 50% for class A).

(c) Crossover distortion

One disadvantage of the simple circuit in Fig. 11.11 is that each transistor does not turn on until the input is about 0.6 V. As a result there is a 'dead zone' in the output waveform, shown in Fig. 11.12, producing what is called *crossover distortion.*

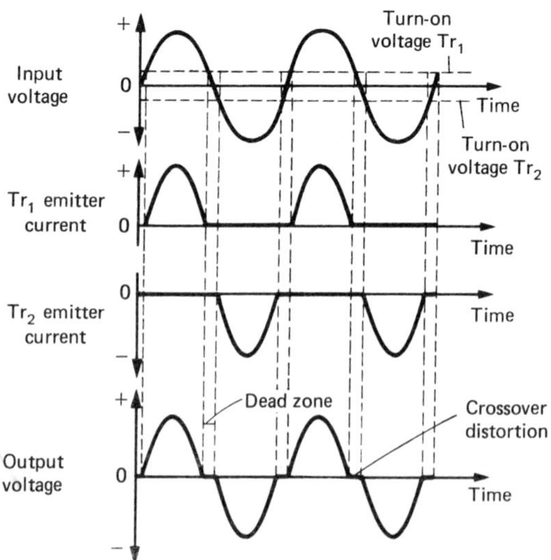

Fig. 11.12

Matters are improved by applying some forward bias to the base-emitter junctions of both transistors so that small quiescent currents flow and the smallest inputs make them conduct. Both transistors then operate in class A for zero and small inputs; for larger inputs each operates in class B on alternate half cycles of the input. This is called *class AB* operation and is normally used in a.f. power amplifiers. It is a compromise between the low distortion, low efficiency class A amplifier and the higher efficiency, higher distortion class B amplifier.

The bias required can be provided by connecting two resistors R_1 and R_2 of suitable values as in Fig. 11.13 so that in the quiescent state the bases of Tr_1 and Tr_2 are, with respect to their emitters, 0.6 V positive and 0.6 V negative respectively. Both transistors are then forward biased. If R_2 is variable, it can be adjusted for minimum crossover distortion.

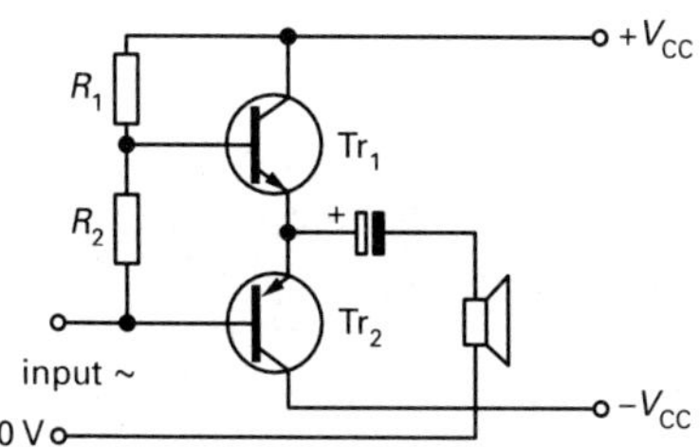

Fig. 11.13

(d) Matched transistors

Tr_1 and Tr_2 should be a 'matched pair' of power transistors, i.e. their characteristics should be identical, otherwise if one has a greater gain than the other, one half of the output waveform will be amplified more and cause distortion.

Matched pairs of both *junction* power transistors and *MOSFET* power transistors are available for various power outputs. In a MOSFET pair, one would be an n-channel device and the other a p-channel type. Power MOSFETs are capable of handling large drain currents at high drain-source voltages and powers of the order of 100 W with less distortion than junction transistors, due to their more linear characteristics.

(e) Maximum output power

The quiescent voltage at the emitter of either of the matched transistors in Fig. 11.13 is $\frac{1}{2}V_{CC}$ where V_{CC} is the d.c. supply voltage. If the alternating input makes the emitter voltage vary from its maximum positive value of V_{CC} (when Tr_1 saturates), the emitter voltage is then a varying direct voltage. The latter has an alternating component with a peak value of $\frac{1}{2}V_{CC}$ and therefore an r.m.s. value (see section 3.4) $V_{r.m.s.} = V_{CC}/(2\sqrt{2})$.

If $I_{r.m.s.}$ is the r.m.s. value of the output current, the maximum output power P_{max} of the amplifier is given by:

$$P_{max} = V_{r.m.s.} \times I_{r.m.s.} = (V_{r.m.s.})^2/Z_L$$

Where Z_L is the loudspeaker impedance (since $Z_L = V_{r.m.s.}/I_{r.m.s.}$).

Hence:

$$P_{max} = \left(\frac{V_{CC}}{2\sqrt{2}}\right)^2 \times \frac{1}{Z_L} = \frac{V_{CC}^2}{8Z_L}$$

For example, when $V_{CC} = 6$ V and $Z_L = 4\ \Omega$, $P_{max} = 36/32 \approx 1$ W. The transistors would of course need to be capable of handling the current and power involved without overheating.

If the input is not large enough to drive the transistors into saturation, the output power is less than P_{max}.

11.9 Decibel scale

(a) Power ratios

To the human ear the change in loudness is the same when the power of a sound increases from 0.1 to 1 W as when it increases from 1 W to 10 W. The ear responds to the *ratio* of the powers and not to their difference.

Now $\log_{10}(1/0.1) = \log_{10} 10 = 1$ and $\log_{10}(10/1) = \log_{10} 10 = 1$. (Logs to base 10 are called common logs; a log button can be found on most calculators). The log to base 10 of the ratio of the powers is therefore the same for each change. This accounts for the way the unit of the *change of power*, called the *bel* (B), is defined. If a power changes from P_1 to P_2, then we define:

$$\text{number of bels change} = \log_{10}\frac{P_2}{P_1}$$

In practice the bel is too large and the *decibel* (dB) is used. It is one-tenth of a bel, hence:

$$\text{number of decibels change} = 10\log_{10}\frac{P_2}{P_1}$$

For example, if the power output from an amplifier increases from 100 mW to 200 mW, the gain in dB = $10\log_{10}(200/100) = 10\log_{10} 2 \approx 10 \times 0.30 = 3$ dB. Fig. 11.14 shows how a very wide range of power ratios is reduced to the more compact range of dBs. Note that a power ratio of 10 (whatever the two powers involved) is a 10 dB change, while a power ratio of 100 is a 20 dB change.

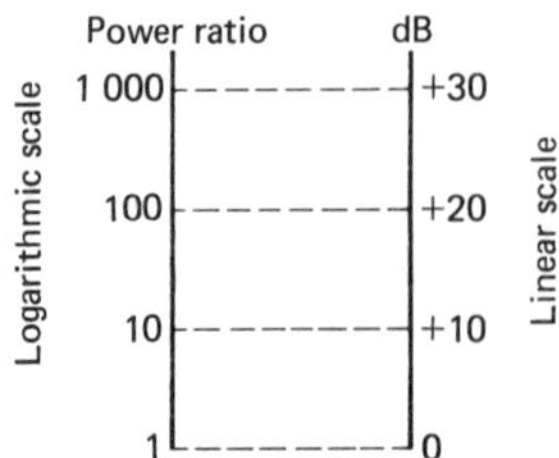

Fig. 11.14

The decibel scale is often used when the frequency response, i.e. the bandwidth (see section 10.6) of a power amplifier is quoted. For example, if a manufacturer states that it is 15 Hz to 35 kHz at the 3 dB level, this means that these are the two frequencies at which the power gain has fallen by 3 dB, i.e. halved, from the maximum gain at the middle frequencies.

Table 11.1 Sound level changes

Sound	dB level
Threshold of hearing	0
Whispering	30
Normal conversation	60
Busy street	70
Noisy factory	90
Jet plane overhead	100
Loud thunderclap	110
Threshold of pain	120

Various sound level changes are given in Table 11.1, taking as zero dB the threshold of hearing, i.e. the sound the average human ear can just detect.

(b) Voltage ratios

It is often more convenient to express the power ratio in terms of the ratio of the corresponding r.m.s. voltages. Suppose these are V_1 and V_2 when the powers are P_1 and P_2. Then, since power is proportional to the square of the voltage (assuming V_1 and V_2 are developed across the same impedance), we can say:

$$\left(\frac{V_2}{V_1}\right)^2 = \frac{P_2}{P_1} \text{ or } \frac{V_2}{V_1} = \sqrt{\frac{P_2}{P_1}}$$

Therefore, if the power gain is halved, i.e. $P_2/P_1 = 1/2$, then $V_2/V_1 = 1/\sqrt{2} \approx 0.7$. We have shown above that halving (or doubling) the power is a 3 dB loss (or gain); we now see that this causes the voltage to decrease to 0.7 (or increase to $\sqrt{2} = 1.4$) of its previous value.

Referring back to section 10.6 you will see that the bandwidth of a voltage amplifier was given with reference to a 0.7 voltage ratio fall, which we have now shown corresponds to a 3 dB power decrease.

Note Since $\log_{10}x^2 = 2\log_{10}x$, we can also write:

$$\text{dB change} = 10\log_{10}\left(\frac{V_2}{V_1}\right)^2 = 10 \times 2\log_{10}\frac{V_2}{V_1}$$

$$= 20\log_{10}\frac{V_2}{V_1}$$

11.10 Noise

Noise in an amplifier or any electronic system is an unwanted voltage or current in the output. It can have any frequency and is heard as a 'hiss' on a loudspeaker. In some cases it originates inside the equipment itself, in other cases it arises externally.

(a) Internal noise

Some of the most common types are:

(i) thermal noise—caused by the haphazard motion of electrons in any conductor, thereby producing very small random voltages which increase with temperature;
(ii) shot noise—caused when charge carriers 'shoot' across a p-n junction, speeding up as they do so;
(iii) partition noise—caused by random variations in the division of current in any device in which a current has to split to reach two (or more) electrodes, as happens in a transistor when the emitter current divides between the base and collector.

(b) External noise

This is mainly picked up by the aerials of radio systems and can have two causes:

(i) noise from equipment—caused by sparking at the contacts of electric motors, thermostats, car ignition systems, etc., or by mains hum from equipment supplied by the a.c. mains or by other radio stations on nearby wavelengths;
(ii) natural noise—caused by lightning flashes (called 'static') that are always occurring somewhere in the world and produce radiation capable of travelling far, or by very short wavelength radiation emitted by stars (called 'cosmic' noise), the sun and the earth itself.

(c) Signal-to-noise ratio

The noise in an electronic system should be a minimum. It will not improve matters if in, say, a radio receiver, we amplify a very weak signal when the noise present with it is strong. The signal-to-noise ratio, S/N, is a useful measure of how 'clean' an output is and is defined by:

$$\text{S/N ratio} = \frac{\text{wanted signal power}}{\text{unwanted noise power}} = \frac{P_S}{P_N}$$

It is often stated in the decibel scale as:

$$\text{S/N ratio} = 10 \log_{10} \frac{P_S}{P_N} = 20 \log_{10} \frac{V_S}{V_N}$$

where V_S and V_N are the signal and noise r.m.s. voltages respectively.

An acceptable S/N ratio in a system depends on what the system is used for. In an a.f. amplifier 60 dB at 1 kHz is a typical value.

11.11 Measuring input and output impedances

(a) Input impedance

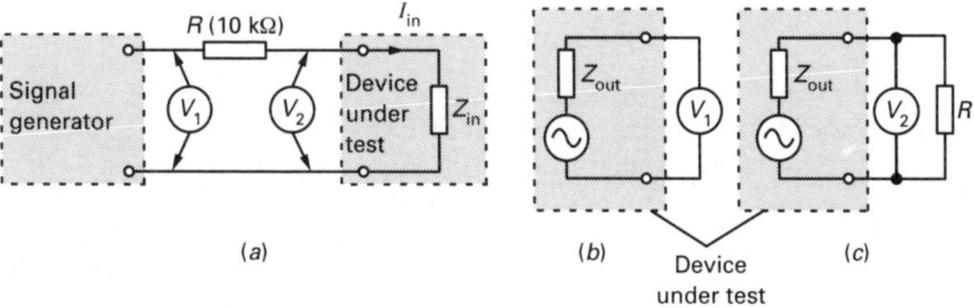

Fig. 11.15

The circuit in Fig. 11.15(*a*) uses a known resistor R joined in *series* with the signal generator supplying an a.c. input signal to the device (e.g. an amplifier) where input impedance Z_{in} is required. The voltages V_1 and V_2 are measured by a CRO or high impedance a.c. voltmeter. Then, if I_{in} is the input current, the voltage across R is:

$$V_1 - V_2 = I_{in}R$$

Hence

$$I_{in} = \frac{V_1 - V_2}{R} \text{ but } Z_{in} = \frac{V_2}{I_{in}}$$

Therefore

$$Z_{in} = \frac{V_2 R}{V_1 - V_2}$$

(b) Output impedance

In this case the a.c. output voltage from the device (e.g. an amplifier getting its input from a signal generator) is measured first on open circuit (V_1), Fig. 11.15(*b*), and then with a known resistance R in *parallel* with the output, to obtain the reduced output voltage (V_2), Fig. 11.15(*c*). The voltage drop (i.e. 'lost' volts) across the output impedance Z_{out} with R connected is:

$$V_1 - V_2 = I_{out}Z_{out} \text{ where } I_{out} \text{ is the output current}$$

But

$$I_{out} = \frac{V_2}{R}$$

Therefore

$$Z_{out} = \frac{(V_1 - V_2)R}{V_2}$$

11.12 Revision questions

1. In the fully stabilized voltage amplifier circuit of Fig. 11.16,
 a) what are the functions of
  (i) R_1 and R_2,
  (ii) R_E,
  (iii) C_E?
 b) why is I made about ten times greater than I_B?
 c) why is R_E chosen so that V_E is usually not more than one-fifth of V_{CC}?

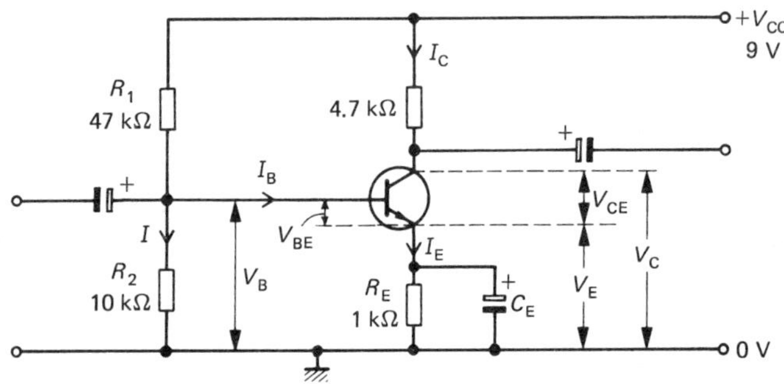

Fig. 11.16

2. Explain the terms
 a) positive feedback,
 b) negative feedback.

3. a) State one disadvantage and four advantages of negative feedback.
 b) Write the negative feedback equation and state the meaning of each symbol.

4. Draw a circuit for
 a) a junction transistor amplifier using *voltage* n.f.b., and
 b) a FET amplifier using *current* n.f.b.

5. a) Explain the terms input impedance and output impedance.
 b) What is the condition for maximum *voltage* transfer between two circuits or devices?

6. a) A signal generator (see section 21.9) is designed to supply an alternating voltage *to* a circuit under test. Why does it have a *low* output impedance (usually less than 100 Ω)?
 b) A cathode ray oscilloscope (see section 21.8) is intended to study voltage *from* a circuit under test. Why is it made with a *high* input impedance (about 1 MΩ)?

7. a) What are the special features of
(i) a Darlington pair,
(ii) an emitter-follower?
b) In the circuit of Fig. 11.17,
(i) why is the output voltage V_o always less than the input voltage V_i?
(ii) why is the input impedance high?
(iii) why does the circuit have a low output impedance?

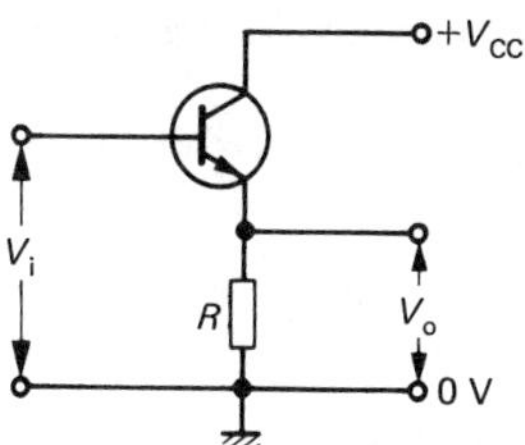

Fig. 11.17

8. a) How does a transistor for use in a power amplifier differ from other types?
b) Why should a power amplifier be matched to its load?
c) What is meant by
(i) class A, and
(ii) class B, operation of a transistor?

9. As applied to a.f. power amplifiers explain the terms
(i) single-ended,
(ii) efficiency,
(iii) heat sink,
(iv) complementary pair,
(v) push-pull.

10. How is matching to the load achieved in
a) a single-ended, and
b) a push-pull, a.f. power amplifier?

11. a) Why does a class B a.f. power amplifier
(i) have a higher efficiency, and
(ii) produce more distortion, than a class A one?
b) What is meant by crossover distortion and how is it reduced?

12. a) What are the advantages of stating power changes on the dB scale?
b) If the power output from an amplifier increases from P_1 to P_2, write down the expression for calculating the power gain in dB.

13. Explain the terms
a) noise,
b) signal-to-noise ratio.

11.13 Problems

1. In the common-emitter amplifier circuit of Fig. 11.16 an n-p-n silicon transistor is used for which $h_{FE} = 100$. Assuming that I is greater than $10I_B$, calculate the quiescent values of
a) V_B, b) V_E, c) I_E, d) V_C, e) V_{CE}, f) I_B.

2. The open-loop gain of an amplifier is 200. What is the closed-loop gain when the feedback factor is 1/50?

3. a) Calculate the feedback factor in the circuit of Fig. 11.4 if $R_L = 1$ kΩ and $R_B = 100$ kΩ.
b) What is the feedback factor in the circuit of Fig. 11.5 when $R_L = 10$ kΩ and $R_S = 1$ kΩ?

4. What is the turns ratio of the transformer required to match a transistor with an output impedance of 400 Ω to a loudspeaker with an impedance of 4 Ω?

5. What is the efficiency of a class A a.f. power amplifier which takes a collector current of 50 mA from a 6 V supply and delivers a.c. power of 60 mW to a transformer-coupled loudspeaker?

6. A push-pull a.f. power amplifier operates from a 9 V supply. What is the maximum a.c. power it can deliver to an 8 Ω load?

7. If the volume control on a radio receiver is turned up so increasing the power output from the loudspeaker from 100 mW to 500 mW, what is the power gain in dB? ($\log_{10} 5 = 0.7$)

8. The bandwidth of a certain a.f. power amplifier is said to be 10 Hz to 50 kHz at the 3 dB level. What does this mean in terms of
a) power output, and
b) voltage output, at these frequencies?

9. a) What is the signal-to-noise ratio of a system in which the signal output power is 1 W and the noise output power is 1 mW?
b) If the input to a radio receiver is 100 μV and the noise level is 1 μV, calculate the signal-to-noise ratio in dB.

UNIT 12

Radio frequency amplifiers and oscillators

12.1 Introduction

Radio frequency (r.f.) or high frequency (h.f.) alternating voltages and currents have frequencies greater than about 30 kHz. They are important in radio, television and other branches of telecommunications and are grouped roughly into bands (see Table 12.1).

Table 12.1 Classification of frequency bands

Frequency band	Classification
30 kHz–300 kHz	low frequency (l.f.)
300 kHz–3 MHz	medium frequency (m.f.)
3 MHz–30 MHz	high frequency (h.f.)
30 MHz–300 MHz	very high frequency (v.h.f.)
300 MHz–3 GHz	ultra high frequency (u.h.f.)
Above 3 GHz	super high frequency (s.h.f.)

Radio frequency amplifiers are similar in many ways to audio frequency amplifiers. They must have a *load* in their output circuit (to convert changing currents to changing voltages) and need to be correctly *biased* (for linear operation). However, they are only required to amplify a band of frequencies and so have to be *selective*. Also, their design, especially for higher frequency work, has to allow for certain capacitance effects (see section 12.5) that can be neglected at audio frequencies.

High power r.f. amplifiers which produce up to hundreds of kilowatts of power use thermionic *valves* since present-day transistors cannot cope with the heat that has to be dissipated for such powers. They are employed mainly in radio and television transmitters. We will not consider them.

Oscillators, both a.f. and r.f., are generators of alternating voltage and current and are basically amplifiers which supply their own input using positive feedback—see section 11.3.

Two circuits that are important in the operation of r.f. amplifiers and oscillators will be discussed first.

12.2 Oscillatory circuit

When a capacitor discharges through an inductor in a circuit of low resistance, an a.c. of constant frequency is produced. This frequency is called the *natural frequency* of oscillation of the circuit.

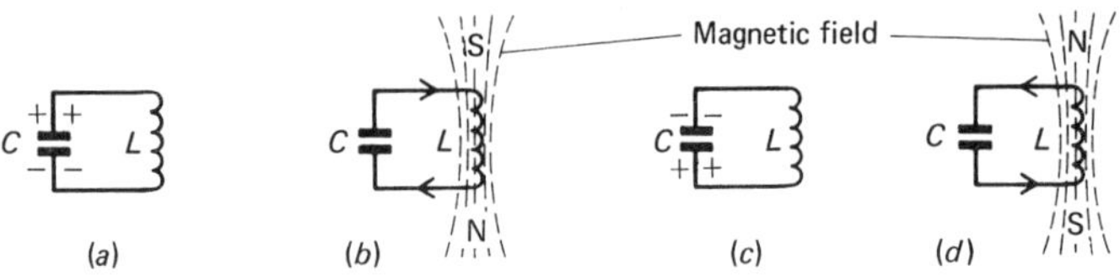

Fig. 12.01

In Fig. 12.01(*a*) a capacitor of capacitance C, charged previously by a power supply, is shown connected across a coil of inductance L. The capacitor starts to discharge immediately, current flows and a changing magnetic field is created which induces a voltage in the coil—see section 6.2. This voltage opposes the current. The capacitor cannot therefore discharge instantaneously and the larger L is, the longer the discharge takes. When the capacitor is completely discharged, the electrical energy originally stored in it has been transferred to the magnetic field round the coil, Fig. 12.01(*b*).

At this instant the magnetic field begins to collapse and a voltage is induced in the coil which tries to maintain the field. Current flows in the same direction as before and charges the capacitor so that its lower plate is positive. By the time the magnetic field has collapsed completely, the energy is again stored in the capacitor, Fig. 12.01(*c*). Once more the capacitor starts to discharge but current now flows in the opposite direction, creating a magnetic field of opposite polarity, Fig. 12.01(*d*). When this field has decayed, the capacitor is again charged with its upper plate positive, as it was at the start. The circuit has produced one electrical oscillation, i.e. one cycle of a.c., at its natural frequency.

The above sequence is repeated several times but energy is gradually lost from the circuit firstly as heat, because the coil has some resistance, and secondly as 'radiation' into space outside the circuit, especially if high frequency a.c. flows. As a result, *damped* oscillations of decreasing amplitude are produced, which eventually die away, Fig. 12.02(*a*). Their frequency, f, can be shown to be given by:

$$f = \frac{1}{2\pi\sqrt{(LC)}}\,\text{Hz}$$

L is in henries and C in farads. To generate r.f. oscillations, L and C must be small; for example if $L = 100\ \mu\text{H} = 100 \times 10^{-6}\ \text{H} = 10^{-4}\ \text{H}$ and $C = 100\ \text{pF} = 100 \times 10^{-12}\ \text{F} = 10^{-10}\ \text{F}$, $f \approx 1.6\ \text{MHz}$.

To obtain undamped oscillations of constant amplitude as in Fig. 12.02(*b*), energy must be fed continuously into the LC circuit to make up for the inevitable losses. This can be done with the help of a transistor as we will see in section 12.6.

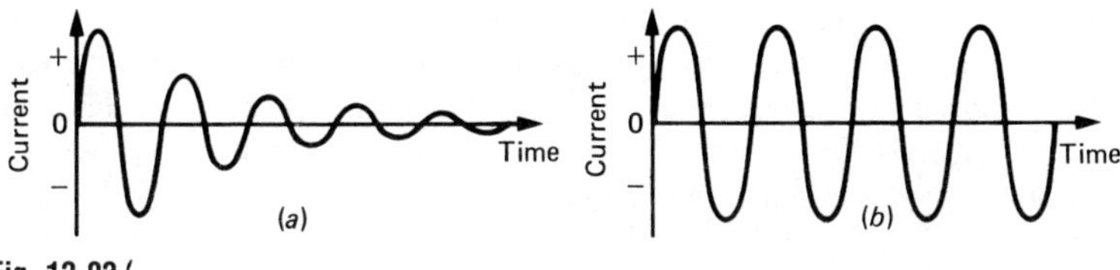

Fig. 12.02

12.3 Tuned circuits

When an alternating voltage is *applied* to a circuit containing a capacitor and a coil, the 'response' of the circuit is a maximum when the frequency of the applied voltage equals the natural frequency of the circuit. Electrical *resonance* occurs in the same way that a child's swing exhibits mechanical resonance and produces large oscillations if pushes are applied at its natural frequency. The form of the electrical 'response' depends on whether the capacitor and coil are in series or in parallel.

(a) Series resonance

In the circuit of Fig. 12.03(*a*) an a.c. voltage V of frequency f is applied to a capacitor C in series with a coil L of small resistance R (shown separately). If $f = 1/[2\pi\sqrt{(LC)}]$, theory and experiment show that resonance occurs with the following results:

(i) the *impedance* Z is a *minimum*, equal to R (i.e. $X_L = X_C$) and so the current I is a maximum, given by $I = V/R$;
(ii) the *voltages* V_L and V_C across L and C are both much *larger* than the voltage V applied to the whole circuit. (This may seem strange but the explanation is that V_L and V_C are in antiphase and so cancel out, leaving $V = V_R$, as shown by the numerical values in brackets in Fig. 12.03(*a*).)

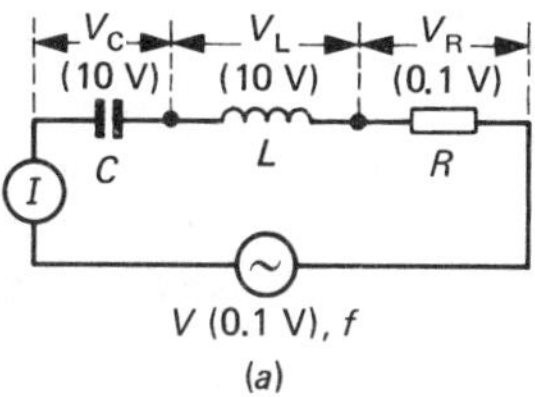

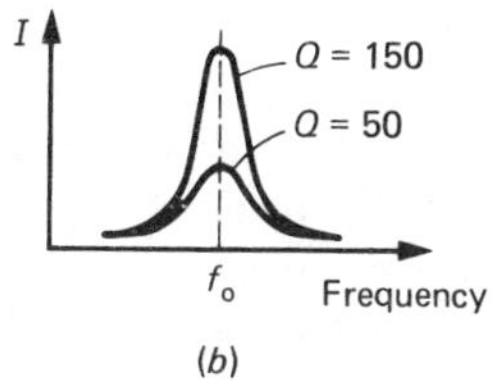

Fig. 12.03

The circuit can thus be used to *select* a voltage at its natural or resonant frequency and *amplify* it. A measure of its ability to do this is its *Q-factor* (*Q* for quality), also called the *magnification factor*, and can be expressed as:

$$Q = \frac{V_L}{V} = \frac{I \times X_L}{I \times R} = \frac{2\pi f L}{R}$$

The higher the ratio L/R (and, it can also be shown, of L/C), the higher is the *Q-factor*, i.e. the coil should have small resistance. The response curve in Fig. 12.03(*b*) of I against frequency shows the effect of the *Q-factor* on selectivity. A high value gives a sharp resonance curve, i.e. a much greater response to the resonant frequency f_0 than to neighbouring frequencies.

(b) Parallel resonance

The circuit is shown in Fig. 12.04(*a*) and again resonance occurs when $f = 1/[2\pi\sqrt{(LC)}]$, if R is small, with two results:

(i) the *impedance* Z is a *maximum*, equal to $L/(CR)$ and the supply current I is a minimum;
(ii) the *voltage* across the circuit is *large*; for the same current at other frequencies it would be much smaller due to Z being smaller.

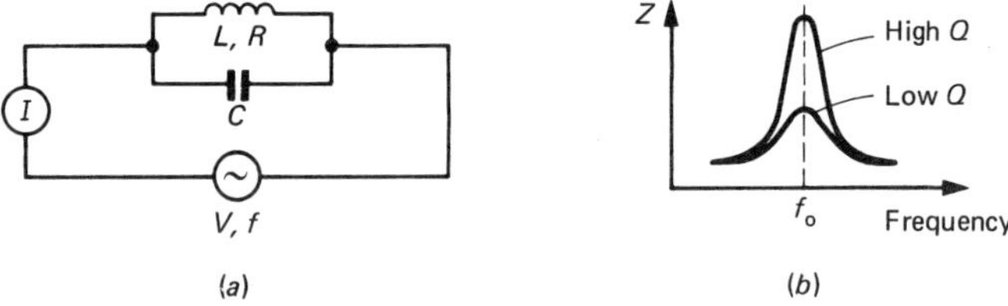

Fig. 12.04

The selectivity is again given by the *Q-factor* ($2\pi fL/R$) and its effect is shown by the *response curve* in Fig. 12.04(*b*) of Z against frequency.

(c) Tuning

If C is a variable capacitor, the resonant frequency of both series and parallel LC circuits can be changed by changing C. This is used in radio communication. For example, in the aerial circuit of a radio receiver (Fig. 12.05) radio signals from different transmitting stations induce voltages of various frequencies in the aerial, e.g. a coil L wound on a ferrite rod, which cause r.f. currents to flow in it. If C is adjusted (tuned) so that the resonant frequency of the LC circuit equals the frequency of the wanted station, a large voltage at that frequency (and no other) is developed across C and can be applied to the next stage of the receiver.

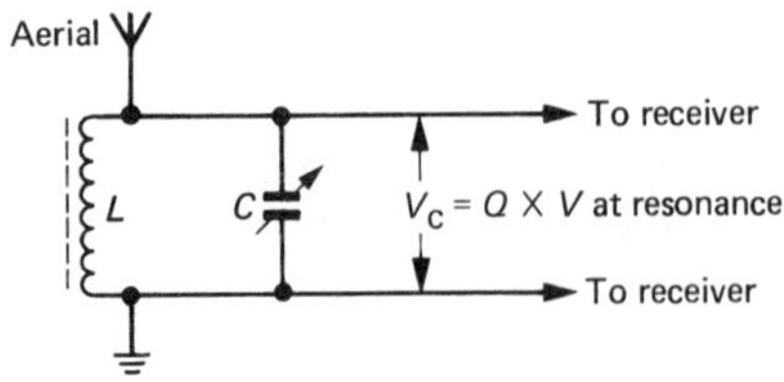

Fig. 12.05

12.4 Radio frequency voltage amplifiers

(a) Single-tuned amplifier

The basic circuit in Fig. 12.06 uses a FET (e.g. 2N3819). The amplifier load is a parallel resonant circuit L_1C_1, tuned to give a high impedance at the selected frequency (by varying C_1). The resulting amplified output voltage developed across L_1C_1 is coupled to the next stage by C_3. The bias components R_S and R_G and the decoupling capacitor C_S function as in an a.f. FET amplifier (Fig. 10.08) but C_S has a smaller value in r.f. amplifiers.

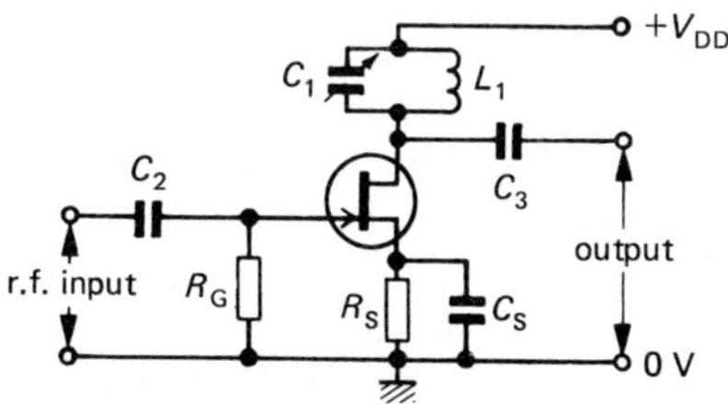

Fig. 12.06

The response curve of a single-tuned amplifier (Fig. 12.04(*b*)) has a sharp peak which falls away rapidly on both sides of the resonant frequency. Radio frequency signals carrying speech, music and vision, require a certain *bandwidth* of frequencies, centred on the resonant frequency, for satisfactory reception, as we will see in section 18.3. The ideal response curve is therefore flat-topped and vertical-sided, Fig. 12.07(*a*). This can be more or less achieved using a double-tuned amplifier.

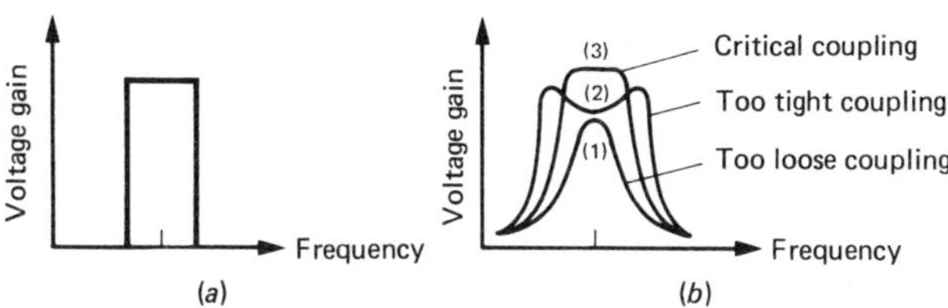

Fig. 12.07

(b) Double-tuned amplifier

A circuit using bipolar transistors (e.g. BF194) is shown in Fig. 12.08. It has two tuned circuits L_1C_1 and L_2C_2, with L_1 and L_2 forming the primary and secondary windings of an r.f. transformer. If the magnetic coupling between L_1 and L_2 is adjusted to its *critical* setting (by altering the position of the iron-dust core in the transformer), a near-ideal response curve like (3) in Fig. 12.07(*b*) is obtained. Curves (1) and (2) show the response for loose and tight coupling respectively.

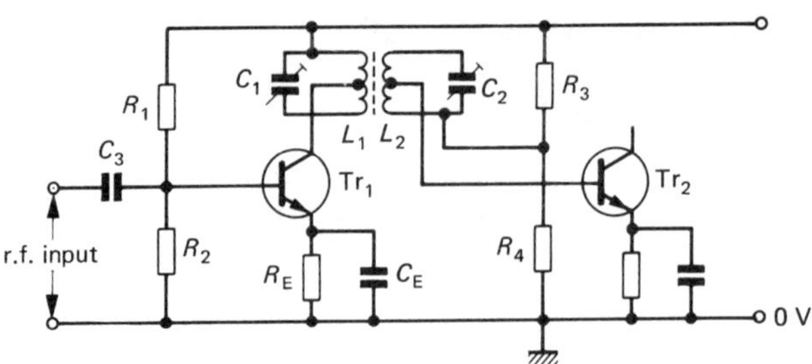

Fig. 12.08

The bias components R_1, R_2, R_E and the decoupling capacitor C_E operate as in an a.f. bipolar amplifier (Fig. 11.01). You should also note that Tr_2 is not connected directly across L_2C_2. If it was, its low input impedance of about 1 kΩ would drastically reduce the *Q-factor* of L_2C_2 and therefore its selectivity. A better impedance match is obtained if the input of Tr_2 is taken from a 'tap' on L_2, as shown. The output impedance of Tr_1 is similarly matched to L_1C_1.

Double-tuned amplifiers are used for fixed frequency applications (hence the trimmers for C_1 and C_2) such as intermediate frequency (i.f.) amplifiers in superhet radio receivers—see section 18.3.

(c) Practical points

Whilst both bipolar transistors and FETs are used for r.f. amplification, FETs generate less 'noise' and have higher input impedances.

A problem which can arise with r.f. amplifiers, for various reasons, is instability. The amplifier then behaves as an oscillator. Careful design, attention to the layout of components, short connections and metal screening cans round coils to shield them from stray fields, all help to give stable operation. The use of crystal or ceramic filters is a further aid.

(d) Crystal and ceramic filters

These are now commonly used as *bandpass* filters in radio receivers instead of *LC* circuits to allow a certain band of frequencies to pass, e.g. in i.f. amplifiers in v.h.f. radios. Their action depends on the reverse piezoelectric effect—see section 9.2.

A thin plate of quartz crystal or a ceramic such as lead zirconium titanate (having a film of conducting material evaporated on two of its faces to act as electrodes) vibrates when an alternating voltage is applied to it. The vibration, which is a maximum at the natural frequency of the crystal, itself generates a 'back' voltage that opposes the applied voltage. In effect, the crystal behaves as a resonant circuit with a very high *Q-factor* (up to 100 000 compared with up to 300 for an *LC* circuit). Resonant frequencies from 10 kHz to about 10 MHz can be obtained, depending on the shape and size of the crystal.

12.5 High frequency amplifiers

When considering the operation of bipolar transistors and FETs at high r.fs., their internal capacitances must be taken into account.

The three capacitances present in a bipolar transistor are shown in Fig. 12.09(*a*). They are partly due to the depletion layer at a p-n junction acting as an insulator that is sandwiched between two conductors (the p- and n-type regions). The base-emitter capacitance C_{BE} is typically of the order of 500 pF.

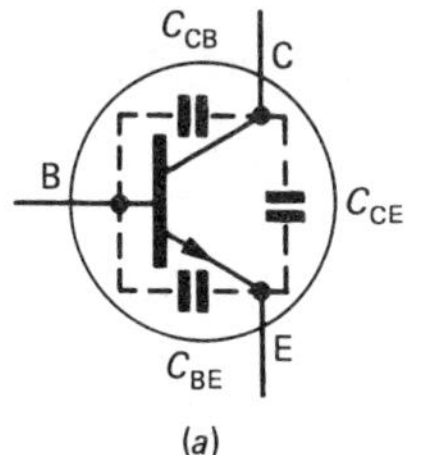

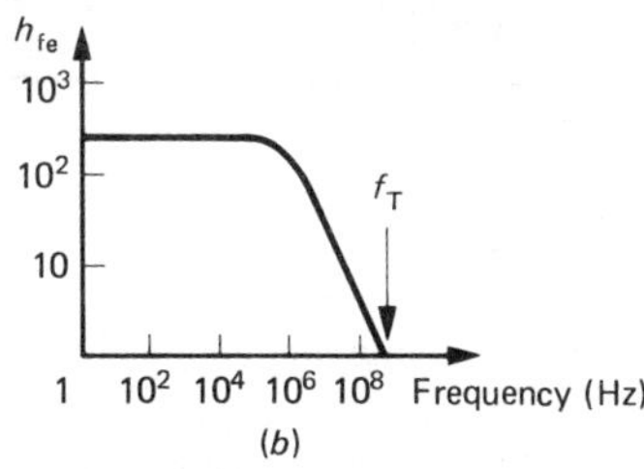

Fig. 12.09

At high frequencies, the reactance ($X_C = 1/(2\pi fC)$) of C_{BE} decreases and becomes comparable with the resistance of the base-emitter function. As a result, the changes of base current due to the high frequency input, instead of controlling the collector current, flow through C_{BE}, causing a drop in the a.c. current gain h_{fe}. C_{CB} and C_{CE} have similar adverse effects on the high frequency performance of the transistor. Fig. 12.09(*b*) shows how h_{fe} varies with frequency; f_T, the *transition frequency*, is the frequency at which $h_{fe} = h_{FE} = 1$ and amplification no longer occurs—see sections 8.4 and 8.5.

The internal capacitances of FETs are much smaller than those of bipolar transistors but they too suffer reduced gain at high frequencies.

Two circuits designed as high frequency amplifiers will now be described.

(a) Dual-gate MOSFET amplifier

This type of amplifier, shown in Fig. 12.10, is used in v.h.f. frequency modulated (FM) radio receivers. A dual-gate MOSFET (e.g. 3N201) has two gates and one channel. Gate G_2 acts as a screen which more or less cuts off the electric field between the drain D and gate G_1, thereby almost eliminating their internal capacitance. Also, by applying the input signal to G_1 and the d.c. bias to G_2, the signal and bias circuits are separated, allowing the best design for each.

The input and output circuits can be tuned by C_1 and C_2 to the v.h.f. radio band (90 to 110 MHz). L_1 and L_2 are 6-turn 1 cm diameter coils. R_1, R_2 and R_S provide bias, C_3 and C_S decouple r.f.

(b) Wide-band amplifier

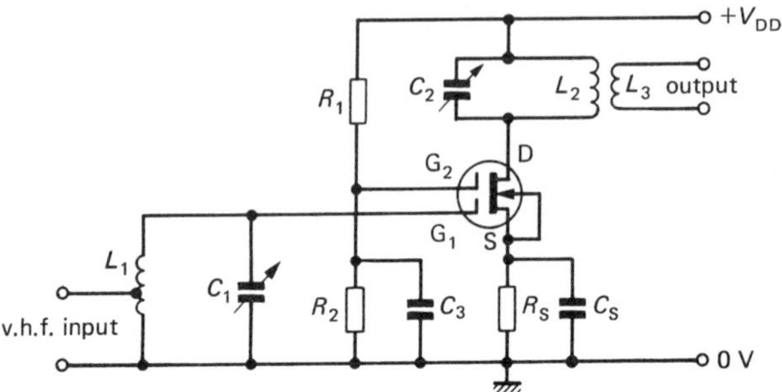

Fig. 12.10

Sometimes a very wide range of frequencies has to be amplified. For example, in a video amplifier handling u.h.f. band (400 to 900 MHz) TV picture signals, a bandwidth of 5.5 MHz is required; and the Y-amplifier in an oscilloscope (see section 21.8) may need to have a flat frequency response from d.c. to 50 MHz.

In such cases a high-Q tuned circuit would not be appropriate as the output load. A resistor would be suitable, since its resistance does not change with frequency. However, internal and stray circuit capacitances at the output, denoted by C_S in Fig. 12.11(*a*), would markedly reduce the load impedance and gain at high frequencies (since C_S is in effect in parallel with R_L, whose top-end in the diagram is 'ground' (0 V) to a.c.).

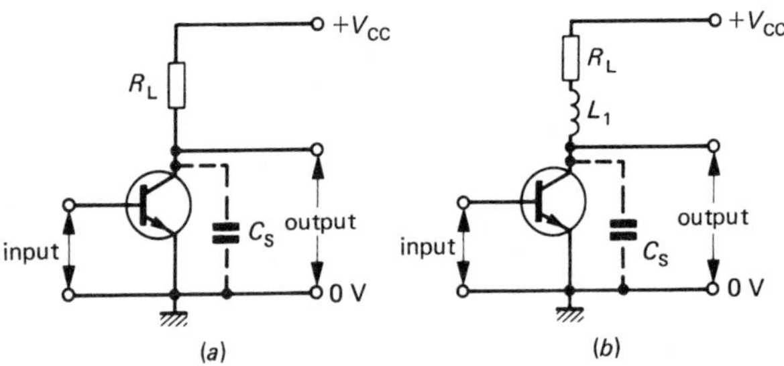

Fig. 12.11

The difficulty is overcome in the wide-band amplifier circuit of Fig. 12.11(*b*) by including a small coil L_1 in series with R_L, as part of the load. L_1 forms a parallel LC circuit with C_S and, as the frequency increases, the reactance of C_S decreases, while that of L_1 increases ($X_L = 2\pi fL$). When the two reactances are equal, resonance occurs and, though damped by R_L, the gain at high frequencies is not reduced.

The value of L_1 may be found experimentally and depends on R_L, C_S and the bandwidth required. Typically it lies between 1 μH and 1 mH.

12.6 Radio frequency oscillators

Radio frequency oscillators are used in radio and television transmitters and receivers, in high voltage power supplies and in test instruments such as r.f. signal generators. They convert d.c. from the power supply into undamped a.c. having a sine, square or some other waveform, depending on which of the many possible circuits is employed.

Most oscillators are amplifiers with a *positive feedback* loop which ensures that the feedback (i) is in phase with the input and (ii) makes good the energy losses in the oscillatory circuit. Sine wave oscillators are described in this section.

(a) *LC* oscillators

The simple circuit of Fig. 12.12 is basically that of an amplifier with d.c. bias and stabilization being provided by R_1, R_2 and R_3 and r.f. decoupling by C_2 and C_3. The tuned circuit L_1C_1 is the collector load (C_1 being variable) and it determines the frequency f of the oscillators, given approximately by $f = 1/[2\pi\surd(L_1C_1)]$.

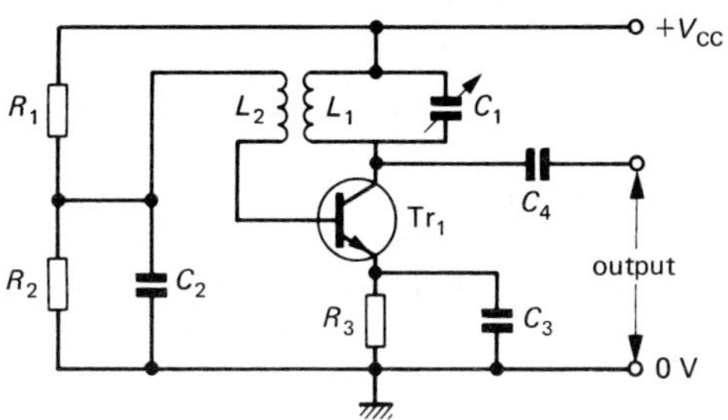

Fig. 12.12

In practice, to start the oscillations, we do not have to apply a sine wave input. Switching on the power supply produces a pulse which charges C_1 and starts oscillations automatically so long as the feedback conditions are met. Feedback occurs by the sinusoidal oscillatory current in L_1 inducing, by transformer action, a voltage of the same frequency in L_2 (coupled magnetically to L_1). This is applied to the base of Tr_1 as the input and is amplified to cause a larger oscillatory current in L_1 and hence a larger voltage in L_2 and so on. The coupling between L_1 and L_2 should be just enough to maintain oscillations.

Note that a single-stage amplifier produces a 180° phase shift between its output and input—see section 10.2. The feedback circuit L_1L_2 must therefore introduce another 180° shift by being connected the 'right way round' to the base of Tr_1.

The frequency of the output, which may be taken from the collector via C_4, is required to be constant in many applications. It depends mainly on the values of L_1 and C_1 but it is affected by other factors such as changes in (i) L_1 and C_1 due to temperature variations, (ii) transistor parameters (e.g. h_{fe}) due to supply voltage variations and (iii) the external load to which the r.f. power is to be supplied. Steps can be taken to minimize all of these but when very high frequency stability is needed, crystal oscillators are used.

(b) Hartley and Colpitts oscillators

The feedback circuits in these two popular r.f. oscillators are similar and, as their block diagram in Fig. 12.13 shows, three impedances Z_1, Z_2 and Z_3 are used to provide the 180° phase shift required. It can be proved that to obtain positive feedback and oscillations, Z_1 and Z_2 must both be inductive (or capacitive) and Z_3 capacitive (or inductive).

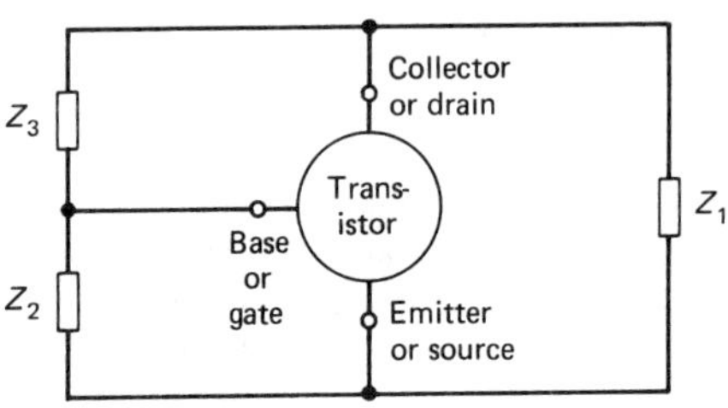

Fig. 12.13

In a Hartley oscillator, shown in Fig. 12.14(*a*), the coil L_0L_1 of the tuned circuit is tapped to provide Z_1 and Z_2. In a Colpitts oscillator, Fig. 12.14(*b*), the capacitor (remember 'C' for Colpitts and capacitor) in the tuned circuit is 'tapped', i.e. two capacitors C_0 and C_1 are used. In both cases R_1, R_2 and R_3 provide bias and the d.c. blocking capacitor C_2 as well as the decoupling capacitor C_3 have negligible reactance at the oscillation frequency f (here fixed).

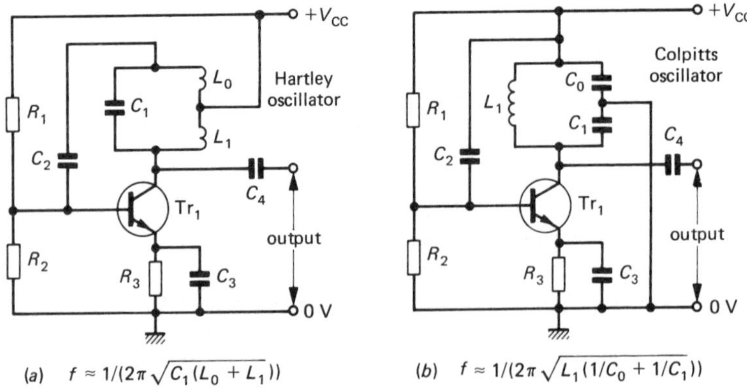

Fig. 12.14

(c) Crystal oscillators

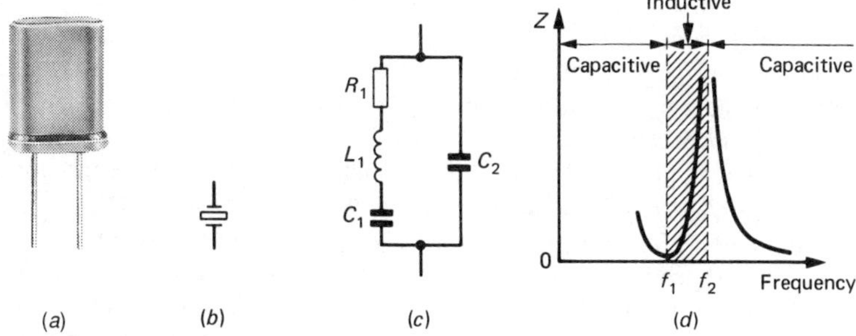

Fig. 12.15

When a sine wave oscillator with a fixed, very stable frequency is required, use is made of a piezoelectric crystal such as quartz. A crystal, enclosed in a metal can with two leads from its electrodes, is shown in Fig. 12.15(*a*) and its symbol in (*b*).

Experiments show that the equivalent electrical circuit of a quartz crystal is a series-parallel arrangement like that in Fig. 12.15(*c*) where, for example, R_1, L_1, C_1 and C_2 might have values 1 kΩ, 100 H, 0.1 pF and 10 pF respectively. At most frequencies the crystal behaves as a capacitor, but at a certain frequency f_1 it exhibits series resonance ($Z \approx 0$) and at a slightly higher frequency f_2, it acts as a parallel resonant circuit (Z very large). At frequencies between f_1 and f_2 it behaves as an inductor. The response curve in Fig. 12.15(*d*) summarizes these properties; they enable us to use crystals for many different feedback loops in oscillators.

The circuit in Fig. 12.16 is a crystal version of a Colpitts oscillator suitable for producing sinusoidal oscillations in the range 1 to 10 MHz. C_1 and C_2 provide the necessary capacitive reactances from the collector and base respectively to the emitter, while X_1 acts as the inductor between collector and base. C_3 allows fine adjustment of the frequency of the oscillations, determined by X_1. R_1 is the collector load from which the output is taken and R_2 provides collector-to-base bias.

The tuned circuit (C_1, C_2, C_3 and crystal) is not in series with the power supply (as in earlier circuits) but is in the alternative 'parallel-fed' arrangement which allows one plate of each of C_1 and C_2 to be connected to ground. Switching on the supply applies a voltage pulse to the crystal, making it vibrate and develop an alternating voltage between its terminals.

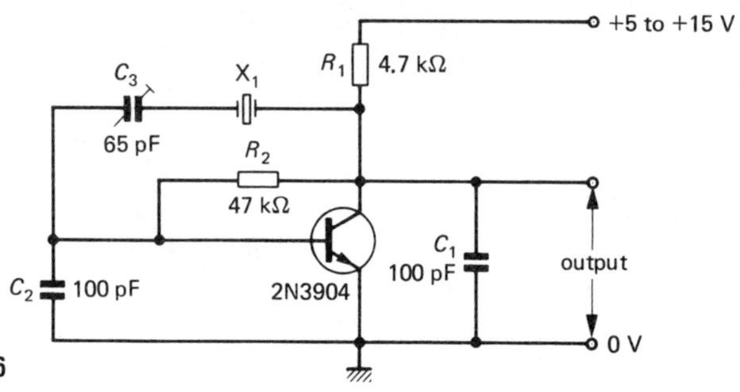

Fig. 12.16

12.7 Audio frequency oscillators

For use in audio frequency sine wave oscillators, coils and capacitors of the required high values are too bulky, as are crystals, which then become expensive and in any case only operate at a fixed frequency. At frequencies up to about 50 kHz, circuits using resistors and capacitors have been developed; beyond this, stray capacitances cause problems.

We saw earlier, in Fig. 5.15, that if a.c. is applied to a resistor and capacitor in series, the voltages developed across them are 90° out of phase. In the Wien a.f. oscillator, a network of two resistors, usually equal, and two capacitors, usually equal, arranged as in Fig. 12.17(*a*), acts as the positive feedback circuit. The network is an a.c. voltage divider and mathematical analysis shows that the output voltage V_o is *in phase* with the input voltage V_i, i.e. the phase shift is zero, at one frequency f, given by:

$$f = \frac{1}{2\pi RC}$$

f is in Hz if R is in ohms and C in farads. At all other frequencies, there is a phase shift. To obtain oscillations the network must therefore be used with a non-inverting amplifier which gives an output to the Wien network that is in phase with its input.

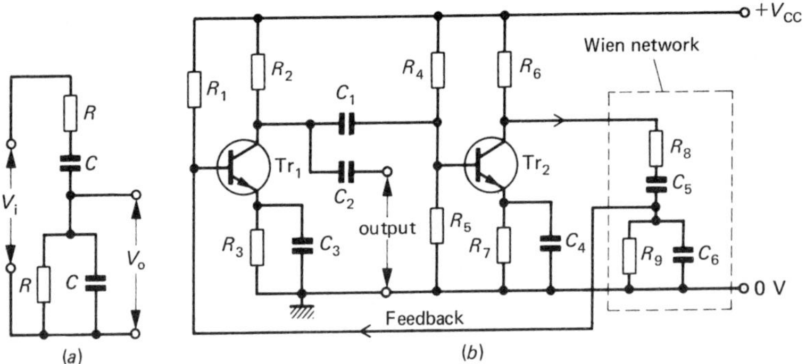

Fig. 12.17

One circuit employs a two-stage transistor amplifier as in Fig. 12.17(*b*), each stage producing a 180° phase shift and so giving an overall shift of 360° (or 0°). The output from the amplifier is applied to the Wien network, R_8C_5, R_9C_6, and part of it is fed back to the input. It can be proved that so long as the voltage gain of the amplifier exceeds three, oscillations will be maintained at the desired frequency f, as determined by the values of R_8, R_9, C_5 and C_6. (Usually $R_8 = R_9$ and $C_5 = C_6$).

To obtain a variable-frequency output, as in an a.f. signal generator, R_8 and R_9 are often variable resistors which are 'ganged', i.e. mounted on the same spindle so that they can be varied simultaneously by one control, also C_5 and C_6 are pairs of capacitors that are switched in for different frequency ranges.

An integrated circuit version of an a.f. oscillator, called a *waveform* or *function generator*, will be considered in section 17.5.

12.8 Relaxation oscillators

The action of a relaxation oscillator depends on the charging of a capacitor followed by a period of 'relaxation' when the capacitor discharges through a resistor. Its output voltage waveform is not a sine wave, like those of the oscillators considered previously, but in two important cases it is a sawtooth wave and a square wave, Figs. 12.18(*a*) and (*b*).

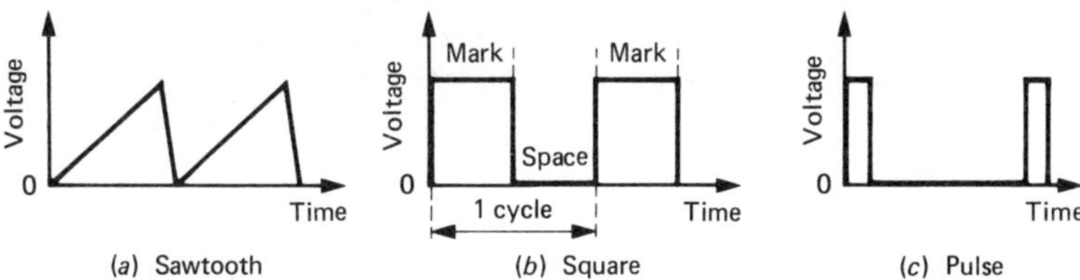

Fig. 12.18

A sawtooth wave oscillator is required in a television receiver and in a cathode ray oscilloscope. Square wave oscillators are used in the operation of digital systems such as computers. They are also used to produce electronic music since square waves are rich in harmonies, i.e. they contain many frequencies; and they are also useful for certain tests on an a.f. amplifier.

The commonest square wave oscillator is the *astable multivibrator*, to be considered in section 14.8. By varying the mark–space ratio of its output, it can become a *pulse generator* which produces a series of very short electrical pulses separated by relatively longer gaps, as in Fig. 12.18(*c*). Square waves can also be obtained from a sine wave oscillator if its output is amplified and then the tops and bottoms of the peaks 'clipped' off (Fig. 10.04(*d*)).

The integrated circuit a.f. oscillator mentioned in the previous section and to be described in section 17.5 generates square, sawtooth and other waveforms.

12.9 Revision questions

1. a) What frequency range is covered by r.f. amplifiers?
b) State two ways in which r.f. and a.f. amplifiers are
(i) similar,
(ii) different.
c) How can an amplifier become an oscillator?

2. Draw an oscillatory circuit. What can you say about
(i) the amplitude,
(ii) the waveform, and
(iii) the frequency, of the oscillations?

3. a) Explain the term electrical resonance.
b) State two properties of
(i) a series resonant circuit,
(ii) a parallel resonant circuit.
c) What is the *Q*-factor of a circuit?

4. a) Draw a circuit for a single-tuned r.f. transistor amplifier and sketch its response curve.
b) Repeat a) for a double-tuned r.f. transistor amplifier.
c) How is the risk of instability reduced in r.f. amplifiers?
d) What advantages do ceramic filters have over *LC* circuits in r.f. amplifiers?

5. a) What effect must be taken into account when designing high frequency r.f. amplifiers?
b) How is this effect almost eliminated in a dual-gate MOSFET FM radio amplifier?
c) How is it used in a wide-band video amplifier?

6. a) State three uses for r.f. oscillators.
b) Draw a circuit for a tuned collector *LC* oscillator explaining
(i) how undamped oscillations are obtained,
(ii) the biasing arrangements,
(iii) the source of the oscillatory power.

7. a) What is meant by the frequency stability of an oscillator?
b) What factors affect it?

8. a) By comparing Figs. 12.13 and 12.14(*a*), state which components provide the positive feedback impedances Z_1, Z_2 and Z_3 in a Hartley oscillator.
b) Repeat a) for a Colpitts oscillator using Figs. 12.13 and 12.14(*b*).

9. a) Draw the electrical circuit equivalent of a piezoelectric crystal and sketch its response curve.
b) Draw a circuit for a crystal-controlled r.f. oscillator and state its advantage over other types of oscillator.

10. Why are *LC* and crystal circuits generally unsuitable for use in a.f. ocillators?

11. Draw a circuit for a Wien oscillator stating
(i) which components determine the frequency of oscillation, and
(ii) why its use is confined mainly to audio frequencies.

12. What kinds of waves are produced by relaxation oscillators? Name one common relaxation oscillator. Draw three waveforms and give examples of their uses.

12.10 Problems

1. Calculate the maximum and minimum frequencies of the oscillations which could be generated by a variable capacitor of maximum capacitance 500 pF and minimum capacitance one-tenth of this value connected to a coil of inductance 100 μH. (1 pF = 10^{-12}F)

2. An *LC* oscillator has a fixed inductance of 50 μH and is required to produce oscillations over the band 1 MHz to 2 MHz. Calculate the maximum and minimum values of the variable capacitor required. (Take $\pi^2 = 10$.)

3. For a *CLR* series circuit in which $C = 100$ pF, $L = 100$ μH and $R = 10\ \Omega$, calculate
   a) the resonant frequency,
   b) the impedance of the circuit at resonance,
   c) the *Q*-factor of the circuit.

4. Repeat question 3 for a parallel circuit with the same values of *C*, *L* and *R*.

5. The voltage gain in dB (see section 11.9) of an r.f. amplifier over a range of frequencies is given in Fig. 12.19. Obtain from it the 3 dB bandwidth, i.e. the range of frequencies within which the gain does not fall below 0.7 of its maximum value.

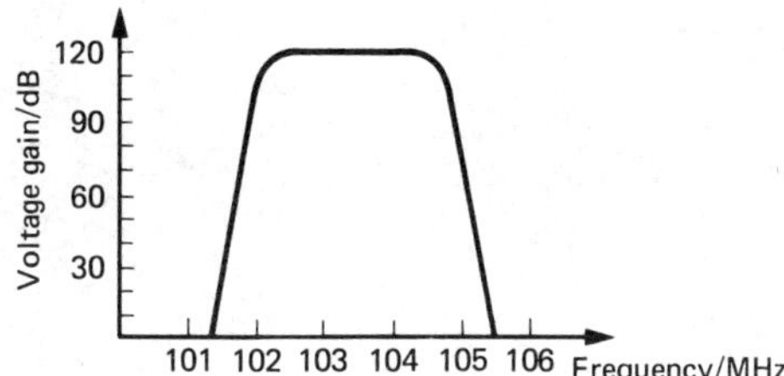

Fig. 12.19

6. In the Wien oscillator circuit of Fig. 12.17(*b*), if $R_8 = R_9 = 100$ kΩ and $C_5 = C_6 = 1$ nF, calculate the frequency of the oscillations. (1 nF = 10^{-9} F)

7. Draw square waves with mark–space ratios of
   a) 1:1,
   b) 2:1,
   c) 3:1.

8. a) For a true *square wave* what is
      (i) the mark–space ratio,
      (ii) the duty cycle?
   (The duty cycle is the ratio of the time the mark lasts to the time the mark and space together last.)
   b) A *rectangular waveform* has a frequency of 100 kHz and is positive for 6 microseconds (6 μs) during each cycle.
      (i) How long does 1 cycle last? (i.e. what is the period?)
      (ii) What is the mark–space ratio?
      (iii) What is the duty cycle?

UNIT 13

Operational amplifiers and linear integrated circuits

13.1 About integrated circuits (ICs)

(a) Description

An IC is a complete electronic circuit containing transistors and perhaps diodes, resistors and capacitors all made from, and on, a chip of silicon about 5 mm square and no more than 0.5 mm thick.

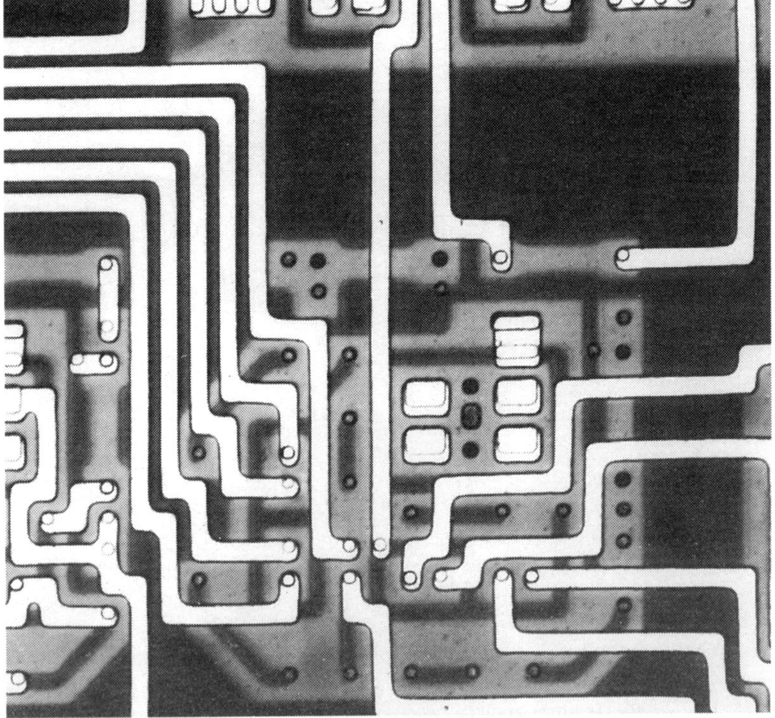

Fig. 13.01

A tiny part of a typical IC is shown much magnified in Fig. 13.01. In Fig. 13.02 it is in its protective plastic case which has been partly removed to reveal the IC ('chip') and the leads radiating from it to the pins that enable it to communicate with the outside world. ICs are packaged in different ways; that shown here is the popular dual-in-line (d.i.l.) arrangement with the pins (often 8, 14, 16 or more) 0.1 inch apart, in two lines on either side of the case. Circular metal packages, similar to those used for some transistors, are also common.

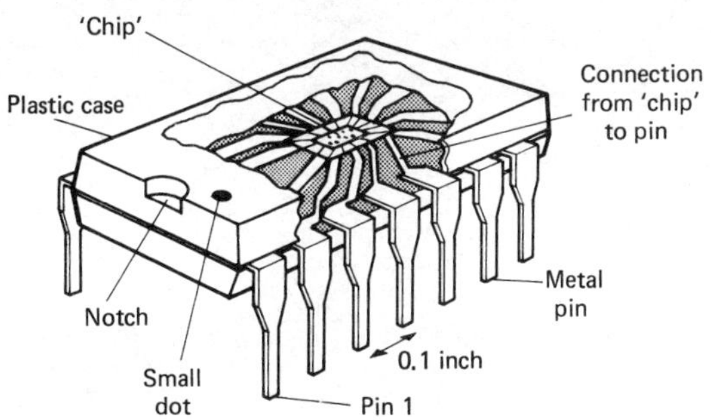

Fig. 13.02

(b) Scales of integration

The first ICs were made in the early 1960s and consisted of fairly simple circuits with fewer than 100 components per chip. They were small-scale integrated (SSI) circuits. The complexity increased rapidly, through medium-scale integrated (MSI) and large-scale integrated (LSI) circuits, until today very-large-scale (VLSI) ones may have millions of components.

(c) Advantages and limitations

Compared with circuits built from separate components ICs are very much smaller, lighter, cheaper and more reliable. However their small size limits the power and voltage (typically 30 V maximum) they can handle. In addition, although silicon is ideal for making diodes and transistors, it is not so good for high value resistors and capacitors (where the present limits are about 50 kΩ and 200 pF respectively) because they need too much space. Also, inductors and transformers cannot be produced on a silicon chip.

(d) Manufacture

Silicon containing no more than 1 in 10^{10} parts of impurity (i.e. 99.999 999 9% pure) is produced chemically from silicon dioxide, the main constituent of sand. It is then melted in an inert atmosphere and crystallization starts when a small crystal of 'pure' silicon (a 'seed') is inserted and slowly withdrawn from it. A cylindrical bar, up to 10 cm in diameter and 1 metre or so long, is formed as a single, near-perfect crystal, i.e. its atoms are arranged almost perfectly regularly throughout. The bar is cut into $\frac{1}{4}$ to $\frac{1}{2}$ mm thick wafers whose surfaces are ground and highly polished, Fig. 13.03.

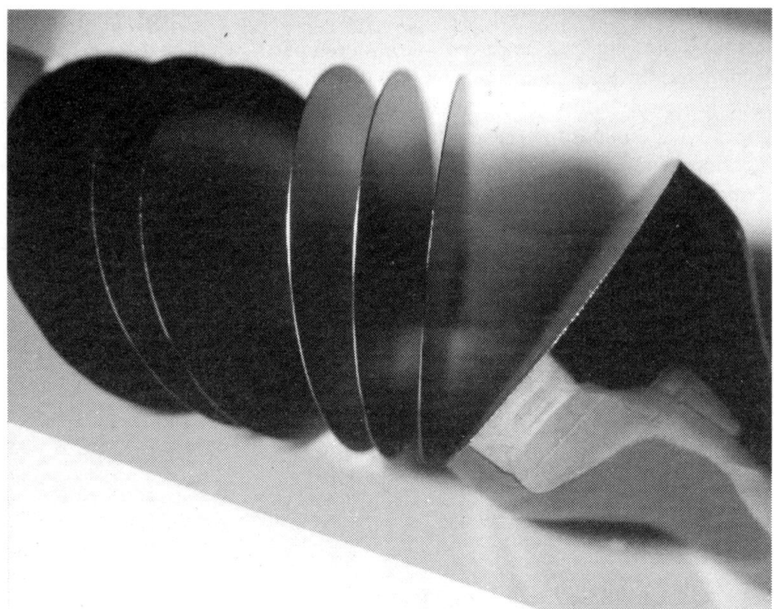

Fig. 13.03

Depending on their size, several hundred identical circuits (the 'chips') may be formed side by side on the surface of one wafer by an extension of the *planar* process used for transistors. This first involves depositing an insulating layer of silicon dioxide on the wafer and then using a pattern of photographic masks, designed from a large drawing of one chip, to create 'windows' in the oxide by exposure to ultraviolet light, followed by developing and etching away with acids, Fig. 13.04(*a*).

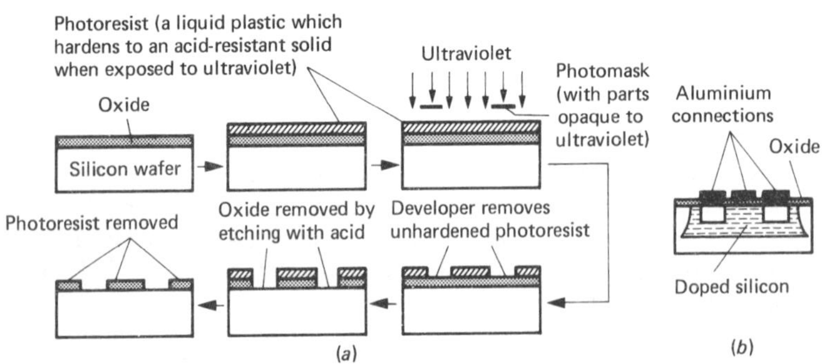

Fig. 13.04

Doping then occurs, in one method by exposing the wafer at high temperature to the vapour of either boron or phosphorus, so that their atoms diffuse through the 'windows' (guided by masks) into the silicon. The p- and n-type regions so produced for the various components are next interconnected to give the required circuit by depositing aluminium, again using masks, Fig. 13.04(*b*). Several layers can be built one on top of the other in this way.

The construction of integrated diodes and transistors is similar to that of their discrete versions. Integrated resistors are thin layers of p-type silicon whose value depends on their length, cross-sectional area and degree of doping. Integrated capacitors are either reverse biased p-n junctions or two conducting areas (e.g. of aluminium or doped silicon) separated by a layer of silicon dioxide as dielectric.

Each chip is tested as shown in Fig. 13.05(*a*) and faulty ones discarded—up to 70% may fail. The wafers are next cut into separate chips as in Fig. 13.05(*b*). Each chip is then packaged and connected (automatically) by gold wires to the pins on the case.

The complete process, which can require up to three months, must be done in a controlled, absolutely clean environment.

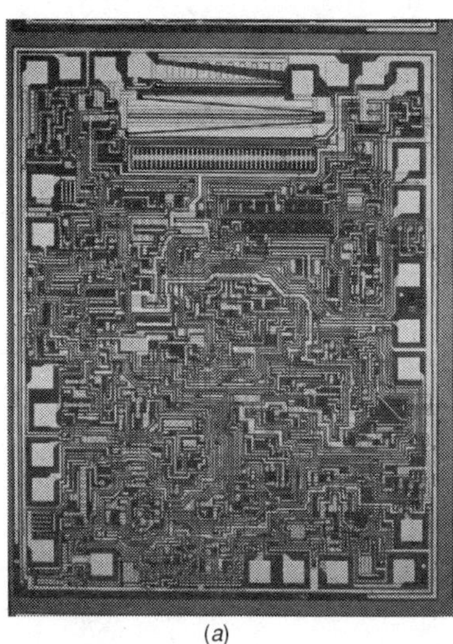

(*a*)

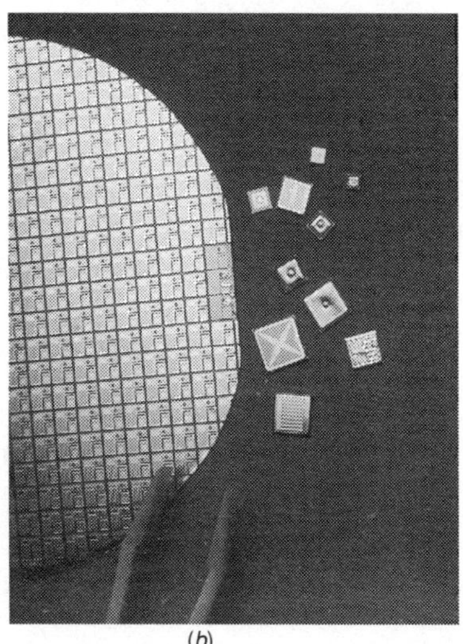

(*b*)

Fig. 13.05

(e) Types

As with discrete-component circuits, there are two broad groups of integrated circuit—linear (or analogue) and digital. The earliest ICs were digital, because they are easier to make and the market for them was larger; they will be considered in Units 15, 16 and 17. Linear ICs are the subject of this Unit, particularly operational amplifiers, which were the first linear type (1964). Most linear ICs are based on bipolar transistors but in some cases FETs are used either exclusively or in addition to bipolar types.

13.2 Operational amplifier—Introduction

Operational amplifiers (op amps) were originally made from discrete components. They were designed to solve mathematical equations electronically, by performing operations such as addition and division in analogue computers—see section 20.10. Nowadays in IC form they have many uses, one of the most important being as high gain d.c. and a.c. voltage amplifiers. A typical op amp contains twenty transistors as well as resistors and small capacitors.

(a) Properties

The chief properties of an op amp are:

(i) a very *high open-loop voltage gain* A_o (see section 11.3) of about 10^5 for d.c. and low frequency a.c., which decreases as the frequency increases;
(ii) a very *high input impedance*, typically 10^6 to 10^{12} Ω, so that the current drawn from the device or circuit supplying it is minute and the input voltage is passed on to the op amp with little loss (see section 11.5);
(iii) a very *low output impedance*, commonly 100 Ω, which means the efficient transfer of its output voltage to any load greater than a few kilohms.

(b) Action

It has one output and two inputs. The *non-inverting* input is marked + and the *inverting* input is marked −, as shown on the amplifier symbol in Fig. 13.06(*a*) and on the pin connections in Fig. 13.06(*b*) for the TL081C and 741 op amps. Operation is most convenient from a dual balanced d.c. power supply giving equal positive and negative voltages $\pm V_S$ in the range ±5 V to ±15 V. The centre point of the power supply, i.e. 0 V, is common to the input and output circuits and is taken as their voltage reference level. (Do not confuse the input signs with those for the supply polarities, which, for clarity, are often left off circuit diagrams.)

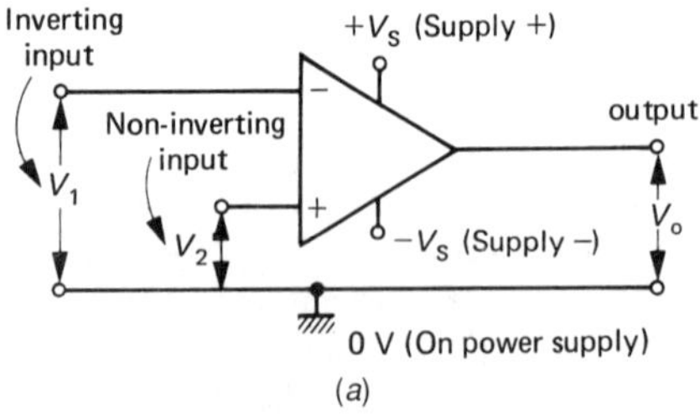

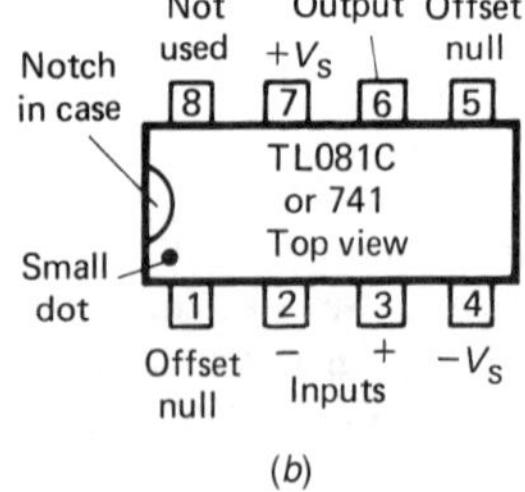

Fig. 13.06

If the voltage V_2 applied to the non-inverting (+) input is positive relative to the other input, the output voltage V_o is positive, similarly if V_2 is negative, V_o is negative, i.e. V_2 and V_o are in phase. At the inverting (−) input, a positive voltage V_1 relative to the other input causes a negative output voltage V_o and vice versa, i.e. V_1 and V_o are in antiphase. Basically an op amp is a *differential amplifier* (see section 13.10), i.e. it amplifies the difference between the voltages V_1 and V_2 at its inputs. There are three cases:

(i) if $V_2 > V_1$, V_o is positive;
(ii) if $V_2 < V_1$, V_o is negative;
(iii) if $V_2 = V_1$, V_o should be zero.

In general, the output is given by:

$$V_o = A_o \times (V_2 - V_1)$$

where A_o is the open-loop voltage gain.

In most applications, but not all (see section 13.8), single-input working is used and the two basic kinds of circuit for this will be considered in sections 13.4 and 13.6.

(c) Transfer characteristic

A typical voltage characteristic showing how the output V_o varies with the input $(V_2 - V_1)$ is given in Fig. 13.07. It reveals that it is only within the very small input range AOB that the output is directly proportional to the input, i.e. when the op amp behaves more or less *linearly* and there is minimum distortion of the amplifier output. Inputs outside the linear range cause *saturation* and the output is then close to the maximum value it can have, i.e. $+V_S$ or $-V_S$.

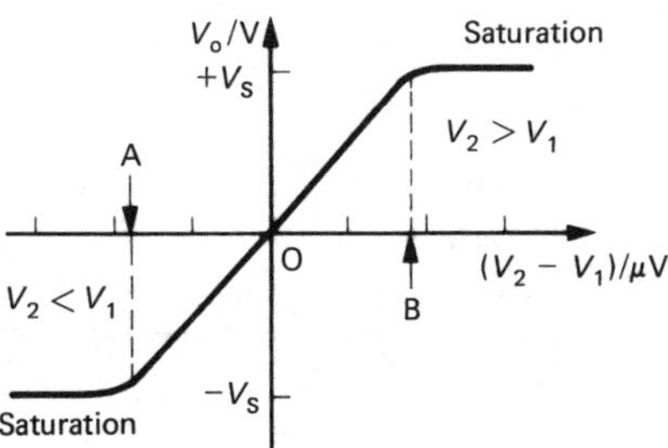

Fig. 13.07

The limited linear behaviour is due to the very high gain A_o and the higher it is, the greater is the limitation. Thus on a ±9 V supply, the maximum output voltage is near ±9 V and, if $A_o = 10^5$, the maximum input voltage swing (for linear amplification) is $\pm 9\ \text{V}/10^5 = \pm 90\ \mu\text{V}$. A smaller value of A_o would allow a greater input.

13.3 Operational amplifier—Practical points

(a) Negative feedback

Op amps almost invariably use n.f.b., obtained by feeding back some of the output to the *inverting* (−) input. The part of the output fed back produces a voltage at the output that opposes the one from which it is taken, thereby reducing the new output of the amplifier, i.e. the one with n.f.b. The resulting closed-loop gain A is then less than the open-loop gain A_0 but a wider range of values of voltages can be applied to the input for amplification. (Feedback applied to the non-inverting (+) input would be positive and would increase the overall output.)

In addition, as stated for discrete component amplifiers, n.f.b. gives greater *stability*, less *distortion* and increased *bandwidth* as shown in Fig. 13.08. Also, the *gain* can be predicted exactly and is almost independent of the particular op amp used, as we will see in section 13.4.

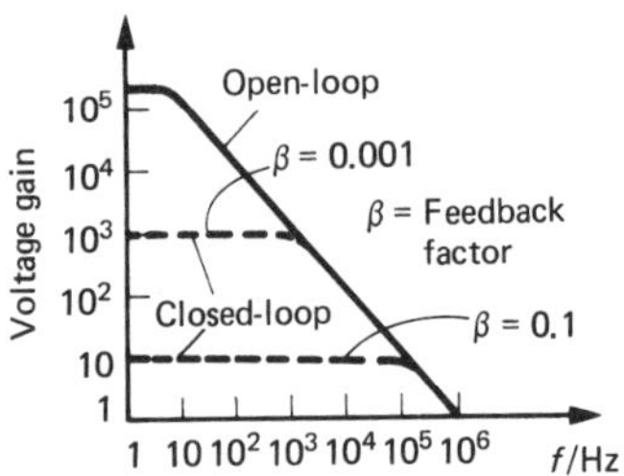

Fig. 13.08

The very high value of A_0 is appreciably greater than is required for most purposes and so it is sensible to sacrifice some of it for the above advantages.

(b) Biasing

To obtain the maximum symmetrical output voltage swing, the quiescent (no input signal) output voltage should be, as in any voltage amplifier, about *half* the supply voltage, i.e. 0 V for a balanced dual supply of $\pm V_S$. As the transfer characteristic of Fig. 13.07 shows, this requires the quiescent voltages at both inputs to be equal (i.e. $V_2 - V_1 = 0$).

In most op amps each input terminal is connected to the base of its own input transistor which must be able to draw the tiny but necessary d.c. bias current to ensure that the input signals operate in a suitable region. This is conveniently achieved by providing both inputs with a d.c. path to ground (i.e. holding the quiescent input voltages near 0 V) and making the emitters of the input transistors 0.6 V negative (for an n-p-n silicon transistor), thus forward biasing the base-emitter junction. The exact arrangements for ensuring a d.c. biasing route to ground depend on how the inputs are coupled to the op amp.

(c) Coupling

In the amplifiers considered in the previous three Units, capacitors were used at the input and output of each stage to couple a.c. signals but to block any d.c. that might upset the operation. However, 'level shifting' circuits are now common (see section 13.10), and these omit capacitors and alter the quiescent output voltage of one stage to a suitable value before applying it to the next. These have been developed firstly because large capacitors cannot be made on ICs, secondly because capacitors reduce signal strengths (due to their impedance) and thirdly because the signals to be amplified are often d.c. voltages. Such circuits are called *direct coupled* amplifiers; normally they can handle a.c. signals as well as d.c. ones. Most linear ICs, including op amps, are direct coupled amplifiers.

In direct coupled applications, the d.c. bias path to ground would be via the signal source or the preceding stage or through feedback circuitry. If these routes do not exist, the inputs must be connected to ground by a high value resistor. Actual examples will arise in the amplifier circuits described in the next few sections.

(d) Offset voltage

In practice, even when the d.c. bias conditions are met, and no input signal is applied to the amplifier, there may still be a small quiescent voltage at the input, called the differential input *offset voltage*. It can arise in several ways (see section 13.10), sometimes being of the op amp's own making. In the 741 op amp it is about 1 mV and in a circuit with a gain of 1000 it would produce 1 V d.c. at the output.

For applications requiring d.c. amplification, e.g. analogue computing, having an offset voltage is unacceptable. It is 'removed' as shown in Fig. 13.09 for the 741 op amp by connecting a 10 kΩ variable resistor to the 'offset null' pins (1 and 5) and adjusting it until the output is zero when the input is zero.

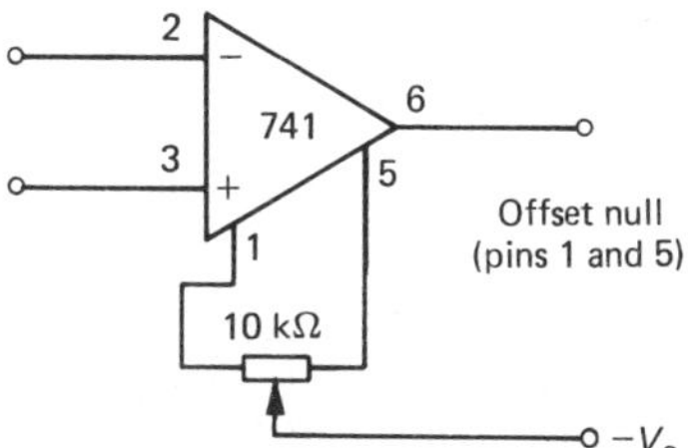

Fig. 13.09

For a.c. operation a coupling capacitor at the output removes any d.c. voltage arising from input offset. Satisfactory amplification then occurs provided the offset has not moved the quiescent output voltage so far from 0 V as to limit the output swing.

(e) Power supplies

An op amp can be operated from a single power supply; the voltage difference available from, for example, a 0 to 18 V supply is the same as that from a +9 V to 0 to −9 V one (i.e. 18 V). However, if the former is used, extra components are required (see section 13.7), which is one reason why a balanced dual supply is usually favoured.

(f) Output current

A typical op amp can supply a maximum output current of about 5 mA. Therefore, the minimum load that can be fed is roughly 2 kΩ (since on a 9 V supply, $V/I = 9\ V/5\ mA \approx 2\ k\Omega$). When greater output currents are required, an emitter-follower output stage (see section 11.6) can be connected to the op amp output or alternatively an op amp current booster can be used.

(g) Slew rate

This is the *maximum rate of change of large amplitude output voltages* that an op amp can allow before it behaves non-linearly. It is measured in volts per microsecond (V/μs) and for the 741 op amp is 0.5 V/μs. This limitation means that op amps should not be used where signals are changing rapidly as in fast-rising digital pulses.

The slew rate limit arises because a rapid rate of change of output voltage causes a rapid rate of charge or discharge of capacitors in the op amp. This in turn requires internal transistors to supply large drive currents which they are not designed to do for power-saving reasons.

Op amps with slew rates greater than 10 V/μs are now available.

13.4 Op amp as an inverting amplifier

(a) Basic circuit

In the circuit of Fig. 13.10 a dual power supply is used and the input voltage (d.c. or a.c.) to be amplified is applied via resistor R_i to the inverting (−) terminal. The output is therefore in antiphase with the input. The non-inverting (+) terminal goes to 0 V. Negative feedback is provided by R_f, called the *feedback resistor*, feeding back a certain fraction of the output voltage, also to the inverting (−) terminal.

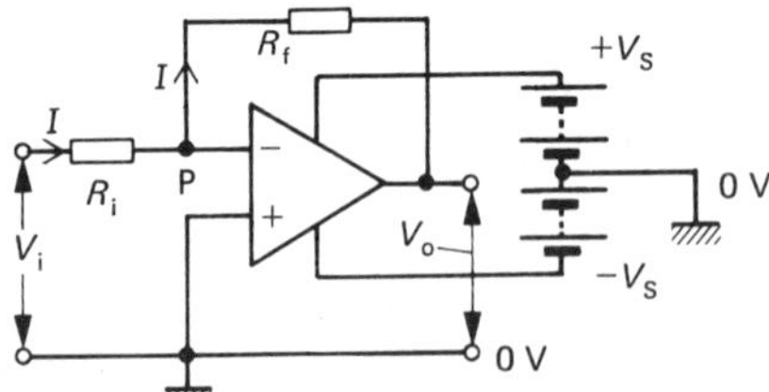

Fig. 13.10

(b) Gain

The voltage at the inverting input (point P) can never be far from zero (in the example in section 13.2(c), its maximum value is ±90 μV) because of the high value of A_o. Therefore, since P is in effect at 0 V, the voltage across R_i equals the input voltage, V_i, and that across R_f equals the output voltage, V_o. P is called a *virtual earth* (or ground) point, though of course it is not connected to ground.

When V_i is positive, current I flows as shown through R_i and then through R_f, only a negligible fraction of I enters the inverting input of the op amp (partly because of its very high input impedance). And so:

$$I = \frac{V_i}{R_i} = \frac{-V_o}{R_f} \qquad (1)$$

The minus sign shows V_i and V_o are in antiphase. Hence from (1), the closed-loop gain A is given by:

$$A = \frac{V_o}{V_i} = \frac{-R_f}{R_i} \qquad (2)$$

For example, if $R_f = 1\ \text{M}\Omega$ and $R_i = 10\ \text{k}\Omega$, $A = -100$ exactly and an input of 0.01 V will cause an output change of 1.0 V.

Equation (2) shows that the gain of the amplifier depends entirely on the values of the discrete resistors R_f and R_i (which can be made accurately) and is independent of the parameters of the op amp (which vary from one sample to another). The required voltage gain can thus be obtained readily and exactly by choosing suitable values for R_f and R_i.

(c) Input impedance

The other versatile feature of this circuit is the way its input impedance can be controlled. Since point P is a virtual earth (i.e. at 0 V), R_i may be considered to be connected between the inverting (−) input terminal and 0 V. The input impedance of the circuit is therefore R_i in parallel with the much greater input impedance of the op amp, i.e. effectively R_i, whose value can be changed.

(d) Further points

In practice the non-inverting (+) input terminal is not connected to 0 V directly (as shown in Fig. 13.10) but via a resistor R_B. If the value of R_B is correctly chosen, each input terminal will 'see' roughly the same resistance to ground. Otherwise the two input bias currents create different voltages at the inputs, resulting in a differential input *offset voltage*—see section 13.3. This produces a steady d.c. voltage at the output (even when there is no external input voltage), which may, in some applications, be undesirable. The value of R_B for correct operation should be about the same as that of R_f and R_i in parallel since some of the bias current for the inverting input is drawn from the output through R_f and the rest through R_i.

When used as an a.c. amplifier, a coupling capacitor C_1 can be included in the input (to block any unwanted d.c. which might overload the amplifier), as shown in the circuit of Fig. 13.11. Both inputs still have the necessary d.c. bias path to ground (via R_f and R_B).

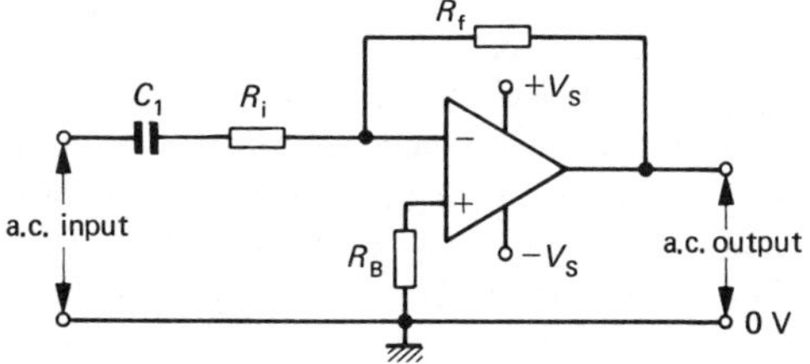

Fig. 13.11

13.5 Op amp as a summing amplifier

When connected as a multi-input inverting amplifier, an op amp can be used to add a number of voltages (d.c. or a.c.) because of the existence of the virtual earth point P (which in turn arises as a consequence of the high value of A_o). Such circuits are employed as 'mixers' in audio applications to combine the outputs of microphones, synthesizers, electric guitars, special effects, etc. They are also used to perform the mathematical process of addition in analogue computing.

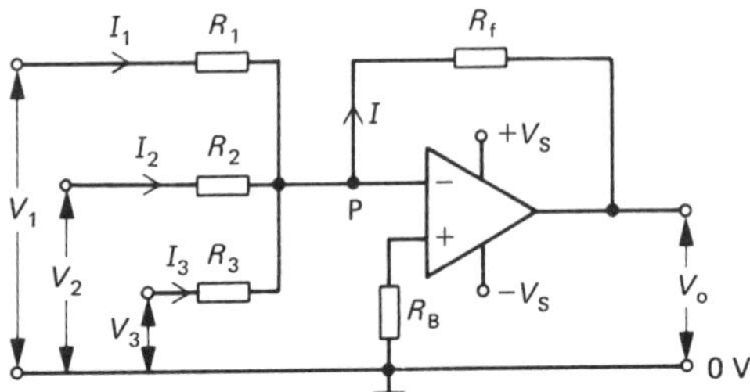

Fig. 13.12

In the circuit of Fig. 13.12 three input voltages V_1, V_2 and V_3 are applied via input resistors R_1, R_2 and R_3 respectively. Assuming that the inverting terminal of the op amp draws no input current, all of it flowing through R_f, then:

$$I = I_1 + I_2 + I_3$$

Hence, since P is a virtual earth point (i.e. at 0 V):

$$\frac{-V_o}{R_f} = \frac{V_1}{R_1} + \frac{V_2}{R_2} + \frac{V_3}{R_3}$$

Therefore:

$$V_o = -\left(\frac{R_f}{R_1} \times V_1 + \frac{R_f}{R_2} \times V_2 + \frac{R_f}{R_3} \times V_3\right)$$

The three input voltages are thus added and amplified if R_f is greater than each of the input resistors.

If $R_1 = R_2 = R_3 = R_i$ (say), the input voltages are amplified equally and

$$V_o = -\frac{R_f}{R_i}(V_1 + V_2 + V_3)$$

If $R_1 = R_2 = R_3 = R_f$ then:

$$V_o = -(V_1 + V_2 + V_3)$$

In this case the output voltage is the sum of the input voltages but is of opposite polarity.

Point P is also called the *summing point* of the amplifier. It isolates the inputs from one another so that each behaves as if none of the others existed and none feeds any of the other inputs even though all the resistors are connected at the inverting input. Also we can select the resistors to produce the best impedance matching with the transducer supplying its input, although compromise may be necessary to obtain both the gain required and the correct input impedance.

13.6 Op amp as a non-inverting amplifier

(a) Basic circuit

The input voltage (d.c. or a.c.) is applied to the non-inverting (+) terminal of the op amp as in Fig. 13.13. This produces an output voltage that is in phase with the input. Negative feedback is obtained by feeding back to the inverting (−) terminal the fraction of the output voltage developed across R_i in the voltage divider formed by R_f and R_i across the output.

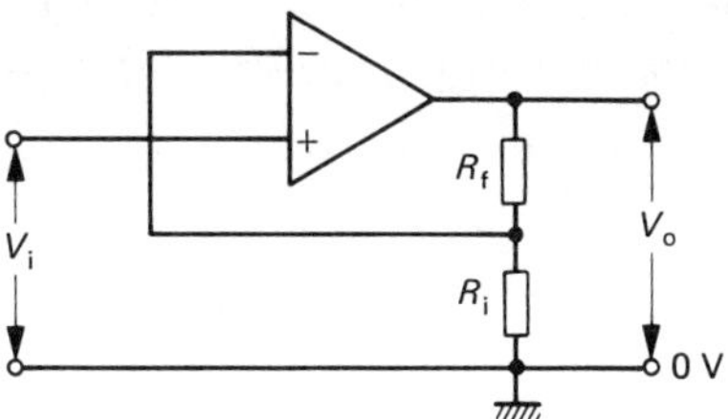

Fig. 13.13

(b) Gain

For the feedback factor β we can write:

$$\beta = \frac{R_i}{R_f + R_i}$$

We saw in section 11.3 that for an amplifier with open-loop gain A_o, the closed-loop voltage gain A is given by:

$$A = \frac{A_o}{1 + \beta A_o}$$

For a typical op amp $A_o = 10^5$, so βA_o is large compared with one in most cases and we can say:

$$A = \frac{A_o}{\beta A_o} = \frac{1}{\beta}$$

Hence:

$$A = \frac{V_o}{V_i} = \frac{R_i + R_f}{R_i} = 1 + \frac{R_f}{R_i}$$

For example, if $R_f = 100\ \mathrm{k}\Omega$ and $R_i = 10\ \mathrm{k}\Omega$, then $A = 110/10 = 11$. As with the inverting amplifier, the gain depends only on the values of resistors R_f and R_i and is independent of the open-loop gain A_o of the op amp.

(c) Input impedance

Since there is no virtual earth at the non-inverting (+) terminal, the input impedance is much higher (typically 50 MΩ) than that of the inverting amplifier. Also, it is unaffected if the gain is altered by changing R_f and /or R_i (but see section (d) below). This circuit gives good matching when the input is supplied by a high impedance source such as a crystal microphone.

(d) Bias

Sometimes the source of the input may not provide the necessary d.c. path to ground for the non-inverting (+) input bias current. On other occasions, such as a.c. operation, an input coupling capacitor might be used. In these cases, a suitable resistor R_B must be connected from the non-inverting input to ground to give the op amp d.c. stability and prevent it from saturating. However, with this modification to the circuit, the high input impedance of the non-inverting amplifier is lost because the shunting effect of R_B presents the input with an alternative path.

The input impedance of the a.c. amplifier in Fig. 13.14 is only about the value of R_B, e.g. 100 kΩ.

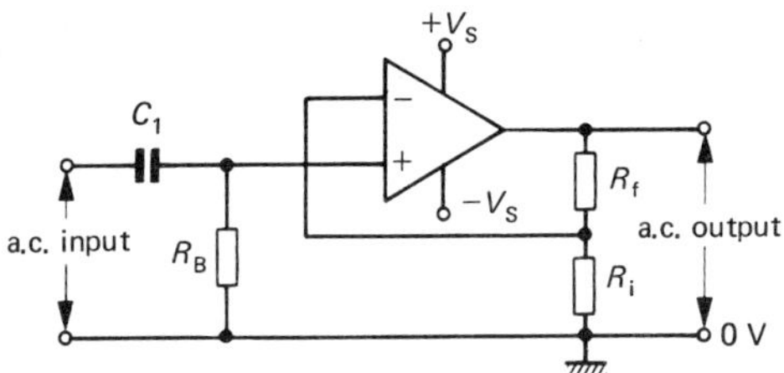

Fig. 13.14

(e) Voltage-follower

This is a special case of the non-inverting amplifier in which 100% negative feedback is obtained by connecting the output directly to the inverting terminal as shown in Fig. 13.15. Thus, $R_f = 0$ and R_i is infinite. Because all of the output is fed back, $\beta = 1$. Therefore, since A_o is very large, $A = 1/\beta = 1$. The voltage gain is therefore one, and the output voltage is the same as the input voltage to within a few millivolts. The circuit is called a *voltage-follower* because the output follows the input. You may wonder what practical use a circuit has that does not give any voltage amplification. However, it has an extremely high input impedance and a low output impedance and, like the emitter- (and source-) follower discussed in section 11.6, the main application is to match a high impedance to a low impedance load, i.e. to act as a *buffer amplifier*.

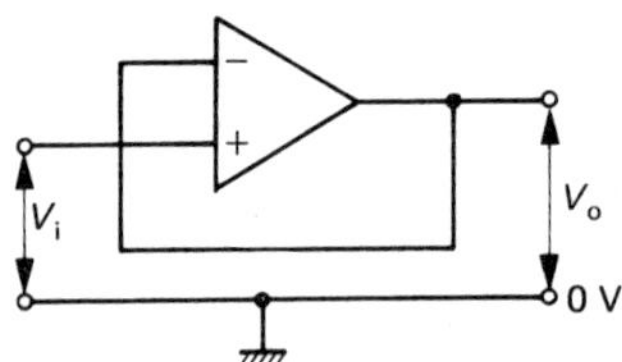

Fig. 13.15

For example, it is used as the input stage of a voltmeter where the highest possible input impedance is required (so as not to disturb the circuit under test) and the output voltage is measured by a relatively low impedance moving-coil meter. In such op amps, FETs replace bipolar transistors in the early stages of the IC.

13.7 Op amp with single power supply

In the quiescent state (no signal input), the 'level shifting' circuitry in an op amp (see section 13.10) ensures that the output voltage is approximately equal to the input voltages. With a balanced dual power supply all are at 0 V, as we have seen. If the input voltages both have some other quiescent value, say +9 V, the quiescent output voltage will also be +9 V.

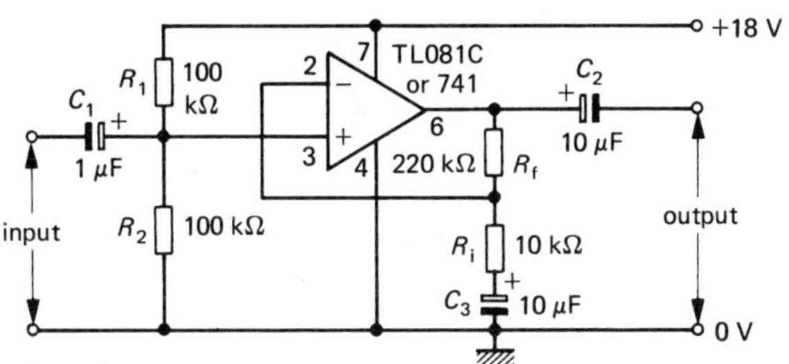

Fig. 13.16

In the non-inverting a.c. amplifier circuit of Fig. 13.16, a single 0 to 18 V power supply is used. Two resistors R_1 and R_2 of equal value are connected across it and act as a voltage divider which provides the non-inverting (+) input of the op amp with a quiescent d.c. bias voltage of +9 V, i.e. *half* the supply voltage. Negative feedback is applied via R_f to the inverting (−) input and blocking capacitor C_3 ensures that the quiescent voltage at this input is the same as the output, i.e. +9 V. Correct biasing is thus obtained. The amplifier is then able to handle both positive- and negative-going input signals and the output has a maximum possible swing capability of ±9 V, i.e. from 0 to 18 V.

From section 13.6 we know that the voltage gain A is given by:

$$A = 1 + \frac{R_f}{R_i}$$

The coupling capacitors C_1 and C_2 prevent the quiescent (d.c.) input and output voltages (9 V) upsetting neighbouring circuits or devices; they also ensure that they in turn do not disturb the biasing of the amplifier.

The circuit could be used as the preamplifier for a crystal microphone.

13.8 Op amp as a voltage comparator

If both inputs of an op amp are used simultaneously, then, as we saw in section 13.2, the output voltage V_o is given by:

$$V_o = A_o \times (V_2 - V_1)$$

where V_1 is the inverting input, V_2 the non-inverting input and A_o the open-loop gain. The voltage difference between the input terminals, i.e. $(V_2 - V_1)$, is amplified and appears at the output, Fig. 13.17.

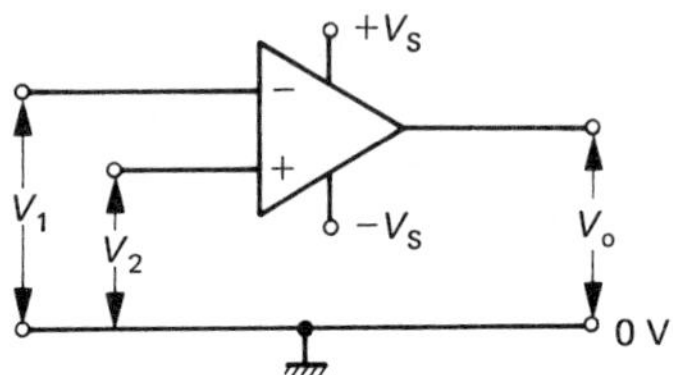

Fig. 13.17

However, A_o is so large that if $(V_2 - V_1)$ exceeds about 100 μV, the op amp saturates. If $V_2 > V_1$, V_o rises to a steady value close to the positive supply voltage $+V_S$. When $V_1 > V_2$, V_o falls to near $-V_S$. The op amp is then behaving as a two-state, digital device with V_o switching from 'high' to 'low'. In this form it is used as a *comparator* to compare voltages, a job it does in an electronic digital voltmeter—see section 21.7.

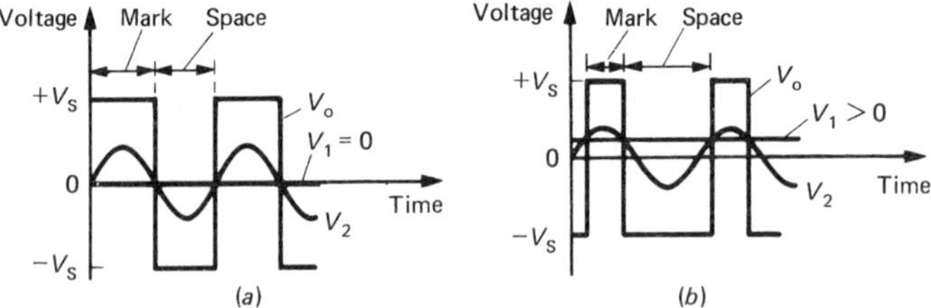

Fig. 13.18

The waveforms in Fig. 13.18 show what happens if input V_2 is an alternating voltage (sufficient to saturate the op amp i.e. greater than 100 μV or so). In Fig. 13.18(*a*), $V_1 = 0$ and $V_o = +V_S$ when $V_2 > V_1$ (i.e. positive) and $V_o = -V_S$ when $V_2 < V_1$ (i.e. negative). In Fig. 13.18(*b*), $V_1 > 0$ and the switching occurs when $V_2 = V_1$ (approx.), giving a mark–space ratio which is no longer one. In effect, the op amp in its saturated condition is converting a continuously varying analogue signal to a two-state digital one.

13.9 Other linear integrated circuits

Although op amps are very versatile amplifiers which can be used for many different purposes, there are some applications where linear ICs having special features are needed. These specialized ICs are also often designed to operate from a single-ended power supply and to require the minimum number of external components. Three types will be considered.

(a) Audio preamplifiers

These are used to give voltage amplification of signals from tape decks, CD players and radio tuners. Their important characteristics are listed below, with data for the 14-pin d.i.l. dual preamplifier LM381 being given as typical.

Open-loop voltage gain	320 000 (110 dB)
Output voltage (peak-to-peak)	$(V_S - 2)$ V
Bandwidth (3 dB)	75 kHz
Input impedance	100 kΩ
Output impedance	150 Ω

For this IC the maximum supply voltage (V_S) is given as 40 V and the input voltage for linear swing as 300 mV.

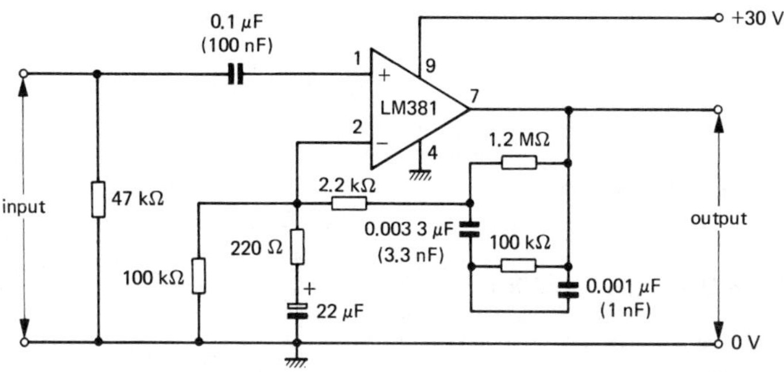

Fig. 13.19

A circuit in which the LM381 is used as a non-inverting preamplifier to amplify the small input voltage (5 to 10 mV) from a dynamic microphone is shown in Fig. 13.19. Some of the external components provide negative feedback (compare Fig. 13.14), other ensure high quality, faithful amplification.

(b) Audio power amplifiers

Like the discrete circuit versions, IC power amplifiers supply power to a loudspeaker. While they have all the advantages of being in integrated form, the problem of dissipating (i.e. getting rid of) unwanted heat makes their design more difficult. As with power transistors, the solution lies in mounting a heat sink (Fig. 11.10) on the IC according to the manufacturer's recommendations.

The LM380 is a 14-pin d.i.l. power amplifier with a typical output of 2 W into a 8 Ω loudspeaker on a supply of 20 V and using as a heat sink about six square centimetres of metal strip connected to the centre pins on its d.i.l. package. It also contains protection circuits which operate if the IC is overloaded (e.g. due to using a speaker of too small impedance or a large input) or its temperature becomes too high.

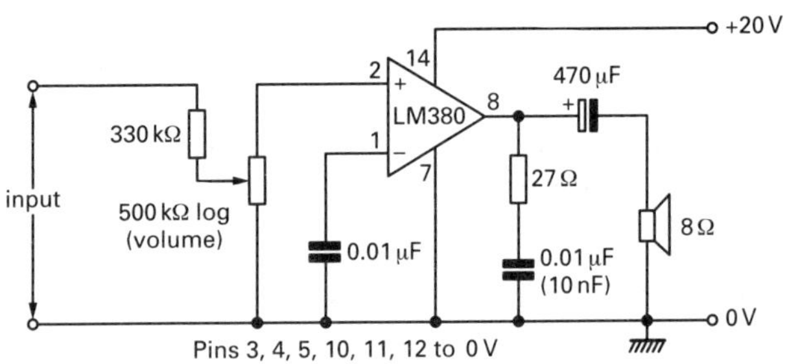

Fig. 13.20

The circuit in Fig. 13.20 uses the LM380 as a power amplifier; it has a fixed-loop gain of 50 and a bandwidth of 100 kHz. (The non-inverting input is connected internally to ground via a 150 kΩ resistor.) The input may be taken directly from a microphone or from the output of the LM381 preamplifier circuit of Fig. 13.19. The few extra components are needed to produce uniform amplification over a wide range of frequencies.

(c) High frequency amplifiers

A wide range of these is available in IC form.

The ZN414Z, Fig. 13.21(*a*), contains ten transistors, fifteen resistors and four capacitors and acts as a radio frequency (r.f.) amplifier with automatic gain control (a.g.c.) and as a detector—see section 18.3. Its main properties are listed below.

Supply voltage	1.2 to 1.6 V (usually 1.3 V)
Supply current	0.3 to 0.5 mA
Frequency range	150 kHz to 3 MHz
Input impedance	4 MΩ
Power gain	72 dB
Output	30 mV (approx.)

In the circuit of Fig. 13.21(*b*) the ZN414Z is the first stage of a simple radio receiver. For good selectivity and sensitivity the tuning circuit should have a high *Q*—see section 12.3.

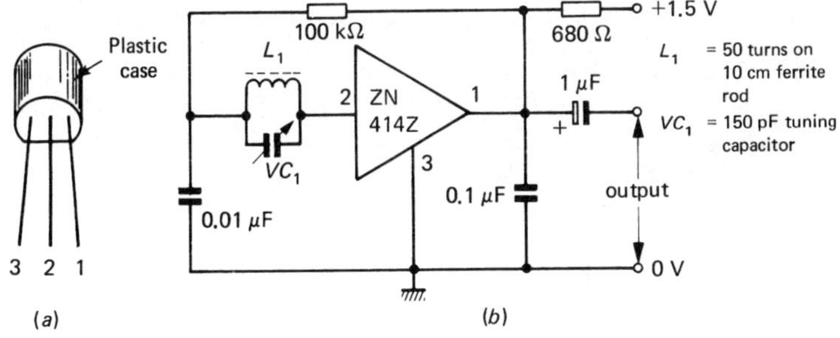

Fig. 13.21

The TAD 100 is intended for superhet radio receivers and incorporates a mixer, oscillator, intermediate frequency (i.f.) amplifier, a.g.c. and audio preamplifier stages. Other linear ICs are used as amplifiers in frequency modulated (FM) radio receivers and as video amplifiers with frequency responses up to several hundred megahertz in television receivers.

13.10 How linear integrated circuits work

Most linear ICs, especially op amps, consist of four stages as shown in the block diagram of Fig. 13.22. The first stage is a differential amplifier, the second provides further voltage amplification, the third is a level shifter and the fourth is usually a complementary emitter-follower output stage. The operation of the first and third stages has not been explained earlier and will now be outlined.

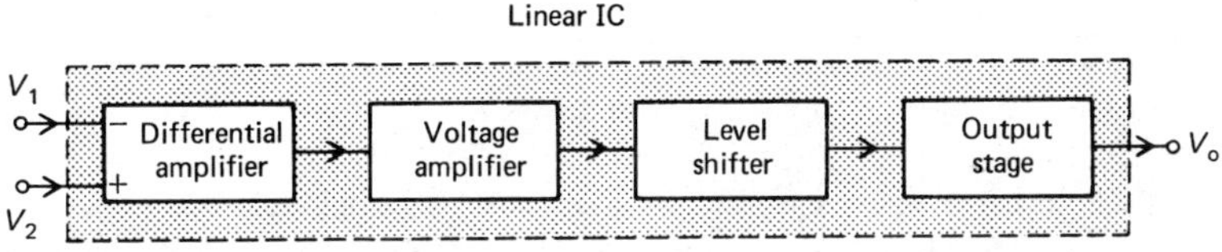

Fig. 13.22

(a) Differential amplifier

The basic arrangement is given in Fig. 13.23. It contains two n-p-n transistors Tr_1 and Tr_2, working from a balanced 9 V dual power supply. R_1 and R_2 are the collector loads and R_E is a common emitter resistor, called the 'tail' resistor, which gives the circuit its alternative name of the 'long-tailed pair' because of the larger-than-usual value (i.e. 4.7 kΩ) of R_E. There is a directly coupled input to the base of each transistor but just one output, i.e. from the collector of Tr_1.

In the quiescent state the bases of Tr_1 and Tr_2 are held at 0 V by their connection to ground via R_3 and R_4 respectively. The emitters are at −0.6 V because the combined quiescent emitter current I_T ($= I_1 + I_2$, ignoring base currents) causes a voltage drop of 8.4 V across R_E. The base-emitter junctions are therefore forward biased, i.e. the bases are positive with respect to the emitters. Tr_1 and Tr_2 collectors both have a d.c. voltage of about +4.5 V.

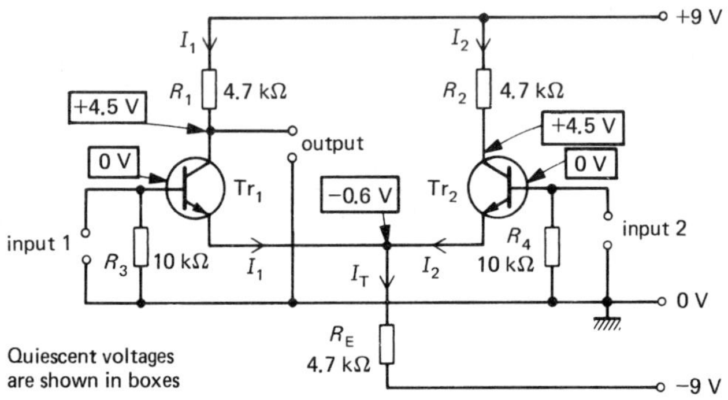

Fig. 13.23

Suppose a signal V_1 (d.c. or a.c.) is now applied to input 1, input 2 being at 0 V. When V_1 is positive, I_1 increases and the output voltage *decreases* (due to the increased voltage drop across R_1). Input 1 is therefore an inverting (−) input. An increase in I_1 also increases the voltage across R_E, i.e. the voltage at the top end of R_E rises, since the lower end is fixed at −9 V. As a result, Tr_2 base-emitter junction voltage falls, causing a decrease in I_2, which, if Tr_1 and Tr_2 are matched and the circuit is well balanced, will be by the same amount as the increase in I_1, i.e. I_T *remains constant*. The voltage across R_E therefore returns to its original value and V_1 has simply caused I_1 to increase at the expense of I_2.

When a positive signal V_2 is applied to input 2, input 1 being at 0 V, I_2 increases and makes (for the reasons given above) I_1 decrease by the same amount. As before, I_T and so also the voltage across R_E are unchanged. The output *increases* in this case (due to the decreased voltage drop across R_1), i.e. input 2 is a non-inverting (+) input.

When the same signal is applied simultaneously to both inputs, neither I_1 nor I_2 changes and there is no output voltage. An output is obtained only when the inputs differ and collector current is diverted from one transistor to the other.

To sum up, a differential amplifier only amplifies the difference between its two inputs and not any signal common to both inputs. This latter property, called *common-mode rejection*, makes the differential amplifier very useful in cases where noise, hum and other unwanted electrical interference is picked up in the leads to both inputs. One application of this is in the electrocardiograph (ECG) where the tiny electrical signals produced by the heart are wanted and are obtained as a voltage between two electrodes on the patient's chest. The large unwanted 50 Hz signals picked up by the body (and connecting leads) from the a.c. mains supply appear at both inputs and are rejected.

A problem with directly coupled amplifiers is caused by temperature changes altering the d.c. operating conditions and producing an amplifier output voltage just as an input signal does. This is called 'drift' and a further advantage of the differential amplifier is that it minimizes the effect because of common-mode rejection whereby voltage changes in one transistor are balanced by opposite changes in the other.

Differential amplifiers in IC form are much superior to those made from discrete components because the characteristics of the transistors match much better— first, because they are made under the same conditions and second, being close together on the same silicon chip, they are more likely to remain matched if the temperature changes.

(b) Level shifter

To obtain the maximum swing capability for the output from a linear IC (on a dual power supply) its quiescent output voltage should be 0 V when the input to the IC is zero. However, the output from the previous stage may, as in the differential amplifier of Fig. 13.23, have a higher quiescent value, e.g. +4.5 V.

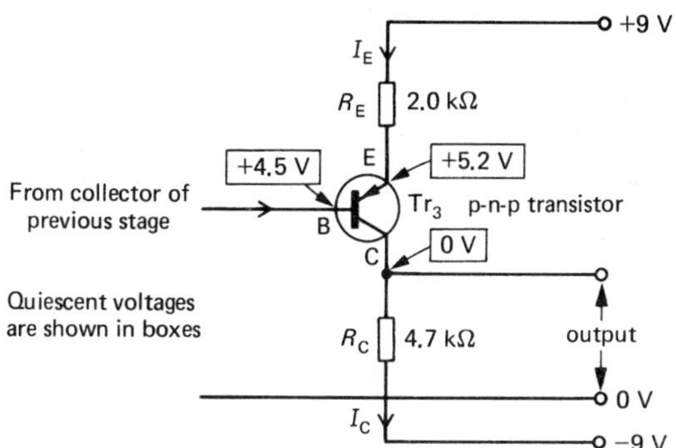

Fig. 13.24

The circuit of Fig. 13.24 shows how 'level shifting' can be achieved using a p-n-p transistor Tr_3. The base Tr_3 is connected to +4.5 V, e.g. to the quiescent collector voltage of Tr_1 in the differential amplifier (or any intermediate voltage amplifier). The level shifter output is taken from the collector of Tr_3 and we want it to have a quiescent value of 0 V. This means that the voltage drop across R_C must be 9 V (under no-signal conditions). Hence $I_C = 9\ V/R_C = 9\ V/4.7\ k\Omega \approx 1.9\ mA$. Ignoring base currents, $I_E = I_C \approx 1.9\ mA$, therefore the voltage drop across $R_E \approx I_E \times R_E \approx 1.9\ mA \times 2.0\ k\Omega \approx 3.8\ V$. And so the emitter voltage = + 9 − 3.8 = + 5.2 V. Since Tr_3 is a p-n-p transistor it is thus forward biased, i.e. its base is about 0.7 V negative with respect to the emitter, i.e. (5.2 − 4.5) V; also, the collector is 4.5 V negative with respect to the base, i.e. (4.5 − 0) V and is reverse biased. Therefore, direct coupling between the output of the previous amplifying stage and the level shifter is possible and the output of the latter sits at 0 V when the input to the IC is zero.

13.11 Revision questions

1. a) State the advantages of integrated circuits over discrete component circuits.
b) Name the two broad groups of ICs and say what each does.

2. a) What are the main properties of operational amplifiers?
b) Explain the terms 'inverting' and 'non-inverting' inputs with reference to an op amp.
c) As amplifiers, why are op amps almost always used with negative feedback?

3. a) What kind of amplifier is represented by the circuit of Fig. 13.25?
b) Write the expression for the closed-loop gain A in terms of R_f and R_i.

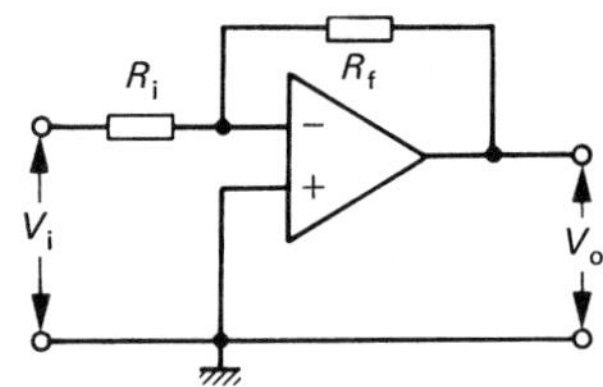

Fig. 13.25

4. Repeat Question 3a) and b) for Fig. 13.26.

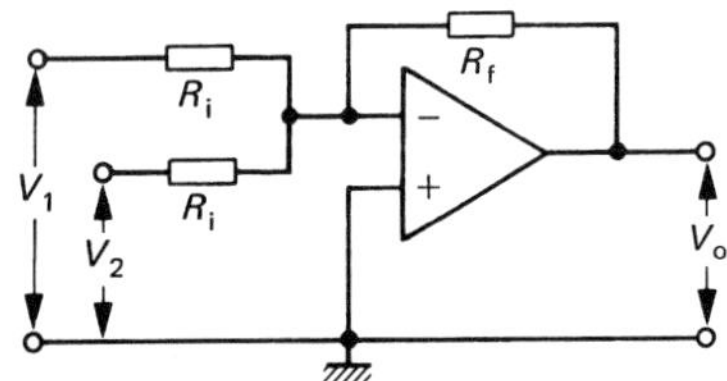

Fig. 13.26

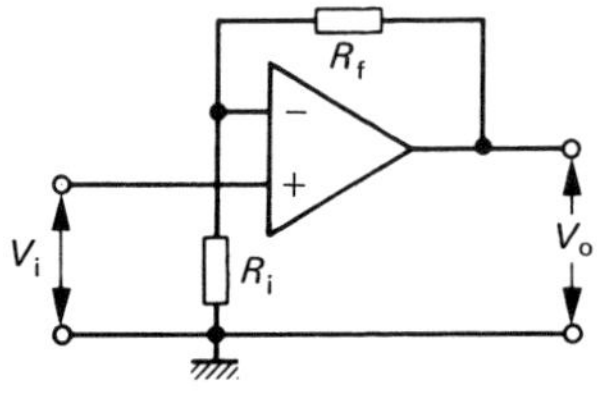

Fig. 13.27

5. Repeat Question 3a) and b) for Fig. 13.27.

6. What is a voltage-follower and what it is used for?

7. How is an op amp used to compare two voltages?

8. Name three specialized types of linear IC.

9. a) Name the four main parts of an op amp.
b) With reference to an op amp explain the terms: direct coupling, offset voltage, common-mode rejection, drift.
c) Why are op amps useful as amplifiers in 'noisy' environments?

13.12 Problems

1. If the open-loop voltage gain A_o of an op amp is 10^5, calculate the maximum input voltage swing that can be applied for linear operation on a ±15 V supply.

2. In Fig. 13.25, if $R_f = 20\ k\Omega$ and $R_i = 10\ k\Omega$, what is
   a) the closed-loop gain A,
   b) the input impedance of the amplifier?

3. Calculate V_o in the circuit of Fig. 13.26 if $R_f = 20\ k\Omega$, $R_i = 10\ k\Omega$, $V_1 = 3$ V, $V_2 = 2$ V and the power supply is ±15 V.

4. In Fig. 13.27, if $R_f = 20\ k\Omega$ and $R_i = 10\ k\Omega$,
   a) calculate the closed-loop gain A, and
   b) say whether the input impedance is greater or less than that of the amplifier in Fig. 13.25.

5. Voltages V_1 and V_2 with waveforms like those shown in Fig. 13.28 are applied simultaneously to the inverting (−) and non-inverting (+) terminals respectively of an op amp on a ±6 V supply. Draw the output waveform V_o.

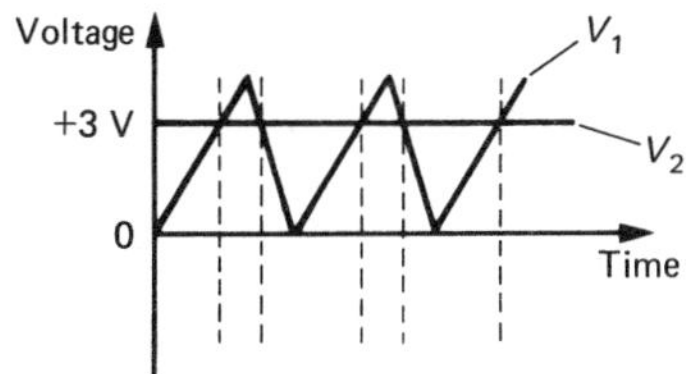

Fig. 13.28

6. If lights falls on the photocell in the circuit of Fig. 13.29, current flows through R_1 and causes a voltage at the + input of the op amp. What happens when this voltage equals the 'reference' voltage (set by the values of R_2 and R_3) at the − input?

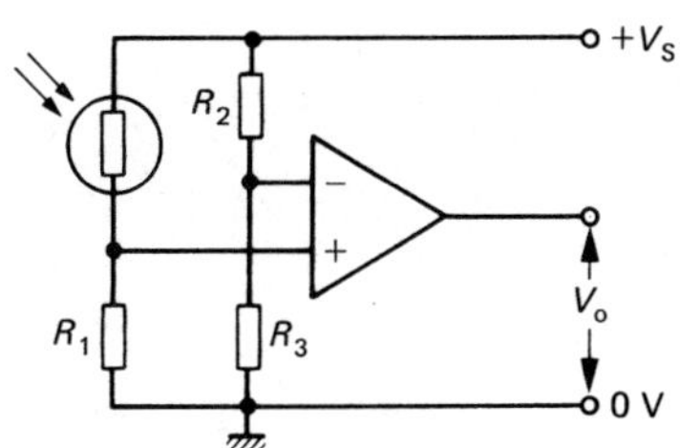

Fig. 13.29

7. A frequency response curve for the 741 op amp is shown in Fig. 13.8 in section 13.3. The device is said to have 'a unit gain bandwidth of 1 MHz'. This means that it has gains of one or more for frequencies up to one million (10^6) hertz. The bandwidth at any gain can be found from:

$$\text{bandwidth} = \frac{10^6}{\text{gain}}$$

What is the bandwidth for a gain of

a) 10,

b) 1 000?

PART FOUR

Digital circuits

UNIT 14

Basic switching circuits

14.1 Introduction

Digital circuits are *switching-type* circuits. Their outputs and inputs involve only two levels of voltage, referred to as 'high' and 'low'. 'High' is near the supply voltage e.g. +5 V, 'low' is near 0 V and (in the scheme known as positive logic), 'high' is referred to as logic level 1 (or just 1) and 'low' as logic level 0 (or just 0). They handle electrical *pulses*.

Digital circuits are *two-state* circuits that are used in computers, in telecommunications, in counting devices such as calculators and electronic watches, in control systems whether they be for domestic appliances such as washing machines or for production processes in industry, and in electronic games.

In this Unit we will first see how a transistor can be operated as a switch and then study some basic circuits where it is used as one. Today these, as well as other digital circuits, are made as integrated circuits and are more complex than the discrete component circuits from which they developed. However, the latter will be discussed in this Unit since they provide a useful introduction to the former and to the subject of digital electronics. Digital integrated circuits are the subject of Units 15, 16 and 17.

14.2 Transistor as a switch

Compared with other electrically operated switches such as relays, transistors have many advantages, whether in discrete or IC form. They are small, cheap, reliable, have no moving parts, their life is almost indefinite (in well designed circuits) and they can switch millions of times a second.

The perfect switch would have no resistance when 'on', infinite resistance when 'off' and would change instantaneously from one state to the other. It would therefore use no power.

(a) Action

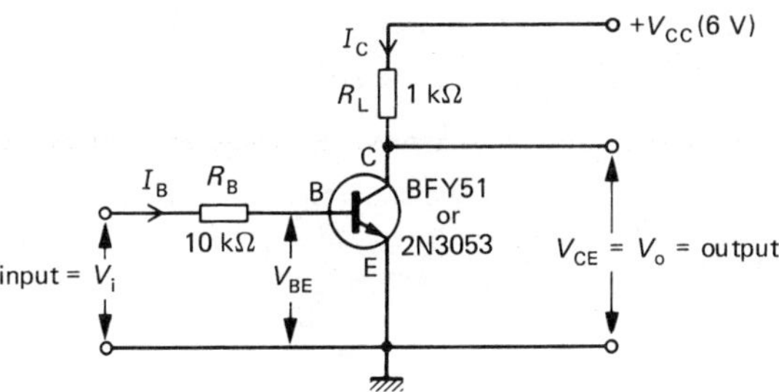

Fig. 14.01

The basic circuit for an n-p-n transistor in common-emitter connection is shown in Fig. 14.01. R_L is the collector load and $\boldsymbol{R_B}$ **prevents excessive base currents which might destroy the transistor**. If the d.c. input voltage V_i is gradually increased from 0 to about +6 V and the corresponding output voltages V_o measured, an input–output voltage characteristic curve like that in Fig. 14.02 can be drawn. It shows that:

(i) along AB, when V_i increases from 0 to about 0.5 V, V_o remains almost equal to the supply voltage V_{CC} (6 V);
(ii) along BC, when V_i increases from about 0.5 V to 1 V, V_o falls rapidly;
(iii) along CD, when V_i increases from 1 V to 6 V, V_o is practically 0 V.

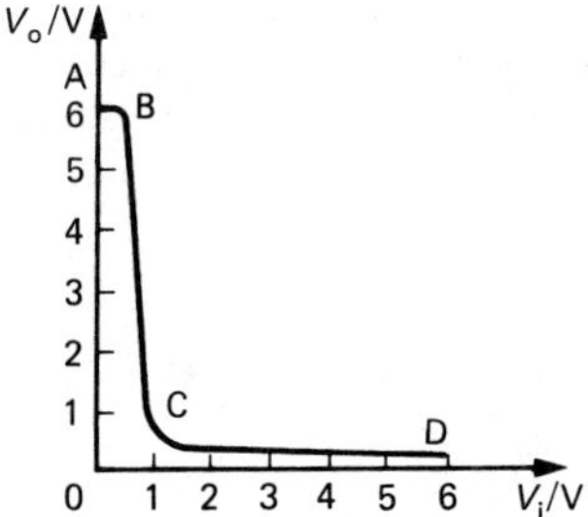

Fig. 14.02

It can also be shown that once V_{BE} reaches 0.6 V it increases very little as V_i is raised to 6 V, i.e. the base-emitter voltage in a forward biased silicon transistor is always close to 0.6 V.

(b) Explanation

For the collector-emitter circuit in Fig. 14.01, since the supply voltage is applied across R_L and the transistor in series, we can say:

$$V_{CC} = I_C \times R_L + V_{CE}$$

I_C is the collector current, which in general depends on the base current I_B, it in turn depending on V_i. Rewriting the equation and putting $V_{CE} = V_o$ (since the collector-emitter voltage is the output voltage), we get:

$$V_o = V_{CC} - I_C \times R_L \qquad (1)$$

That is, the output voltage V_o is less than the supply voltage V_{CC} by the voltage drop across R_L i.e. $I_C \times R_L$. Consider two cases.

When $V_i = 0$, i.e. is 'low', $I_C = 0$ and the transistor, being cut off, behaves like a very high resistance. Since $I_C \times R_L = 0$ when $I_C = 0$, it follows that none of the supply voltage is dropped across R_L. Hence $V_o = + V_{CC}$, i.e. is 'high' (6 V), as equation (1) confirms.

When $V_i = + V_{CC}$, i.e. is 'high', I_B makes I_C, and so also $I_C \times R_L$, large. From (1), we see that the greatest value $I_C \times R_L$ can have is V_{CC}. Then $V_o = 0$, i.e. is 'low' and all the supply voltage is dropped across R_L and none across the transistor which behaves as if it has zero resistance. I_C has its maximum possible value, given by $I_C = V_{CC}/R_L$, and any further increase in I_B does not increase I_C. The transistor is then said to be *saturated* or *bottomed.*

Summing up, when $V_i = 0$, $V_o = + V_{CC}$ and when $V_i = + V_{CC}$, $V_o = 0$. Thus depending on which of the two values V_i has, V_o switches between the two voltage levels $+ V_{CC}$ ('high') and 0 V ('low').

It is also worth noting that I_B is given by:

$$V_i = I_B \times R_B + V_{BE}$$

Thus:

$$I_B = \frac{V_i - V_{BE}}{R_B}$$

(c) Power considerations

The power P used by a transistor as a switch should be as small as possible. It is given by $P = V_o \times I_C$. In practice, at cut-off, I_C is nearly but not quite zero due to leakage current and so $P \approx 0$. At saturation, V_o is typically 0.2 V (for the ideal switching transistor it should be zero) and again $P \approx 0$ but it is essential to ensure the transistor is saturated. This is achieved if I_B is large enough to make I_C/I_B about five times less than h_{FE} (normally $I_C/I_B = h_{FE}$). For example, if $V_{CC} = 6$ V, $R_L = 1$ kΩ and $h_{FE} = 100$, then $I_C(\max) = V_{CC}/R_L = 6$ V/1 kΩ = 6 mA. Hence since 20 is one-fifth of 100, if $I_C/I_B = 20$, i.e. $I_B = I_C/20 = 6/20 = 0.3$ mA, then satisfactory saturation (called 'hard' bottoming) occurs.

Power is also used if the switch-over from cut-off to saturation happens slowly. The larger R_B is, the slower is the switching rate and the greater the value of V_i at which it occurs. Fast-switching is desirable to avoid overheating and damage to the transistor. For this reason n-p-n types are preferred because their majority carriers (electrons) travel faster than the majority carriers (holes) in p-n-p types.

(d) Comparison of transistor as an amplifier and as a switch

In both cases a 'load' (e.g. a resistor) is required in the collector circuit but the base bias is different. As an amplifier, the bias has to cause a collector current which makes the quiescent value of $V_{CE} \approx \frac{1}{2} V_{CC}$. As a switch, the bias has to make the collector current either zero or a maximum. In the first case the transistor is cut off and $V_{CE} \approx V_{CC}$, in the second case the transistor is saturated and $V_{CE} \approx 0$. This is due to the collector-base junction becoming forward biased, the voltage across it being opposite and nearly equal to that of the base-emitter junction, so making V_{CE} almost zero. Table 14.1 sums up these facts.

Table 14.1 Comparison of transistor applications

Application	V_{CE}
Amplifier	$\frac{1}{2} V_{CC}$
Switch-on	0
Switch-off	V_{CC}

You should also note that a transistor as an amplifier acts *linearly* (i.e. the output is more or less directly proportional to the input) because it works on the straight part BC of Fig. 14.02. By contrast, a transistor as a switch operates between cut-off and saturation, its behaviour is non-linear, the output having one of two values.

(e) FETs as switches

MOSFETs can also act as switches. Compared with junction transistors, the input current (to the gate) is much smaller but when switched 'on' their resistance tends to be higher initially; however their switching speed is about ten times greater. They are popular for use in power circuits (see section 8.8).

As a switch a JUGFET has a higher 'on' resistance than a junction transistor and its switching speed is lower.

14.3 Alarm circuits

Alarm circuits can be designed in which an appropriate transducer controls the operation of a transistor as a switch.

(a) Light-operated

A simple circuit which switches on a lamp L when it gets dark is shown in Fig. 14.03. The resistor *R* and the light dependent resistor (LDR) form a voltage divider across the supply. The input, applied between the base and emitter of the transistor, is the voltage across the LDR and depends on its resistance.

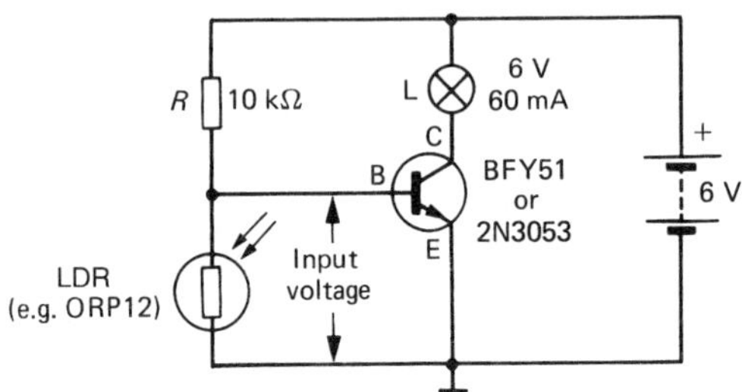

Fig. 14.03

In bright light, the resistance of the LDR is low (e.g. 1 kΩ) compared with *R* (10 kΩ). Most of the supply voltage is dropped across *R* and the input is too small to turn on the transistor (for silicon about 0.6 V is needed). In the dark the LDR has a much greater resistance (e.g. 10 MΩ) and more of the supply voltage is dropped across it and less across *R*. The voltage across the LDR is then enough to switch on the transistor and produce a collector current sufficient to light L. If *R* is replaced by a variable resistor, the light level at which L comes on can be adjusted.

When *R* and the LDR are interchanged, L is on in the light and off in the dark and the circuit could form the basis of an intruder alarm.

(b) Temperature-operated

In the circuit of Fig. 14.04 an n.t.c. thermistor (see section 9.9) and resistor *R* form a voltage divider across the supply and the input to the transistor is the voltage across *R*. When the temperature of the thermistor rises, its resistance decreases, causing less of the supply voltage to be dropped across the thermistor and more across *R*. When the latter exceeds 0.6 V (for a silicon transistor), the transistor switches on and collector current (too small to ring the bell directly) flows through the coil of the relay. As a result the 'normally open' contacts close, enabling the bell to obtain directly from the supply the larger current it requires. The purpose of the diode across the relay coil was explained in section 9.11.

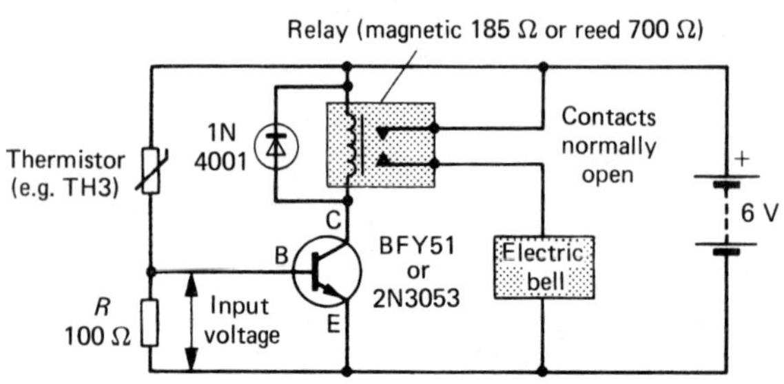

Fig. 14.04

14.4 Logic gates—Types

The various kinds of logic gate are switching circuits that 'open' and give a 'high' output depending on the combination of signals at their inputs of which there is usually more than one. The behaviour of each kind is summed up in a *truth table* showing, in terms of 1s ('high') and 0s ('low') what the output is for all possible inputs.

(a) NOT gate or inverter

This is the simplest gate, with one input and one output. Its circuit is that of the transistor switch discussed in Unit 14.2 and shown again here as Fig. 14.05, along with its British and American symbols and its truth table.

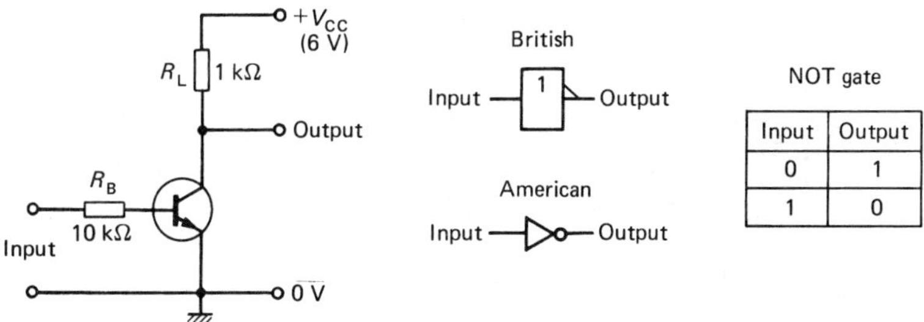

Input	Output
0	1
1	0

Fig. 14.05

As explained before, its action is to produce a 'high' output (e.g. +6 V) if the input is 'low' (e.g. 0 V), and vice versa. That is, the output is 'high' if the input is *not* 'high' and whatever the input, the circuit 'inverts' it.

(b) NOR gate

The circuit is similar to that of the NOT gate but it has two (or more) inputs, A and B in Fig. 14.06. If either or both of A and B are 'high' (e.g. connected to +6 V), the transistor is switched on and saturates and the output F is 'low'. When A and B are both 'low' (e.g. connected to 0 V), i.e. neither A *nor* B is 'high', the output is 'high'.

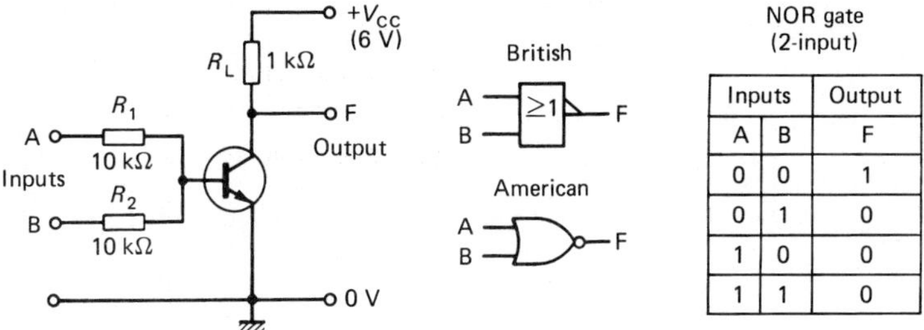

Inputs		Output
A	B	F
0	0	1
0	1	0
1	0	0
1	1	0

Fig. 14.06

The symbols and truth table for a two-input gate are also given in Fig. 14.06. With more inputs the same happens, i.e. if any input is 'high', the output is 'low'. A NOT gate is a NOR gate with one input.

The circuits of Figs. 14.05 and 14.06 contain only resistors and a bipolar transistor. They use a type of circuitry called resistor-transistor logic (shortened to RTL).

(c) OR gate

This is a NOR gate followed by a NOT gate; it is shown with two inputs in symbol form in Fig. 14.07. The truth table can be worked out from that of the NOR gate by interchanging 0s and 1s in the output. The output F is 1 when either A *or* B *or* both is a 1, i.e. if any of the inputs is 'high', the output is 'high'.

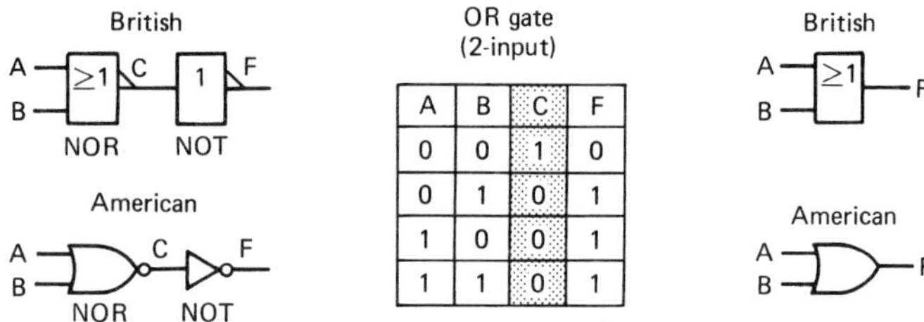

A	B	C	F
0	0	1	0
0	1	0	1
1	0	0	1
1	1	0	1

Fig. 14.07

(d) NAND gate

The symbols and truth table for a two-input NAND gate are given in Fig. 14.08. If A *and* B are both 'high', the output is *not* 'high'. (The gate gets its name from this AND NOT behaviour.) Any other input combination produces a 'high' output.

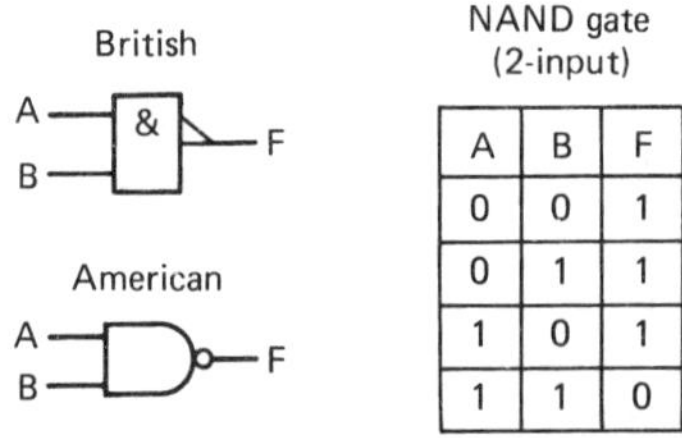

A	B	F
0	0	1
0	1	1
1	0	1
1	1	0

Fig. 14.08

(e) AND gate

This is a NAND followed by a NOT; it is shown with two inputs in symbol form in Fig. 14.09. The truth table may be obtained from that of the NAND gate by changing 0s to 1s and 1s to 0s in the output. The output is 1 only when A *and* B are 1.

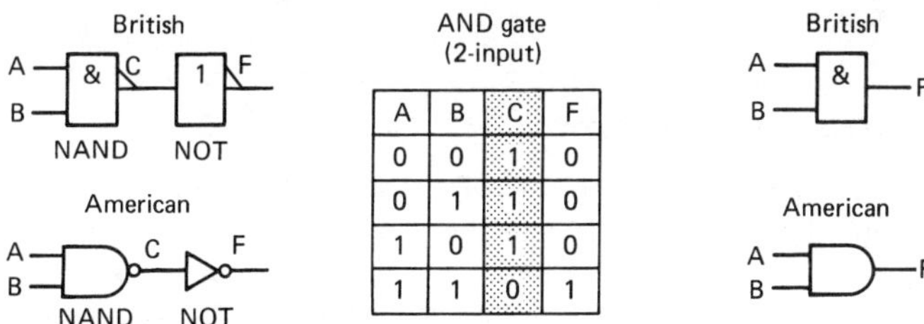

A	B	C	F
0	0	1	0
0	1	1	0
1	0	1	0
1	1	0	1

Fig. 14.09

(f) Exclusive-OR gate

This is an OR gate with only two inputs which gives a 'high' output when either input is 'high' but not when both are 'high'. Unlike the ordinary OR gate (sometimes called the *inclusive*-OR gate) it excludes the case of both inputs being 'high' for a 'high' output. It is also called the *difference* gate because the output is 'high' when the inputs are different.

The symbols and truth table are shown in Fig. 14.10.

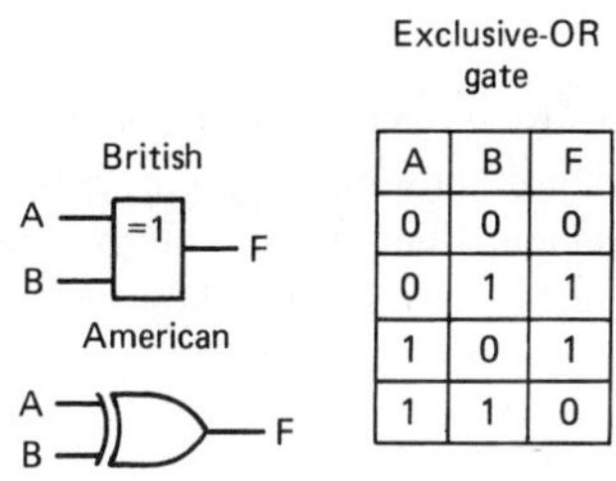

A	B	F
0	0	0
0	1	1
1	0	1
1	1	0

Fig. 14.10

14.5 Logic gates—Further points

(a) Floating inputs

The input to a logic gate should not be left unconnected, i.e. floating, since its apparent voltage level might not be the one expected. For example, a floating input in the NOR gate RTL circuit of Fig. 14.06 behaves as if it were 'low'. In practice, depending on the voltage level required, an input should be connected to either $+V_{CC}$ or to 0 V.

(b) Sources and sinks of current

In Fig. 14.06, if another circuit (e.g. a logic gate) is providing an input to the NOR gate, the output of the circuit when it is 'high' must be able to *supply* the necessary base current to saturate the transistor, i.e. it must be a current *source*.

On the other hand, the output of any circuit connected to the input of a gate must when it is 'low' be able to *accept* current, i.e. it must be a current *sink*.

(c) Fan-out

The ability of the output of, say a logic gate to source or sink current is expressed by its *fan-out*. This is the number of inputs of similar logic gates which can be connected to its output without seriously changing its voltage levels.

(d) Logic families

Various types of circuitry, called logic families, are used to construct logic gates. The first semiconductor gates used resistor-transistor logic (RTL), to be followed later by diode-transistor logic (DTL). Both of these are now obsolete, and in integrated circuit logic gates more complex circuitry based on two transistors is used to avoid problems that arose with the earlier simple gates.

(e) Basic gates

These are usually considered to be the NOT, NOR and NAND gates, for two reasons. First, they are the most useful for building more complex digital circuits. Second, in more advanced treatments they enable us to put into practice mathematical rules, called Boolean algebra, that describe how logic circuits behave (see section 19.2).

(f) Decision-making

Logic gates can make logical decisions in the form of 'yes'/'no' answers, based on information (also in 'yes'/'no' form) supplied to their inputs, 'yes' and 'no' being represented by 'high' and 'low' voltages respectively. A simple example of decision-making by a logic gate is given in section 14.13, Problem 5.

14.6 Multivibrators

Multivibrators are two-stage transistor switching circuits in which the output of each stage is *fed back* to the input of the other by coupling resistors or capacitors. As a result the transistors are driven alternately into saturation and cut-off and whilst the output from one is 'high', the other is 'low'; we say their outputs are *complementary*.

The switch-over in each transistor from one output level or state to the other is so rapid that the collector voltage waveforms are almost 'square'. The term 'multivibrator' refers to this since a square wave can be analysed into a large number of sine waves with frequencies that are multiples (harmonics) of the fundamental.

Multivibrators are of three types.

(i) The bistable or 'flip-flop' This has two stable states. In one, the output of the first transistor is 'high' and of the second 'low'. In the other state the opposite is the case. It will remain in either state until a suitable external trigger or clock pulse makes it switch. Bistables are *memory-type* circuits that are used in digital systems, such as computers, to store the binary digits 0 ('low' output) and 1 ('high' output). They are also at the heart of binary counters.

(ii) The astable or free-running multivibrator This has no stable states. It switches from one state to the other automatically at a rate determined by the circuit components. Consequently it generates a continuous stream of almost square wave pulses, i.e. it is a square wave oscillator and belongs to the family of *relaxation* oscillators—see section 12.8. One of the most important of its many uses is to produce timing pulses for keeping the different parts of a digital system, such as a computer, in step—it is then known as the 'clock'.

(iii) The monostable or 'one-shot' This has one stable state and one unstable state. Normally it rests in its stable state but can be switched to the other state by applying an external trigger pulse. It stays in the unstable state for a certain time before returning to its stable state. It converts a pulse of unpredictable length (time) from a switch into a 'square' pulse of predictable length and height (voltage) which may be used to cause a 'delay' in a circuit or to act as a 'gate' for another circuit and allow a number of timing pulses to pass for a certain time.

Multivibrators can be constructed using junction or field effect transistors, operational amplifiers (see section 17.4) and logic gates (see sections 15.4 and 16.2).

14.7 Bistable or flip-flop

The basic circuit for a bistable or flip-flop is shown in Fig. 14.11. The collector of each transistor is coupled to the base of the other by a resistor R_1 or R_2. The output of Tr_1 is thus fed to the input of Tr_2 and vice versa.

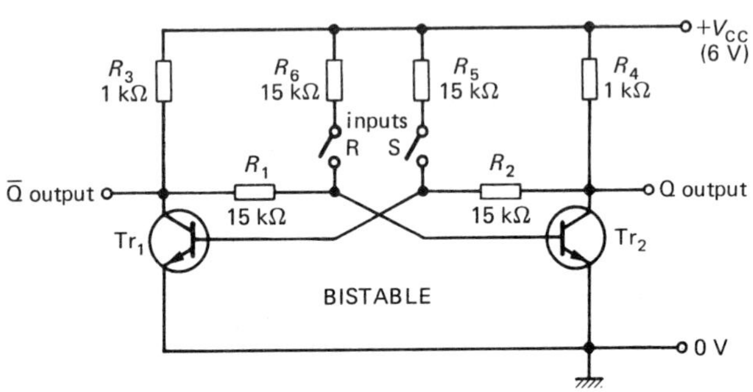

Fig. 14.11

(a) Action

When the supply is first connected, both transistors draw base current, but because of slight differences (e.g. in h_{FE}), one conducts more than the other. A cumulative effect occurs and as a result one, say Tr_1, rapidly saturates while the other, Tr_2, is driven to cut-off.

The collector voltage of Tr_1 is therefore 'low' (most of V_{CC} being dropped across R_3) and so no current can flow through R_1 into the base of Tr_2 (since about 0.6 V is required for this in a silicon transistor). Tr_2 remains off, its collector voltage is thus 'high' ($+V_{CC}$) and causes current to flow via R_2 into the base of Tr_1, thus reinforcing Tr_1's saturated condition. The feedback is *positive* and the circuit is in a stable state which it can maintain indefinitely with output Q = 1 (since Tr_2 is cut off) and its complementary output Q = 0 (since Tr_1 is saturated). (Q is pronounced 'not Q'.)

The state can be changed by applying a positive voltage (greater than +0.6 V) to the base of Tr_2 via R_6 (e.g. by temporarily connecting input R to $+V_{CC}$). Tr_2 now draws base current (through R_6) which is large enough to drive it into saturation. Its collector voltage falls from $+V_{CC}$ to near zero, so cutting off the base current to Tr_1. Tr_1 switches off, its collector voltage rises from near zero to $+V_{CC}$ and is fed via R_1 to the base of Tr_2 to keep it saturated. The circuit is in its second stable state but with Tr 1 off (Q = 1) and Tr_2 saturated (Q = 0).

The collector voltage (the output) of each transistor can thus be made to 'flip' to $+V_{CC}$ or 'flop' to zero by an appropriate input voltage. A negative voltage applied to the base of the transistor that is saturated also causes a change of state.

(b) Further points

To ensure the transistors are driven into cut-off and saturation (thereby keeping the power dissipation to a minimum during switching), R_1/R_3 and R_2/R_4 should be less than h_{FE} for the transistors, i.e. in Fig. 14.11 we require $h_{FE} > 15$.

If input R is made 'high' (e.g. at +6 V) and input S is 'low' (e.g. at 0 V) or disconnected, Q = 0, i.e. the bistable has been 'reset'; if input S is 'high' and input R 'low', Q = 1, i.e. the bistable is 'set'. It is called an SR (set-reset) *flip-flop*.

(c) Bistable as a memory

A bistable can 'store' a single binary digit (a 'bit'—see section 15.5) by staying in one of its stable states until there is a switching pulse. Taking the output of the bistable as the collector voltage of Tr_2 (i.e. the Q output in Fig. 14.11), when Q = 1, the bistable stores (remembers) the bit '1' as a 'high' voltage level. It is an electrical memory which stores data presented to its input and makes that data available at its output. When Q = 0, the bit stored is '0'. A separate bistable is needed to store each bit.

14.8 Astable

The circuit of Fig. 14.12 is similar to that of the bistable (Fig. 14.11) but the transistors are coupled by capacitors C_1 and C_2, not by resistors.

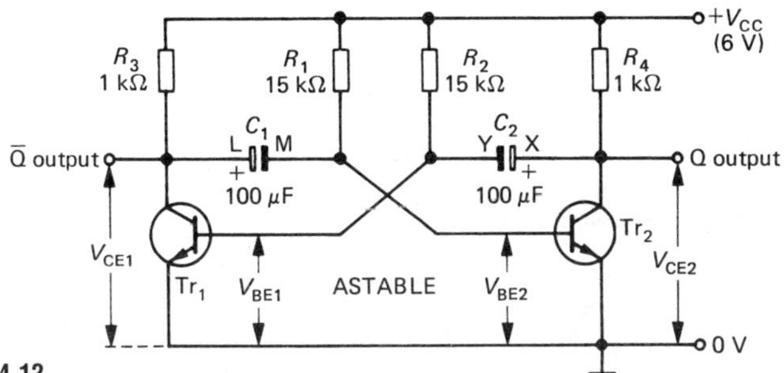

Fig. 14.12

(a) Action

When the supply is connected, one transistor quickly saturates and the other cuts off (as with the bistable). Each then switches automatically to its other state, then back to its first state and so on. As a result, the output voltage, which can be taken from the collector of either transistor, is alternately 'high' (+6 V) and 'low' (near 0 V) and is a series of almost square pulses.

To see how these are produced suppose that Tr_2 was saturated (i.e. on) and has just cut off, while Tr_1 was off and has just saturated. Plate L of C_1 was at +6 V, i.e. the collector voltage of Tr_1 when it was off; plate M was at +0.6 V, i.e. the base voltage of Tr_2 when it was saturated. C_1 was therefore charged with a voltage between its plates of (6 V − 0.6 V) = + 5.4 V. At the instant when Tr_1 suddenly saturates, the voltage of the collector of Tr_1 and so also of plate L, falls to 0 V (nearly). But since C_1 has not had time to discharge, there is still +5.4 V between its plates and therefore the voltage of plate M must fall to −5.4 V. This negative voltage is applied to the base of Tr_2 and turns if off.

C_1 starts to charge up via R_1 (the voltage across it is now 11.4 V), aiming to get to +6 V, but, when it reaches +0.6 V, it turns on Tr_2. Meanwhile plate X of C_2 has been at +6 V and plate Y at +0.6 V (i.e. the voltage across it is 5.4 V). Therefore, when Tr_2 turns on, X falls to 0 V and Y to −5.4 V and so turns Tr_1 off. C_2 now starts to charge through R_2 and when Y reaches +0.6 V, Tr_1 turns on again. The circuit thus switches continuously between its two states.

(b) Voltage waveforms

The voltage changes at the base and collector of each transistor are shown graphically in Fig. 14.13. We see that V_{CE1} (collector voltage of Tr_1) and V_{CE2} (collector voltage of Tr_2) are complementary (i.e. when one is 'high' the other is 'low'). Also, they are not quite square but have a rounded rising edge. This happens because as each transistor switches off and V_{CE} rises from near 0 V to $+V_{CC}$, the capacitor (C_1 or C_2) has to be charged (via the collector load R_3 or R_4). In doing so it draws current and causes a small, temporary voltage drop across the collector load which prevents V_{CE} rising 'vertically'.

(c) Frequency of the square wave

The time t_1 for which Tr_1 is on (i.e. saturated with $V_{CE1} \approx 0$) depends on how long C_1 takes to charge up through R_1 from −5.4 V to +0.6 V (i.e. 6.0 V = V_{CC}) and switch on Tr_2. It can be shown that:

$$t_1 = 0.7C_1 \times R_1$$

That is, it depends on the time constant $C_1 \times R_1$—see section 5.6. If $C_1 \times R_1$ is reduced, Tr_1 is on for a shorter time.

Similarly, the time t_2 for which Tr_2 is on (i.e. the time for which Tr_1 is off) is given by:

$$t_2 = 0.7C_2 \times R_2$$

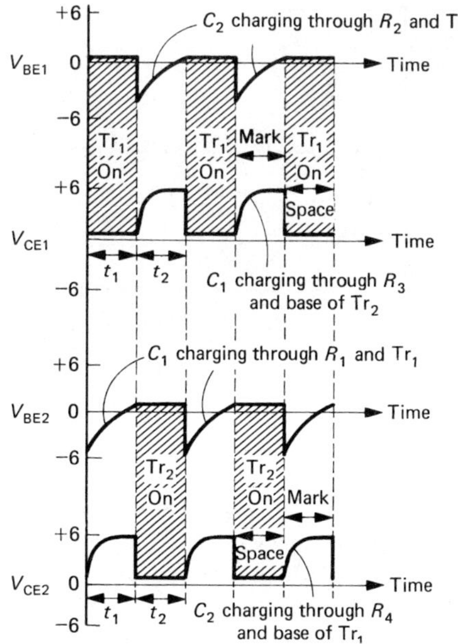

Fig. 14.13

The frequency f of the square wave is therefore:

$$f = \frac{1}{t_1 + t_2} = \frac{1}{0.7(C_1R_1 + C_2R_2)}$$

If the circuit is symmetrical, $C_1 = C_2$ and $R_1 = R_2$ and $t_1 = t_2$, i.e. the transistors are on and off for equal times. Their *mark–space ratio* (see section 12.8) is one. The frequency f in hertz is therefore:

$$f = \frac{1}{1.4C_1R_1} = \frac{0.7}{C_1R_1}$$

C_1 is in farads and R_1 in ohms. For example, if $C_1 = C_2 = 100\mu\text{F} = 100 \times 10^{-6}$ F $= 10^{-4}$F and $R_1 = R_2 = 15\ \text{k}\Omega = 1.5 \times 10^{-4}\Omega$, then $f = 0.7/(10^{-4} \times 1.5 \times 10^{-4})$ $= 0.7/1.5 \approx 0.5$ Hz.

The mark–space ratio of V_{CE} (the output from either transistor) can be varied by choosing different values for the time constants $C_1 \times R_1$ and $C_2 \times R_2$. However, the base resistors R_1 and R_2 should be low enough to saturate the transistors. The condition for satisfactory saturation and cut-off is (as for the bistable) R_1/R_3 and R_2/R_4 both less than h_{FE}. In Fig. 14.12 the transistors must have h_{FE} greater than 15.

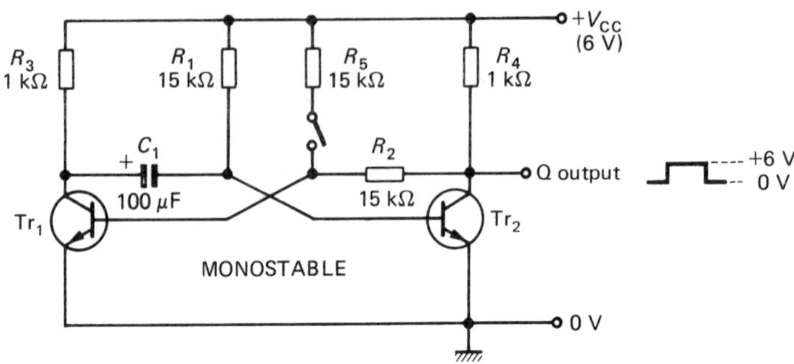

Fig. 14.14

14.9 Monostable

The circuit is shown in Fig. 14.14; one stage is coupled by a capacitor C_1, and the other by a resistor R_2. When the supply is connected, the circuit settles in its one stable state with Tr_1 off and Tr_2 held on (saturated) by resistor R_1. The output voltage is therefore zero (i.e. Q = 0).

A positive pulse, i.e. voltage, applied via R_5 (e.g. by connecting it momentarily to +6 V) to the base of Tr_1 switches it on. The voltage of the right-hand plate of C_1 therefore falls rapidly from +0.6 V to −5.4 V (as explained for the bistable) and switches off Tr_2 making Q go 'high' (+6 V). The monostable is now in its second state but only until the right-hand plate of C_1 charges up through R_1 to +0.6 V. Then Tr_2 is switched on again and Q goes 'low' (0 V).

The time T of the square output pulse is the time for which Q is 'high' and is given approximately by $T = 0.7C_1 \times R_1$.

14.10 Triggered flip-flop

Triggered flip-flops are the building blocks of binary counting circuits. They are similar to the basic SR bistable of Fig. 14.11, with extra components that enable successive pulses, applied to an input called the *trigger*, to make the bistable switch to and fro (or 'toggle') from one state to the other. It is also called a *T-type* (toggling) *flip-flop*.

In Fig. 14.15 the pulse 'steering' is done by connecting between the collector and base of each transistor, a diode (D_1 or D_2) and a resistor (R_5 or R_6), their junctions going via capacitors (C_1 or C_2) to the trigger input. The action is as follows. Suppose Tr_1 is on (saturated) and Tr_2 off. The base of Tr_1 (point N) is therefore at +0.6 V and its collector (point O) near 0 V, so D_1 is on the verge of conducting since it is almost forward biased (the voltage between points N and M being close to 0.6 V). D_2 on the other hand is heavily reverse biased because the collector of Tr_2 (point Z) is at +6 V (and so also is point X because no current flows through R_6) and its base (point Y) is at 0 V.

When square wave pulses (6 V amplitude) are applied to the trigger input, the rapidly rising, positive-going edge of the first plate (i.e. AB) raises the voltage of both L and M to +6 V and of W and X to +6 V and +12 V respectively since the charges on C_1 and C_2, and therefore the voltages across them, cannot change instantly. As a result the voltages at the lower ends of D_1 and D_2 (points M and X)

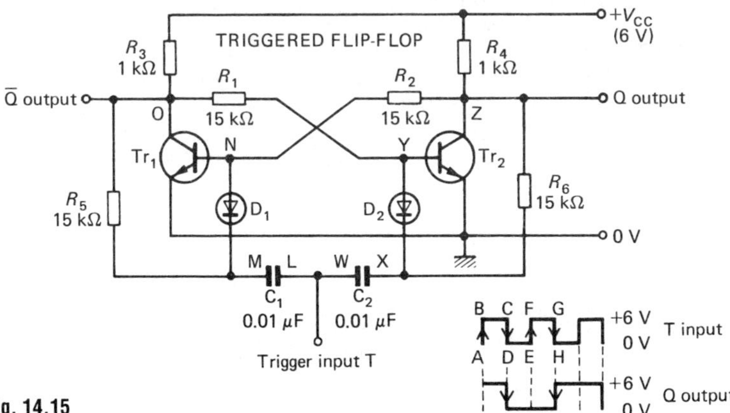

Fig. 14.15

are raised by 6 V, forcing D_1 into reverse bias and making D_2 even more reverse biased than previously. The state of the bistable stays the same (Tr_1 on and Tr_2 off) and the charges on C_1 and C_2 quickly adjust to restore the voltage at M and X to 0 V and 6 V respectively.

The rapidly falling negative-going edge of the first pulse (i.e. CD) suddenly lowers the voltage of L from 6 V to 0 V, of M from 0 V to −6 V and of both W and X from 6 V to 0 V. This means that while D_2 is now on the point of conducting, D_1 becomes forward biased, conducts heavily and in effect short-circuits the base-emitter of Tr_1, diverting the current into C_1 which becomes charged. Tr_1 is cut off, its collector current falls, thereby causing its collector voltage to rise and turn on Tr_2. This switches the bistable into its second state (Tr_1 off and Tr_2 on).

The switching process is repeated and the bistable returns to its first state (Tr_1 on and Tr_2 off) when the negative-going edge (GH) of the second input pulse arrives and so on. The triggered flip-flop is *edge* triggered; the SR flip-flop of section 14.7 is *level* triggered, i.e. it needs a 1 or a 0 to change state.

Table 14.2 shows that each output is a 1 once for every two trigger input pulses. That is, the circuit *divides by two* and this is the basis of binary counting—see section 16.7.

Table 14.2 Output of triggered flip-flop

Trigger input pulse number:	**1**	**2**	**3**	**4**
Q output	0	1	0	1
$\bar{Q}$ output	1	0	1	0

14.11 Schmitt trigger

The Schmitt trigger is a bistable multivibrator in which feedback is via a resistor that is common to the emitter circuits of both transistors, rather than from the collector of one to the base of the other as in an ordinary bistable. It has several uses, to be considered in section 17.6, but one is to convert a sine wave into a square wave—see section 14.13, Problem 8.

(a) Action

To understand its action consider the circuit of Fig. 14.16(*a*), in which R_5 is the common emitter resistor.

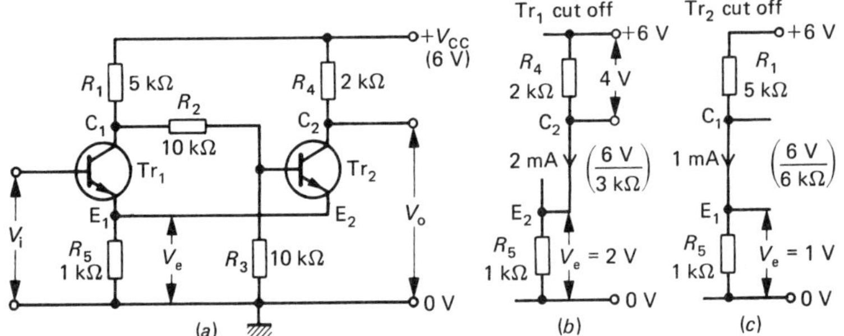

Fig. 14.16

If the input voltage $V_i = 0$, Tr_1 is off and the voltage at the base of Tr_2 due to the voltage divider formed by R_2 and R_3 is enough to saturate Tr_2. The voltage between the collector C_2 and the emitter E_2 of Tr_2 is therefore almost zero and the 6 V supply is dropped across R_4 and R_5 in the ratio of their resistances. R_4 drops 4 V and R_5 2 V as shown in Fig. 14.16(*b*). The common emitter voltage V_e is 2 V,as is that of Tr_2 collector, i.e. the output voltage $V_o = 2$ V and will always be so provided V_i is less than 2 V. This is the 'low' state of the circuit.

If V_i is increased above 2 V, the base-emitter junction of Tr_1 becomes forward biased and it conducts. Tr_1 collector voltage falls (due to the collector current now flowing in R_1), reducing the base current to Tr_2 which comes out of saturation. The resulting drop in Tr_2 emitter current reduces V_e and turns on Tr_1 faster (since the voltage between its base and emitter increases due to V_e decreasing). This positive (regenerative) feedback causes Tr_1 collector voltage to fall faster still and this continues until Tr_1 saturates and Tr_2 cuts off. V_o then jumps to 6 V because there is no current through R_4 and so no voltage drop across it. This is the 'high' state of the circuit which is retained even if V_i is increased further. As Fig. 14.16(*c*) shows, with Tr_1 saturated, V_e is now only 1 V.

To return the circuit to the 'low' state, V_i must be reduced so that Tr_1 comes out of saturation into cut-off and Tr_2 does the opposite. As V_i is decreased, Tr_1 collector current decreases and its collector voltage increases, causing Tr_2 to conduct. This increases V_e and so Tr_1 collector current decreases further (since its emitter voltage has risen) until the regenerative action cuts off Tr_1 and saturates Tr_2. The position is then as in Fig. 14.16(*b*) again, with $V_e = 2$ V.

(b) Characteristic

The characteristic of a Schmitt trigger is shown in Fig. 14.17. The value of V_i which triggers the circuit and makes V_o jump from 'low' to 'high' is called the *upper trip point* (UTP), whilst the *lower trip point* (LTP) is the value of V_i which causes it to jump from 'high' to 'low'.

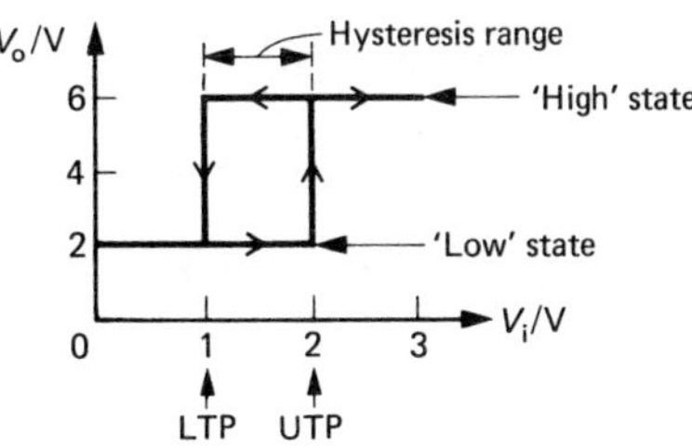

Fig. 14.17

In the circuit of Fig. 14.16(*a*), the UTP is about 2 V and the LTP about 1 V. Since the change from the 'high' to the 'low' state occurs at a lower value of V_i than does the reverse action, the circuit exhibits a 'lagging' or *hysteresis* effect in its switching action. The values of the UTP and LTP can be altered by making the emitter resistor R_5 variable thus giving a different hysteresis range or dead band.

14.12 Revision questions

1. Which of the following are two-state devices:
a) the on-off switch on a radio,
b) the volume control on a TV set,
c) a door lock,
d) an electric iron thermostat,
e) the gear box in a car?

2. How many states does a set of British traffic lights have?

3. a) What is meant by saying that a transistor is
(i) saturated or bottomed,
(ii) cut-off?
b) When a transistor is saturated, what is the effective value of the collector-emitter
(i) resistance,
(iii) power consumption?
(ii) voltage,
c) Repeat b) for a transistor in the cut-off state.
d) What is the condition for 'hard' bottoming?
e) At saturation what two factors determine the maximum possible value of the collector current?
f) The collector-emitter voltage at saturation for transistor A is 1.0 V and for transistor B 0.15 V. Which one would you use as a switch and why?

4. Copy and complete the truth table for each of the two-input logic gates.

Input A	Input B	NOR output	OR output	AND output	NAND output	Exclusive-OR output
0	0					
1	0					
0	1					
1	1					

5. Which type of logic gate is represented by each of the electrical circuits in Figs. 14.18(*a*) and (*b*)?

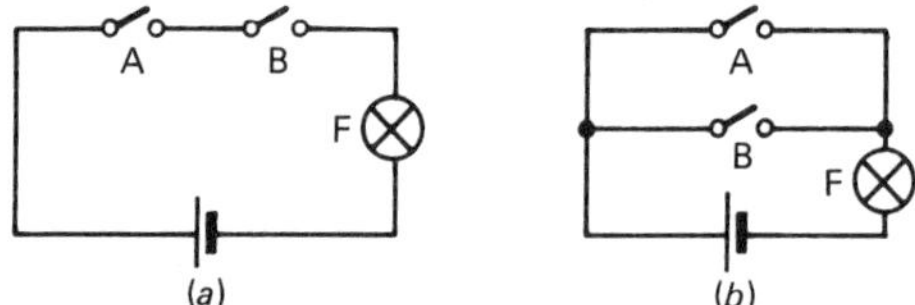

Fig. 14.18 (*a*) (*b*)

6. What do the symbols represent in Figs. 14.19(*a*), (*b*), (*c*), (*d*), (*e*) and (*f*)?

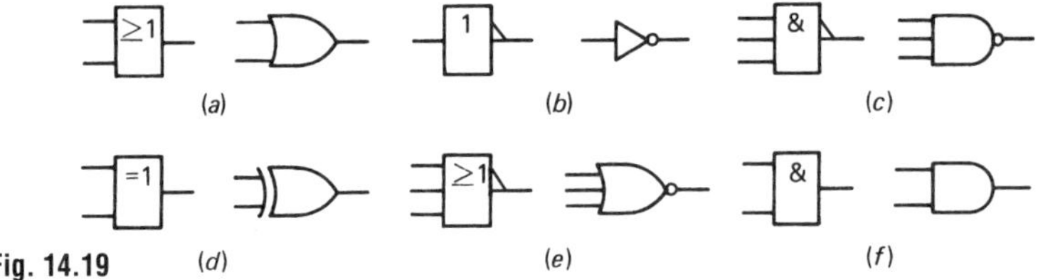

Fig. 14.19 (*a*) (*b*) (*c*) (*d*) (*e*) (*f*)

7. Explain the terms:
a) floating input,
b) fan-out,
c) logic family.

8. Basic circuits for the three kinds of multivibrator are shown in Figs. 14.20(*i*), (*ii*) and (*iii*).

a) Which circuit is a bistable and why is it so called? In the bistable, if output Q is a 1, is Tr_2 cut off or saturated? How would you change the state of the bistable to make Q a 0? Why should the transistors be driven into either cut-off or saturation? How is this achieved in practice? What is another name for a bistable? State two uses of bistables.

b) Which circuit is an astable and why is it so called? Draw the waveform of the output at Q. How would you increase the frequency of the output? How would an astable giving a symmetrical waveform (i.e. mark–space ratio of 1) differ from one giving a non-symmetrical one? How does increasing V_{CC} affect the output? State two uses of astables.

c) Which circuit is a monostable and why is it so called? Is the Q output normally 'high' or 'low' and how is it changed from one to the other? What is the effect of increasing R_1? State one use for a monostable.

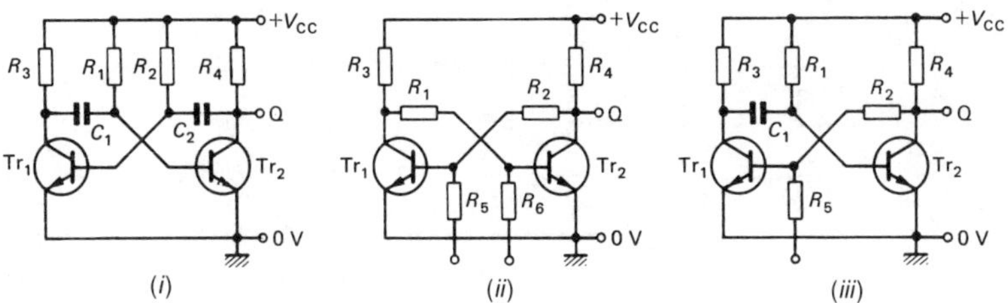

Fig. 14.20

9. Redraw the bistable circuit of Fig. 14.20 so that it is a triggered flip-flop. Why is it called a 'divide by two' circuit?

10. What is a Schmitt trigger? What does it do? State a major use for it. Explain the following terms: upper trip point, lower trip point, hysteresis.

14.13 Problems

1. In the circuit of Fig. 14.21, what is the voltage at A when
   a) S is open, b) S is closed?

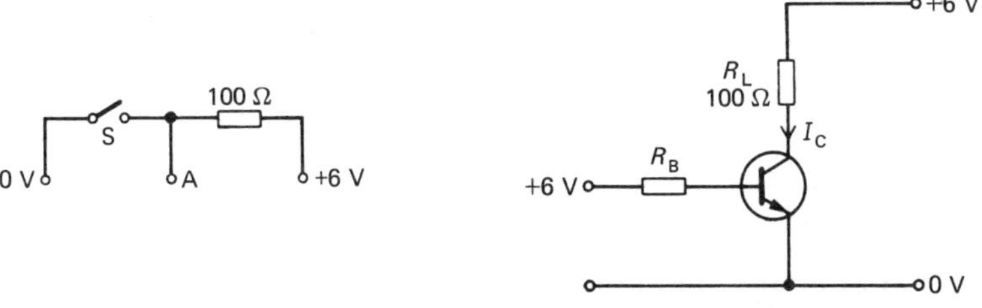

Fig. 14.21 **Fig. 14.22**

2. The silicon transistor in the circuit of Fig. 14.22 is saturated.
   a) Calculate the maximum possible value of collector current.
   b) If $h_{FE} = 150$ what is the value for hard bottoming of
      (i) the base current, (ii) the base resistor R_B?
   Assume the base-emitter voltage is 0.6 V.

3. Write down the truth table for the circuit (given in both British and American symbols) in Fig. 14.23.

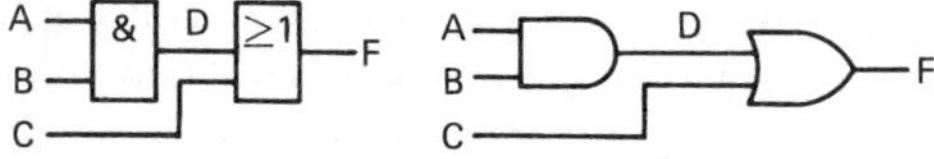

Fig. 14.23

4. The waveforms in Fig. 14.24 are applied one to each input of a two-input
a) AND gate, and
b) NOR gate.
Draw the output waveform for each case.

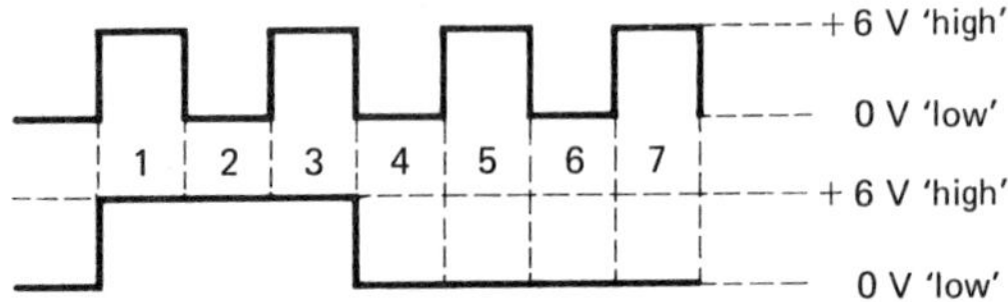

Fig. 14.24

5. The gas central heating system represented by the block diagram in Fig. 14.25 has digital electronic control. The gas valve turns on the gas supply to the boiler when the output from the logic gate is 'high'. This is so only if both of the following conditions hold:

(i) the output from the *thermostat* is 'high', indicating that the room temperature has fallen below that desired and in effect saying 'yes', the room needs more heat; and

(ii) the output from the *pilot flame sensor* is 'high', meaning 'yes' the pilot is lit.

What type of logic gate is used?

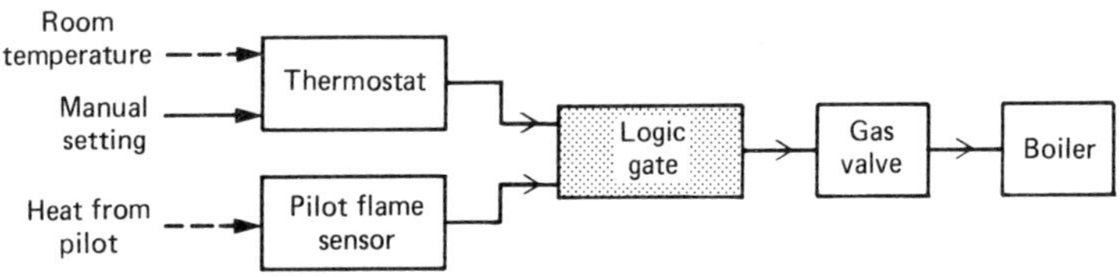

Fig. 14.25

6. A capacitor is charged to 6 V so that one plate, A, is at +6 V and the other plate, B, at 0 V. What will be the voltage of plate B if plate A is suddenly connected to
a) 0 V,
b) +12 V,
c) −6 V?
Assume in each case that initially A is at +6 V and B at 0 V.)

7. In the circuit of Fig. 14.20(*i*), $C_1 = C_2 = 1\ \mu$F and $R_1 = R_2 = 6.8\ \text{k}\Omega$. Calculate the frequency of the output.

8. a) Draw the characteristic for a Schmitt trigger which has the following properties: UTP = 3 V, LTP = 2 V, 'high' state = 9 V, 'low' state = 3 V.
b) Sketch the output voltage waveform from the Schmitt trigger of a) if the input is a sine wave of peak value 5 V.

UNIT 15

Decision-making circuits

15.1 Introduction

In this and the next two Units, the properties and operating principles of the main types of digital IC are explained and their commercial forms described. The aim is to give you a working knowledge and general understanding of what are otherwise 'mysterious little black plastic blocks'. The basic principles involved in digital circuit design are considered in Unit 19. (You will find it useful to revise section 13.1 now.)

The digital ICs dealt with in this Unit are classed as the 'decision-making' kind (see section 14.5) and consist of circuits containing mainly logic gates. They use what is called *combinational logic* because their outputs are always the same for the same combination of inputs. Also, unlike the digital circuits to be considered in Unit 16, their outputs depend only on their *present* inputs and not on any previous ones.

15.2 About digital integrated circuits

(a) Digital logic families

There are two main families available today, each with its own special circuitry or, as it is called, its *logic*. One is TTL (standing for transistor-transistor logic) using bipolar transistors; the other is CMOS (standing for complementary metal oxide semiconductor and pronounced 'see-moss') based on MOSFETS. The basic circuits round which each is built are described in sections 17.8 and 17.9. The performance of TTL and CMOS is much superior to earlier families such as RTL (resistor-transistor logic—see section 14.5) and DTL (diode-transistor logic).

TTL ICs are made by various manufacturers as the '74' series; figures following '74' indicate the nature of the IC, e.g. the kind of logic gate. CMOS ICs belong to the '4000B' series; the numbers after '4' depend on what the IC does. Both types are usually in the form of 14 or 16 pin d.i.l. packages (like many linear ICs), the connecting pins being 0.1 inch apart and pin 1 being identified from a small dot or notch on the case, as was shown in Fig. 13.02 (p. 179).

(b) Properties of TTL and CMOS

Some properties of each family are summarized in Table 15.1.

Note that TTL must have a stabilized (steady) 5 V supply (see section 21.4), while CMOS will work on any d.c. voltage (unstabilized) between 3 V and 15 V and usually requires much less power.

Table 15.1 Properties of TTL and CMOS

Property	TTL	CMOS
Supply voltage	5 V±0.25 V d.c.	3 V to 15 V d.c.
Power used	Milliwatts	Microwatts
Switching speed	Fast	Slow
Input impedance	Low	High

The original TTL and CMOS ranges of ICs have been augmented by newer versions.

The 'Schottky' TTL or *74LS series* is a *lower power* version of the 74 series operating on a 5 V supply. It uses the same identification system as the 74 series. For example, 74LS00 is a quad two-input NAND gate like the 7400.

The *74HC series* is a *higher speed* version of the 4000B CMOS series but operates on a 2 to 6 V supply. The same pin connections and numbering system are used as the equivalent 74LS ICs. For example, the 74HC00 is also a quad two-input NAND gate like the 74LS00.

Eventually the 74HC range may replace the other ranges since it combines the advantages of both TTL and CMOS, i.e. high switching speed and low current.

(c) Some uses

The fast switching speed of TTL makes it suitable for use in large computers containing several million switching circuits. The time for signals to pass through the system then becomes important.

CMOS circuitry lends itself to very-large-scale integration (VLSI) because MOSFETS take up much less room than bipolar transistors (about one-fifth). It is used for pocket calculator and digital watch ICs (where low power consumption is also important) as well as in microprocessors and certain types of memory.

15.3 Digital integrated circuits—Practical points

(a) TTL

Unused inputs behave as if they were connected to +5 V (V_{CC}), i.e. they assume logic level 1 unless connected to 0 V (ground, Gnd). It is good practice to join them to +5 V.

Ideally the 'low' and 'high' input and output voltages which represent *logic levels* 0 and 1 respectively for TTL ICs, are 0 V and +5 V. In fact, due to voltage drops across transistors and resistors inside the IC, these values are not achieved. For example, an input voltage of from 0 to 0.8 V behaves as 'low', while one from 2 to 5 V is 'high' because it can cause an output to change state. Similarly a 'low' output may be 0 to 0.4 V (since the voltage across a saturated transistor is not 0 V) and 'high' is from 2.4 to 5 V since any voltage in this range can operate other TTL ICs.

Table 15.2 summarizes these values. There is, in the worst possible case, a safety margin of 2.4 − 2.0 = 0.4 V when the output of one IC provides the input of another. Typically it is nearer 1 V.

Table 15.2 Values of TTL and CMOS 'low' and 'high' input and output voltages

	TTL		CMOS	
State	**Input**	**Output**	**Input**	**Output**
'Low'	0.8 V	0.4 V	$0.3V_{DD}$(1.5 V)	V_{SS}(0 V)
'High'	2.0 V	2.4 V	$0.7V_{DD}$(3.5 V)	V_{DD}(5.0 V)

The *fan-out* of a TTL IC is ten; this means that up to ten other TTL ICs can take their inputs from one TTL output and switch reliably. If this number is exceeded, the output is overloaded and even the worst-case voltage levels are not obtained.

The 5 V *power supply* used with TTL ICs should be well regulated and adequately decoupled to deal with the brief, but large current 'spikes' which occur when ICs switch from one state to another. It should also be capable of supplying the current required; a simple gate IC takes about 10 mA, a dual flip-flop 25 mA and an MSI package 50 mA or so.

(b) CMOS

Unused inputs must be connected either to supply + or ground (V_{DD} or V_{SS}) depending on the circuit and not left to 'float', otherwise erratic behaviour occurs.

The very high input impedance of a CMOS IC (due to the use of MOSFETs) accounts for its low current consumption but it can allow static electric charges to build up on input pins when, for example, they touch insulators, e.g. plastics or clothes, in warm dry conditions. (This does not happen with TTL ICs because the low input impedance of bipolar transistors ensures that any charge leaks harmlessly through junctions in the IC.)

These static voltages can break down the very thin layer of silicon oxide insulation between the gate and other electrodes of MOSFETs in the input, thereby destroying the IC. CMOS ICs are therefore supplied in antistatic or conductive carriers. Also, protective diode circuits are incorporated which operate when the IC is connected to the power supply.

Logic level voltages are given in Table 15.2 in terms of V_{DD} and V_{SS}; the values for a 5 V supply are in brackets. The 'ideal' output values are more or less attained and the safety margin in the worst-case conditions is $5.0 - 3.5 = 1.5$ V, which is better than for TTL.

The *fan-out* is about 50 for one CMOS IC driving other CMOS ICs, due to their very high input impedance. The problems (and their solution) which arise when CMOS and TTL are used together in the same system will be discussed in section 17.7.

The *power supply* needs of CMOS ICs are much less demanding than those of TTL, batteries (e.g. 9 V) being quite satisfactory in many cases.

15.4 Logic gates

(a) Introduction

The various logic gates are available as ICs in both TTL and CMOS forms. We will see shortly that, although every possible digital logic circuit can be built by combining only NAND gates or only NOR gates, using the complete range sometimes allows more economical circuit designs, i.e. fewer ICs are needed.

Each IC package usually has several gates on the same chip. For example, a quad two-input NAND gate contains four identical NAND gates, with two inputs and one output per gate. Every gate therefore has three pins, making 14 pins altogether on the package. This includes two for the positive and negative power supply connections which are common to all four gates. A hex inverter has six NOT gates.

Truth tables for each gate were given in section 14.4. You should try to remember the following:

NOT: output always *opposite* of input

AND: output 1 *only* when all inputs 1

NAND: output 1 *unless* all inputs 1

NOR: output 1 *only* when all inputs 0

OR: output 1 *unless* all inputs 0

exclusive-OR: output 1 when inputs *different*

Note that the American circuit symbols for gates will be used in this book, since these are favoured by component manufacturers and most journals.

(b) Pin connections

The pin connections (top view) for a selection of TTL and CMOS gates are given in the Appendix. The pin connections for the low power Schottky TTL and the 74HC CMOS series are the same as for the corresponding gates in standard TTL, e.g. 7400, 74LS00 and 74HC00 are identical.

(c) Logic gates from two-input NAND gates

The various arrangements are shown in Fig. 15.01. You can check that each one gives the correct output for the various input combinations by constructing a

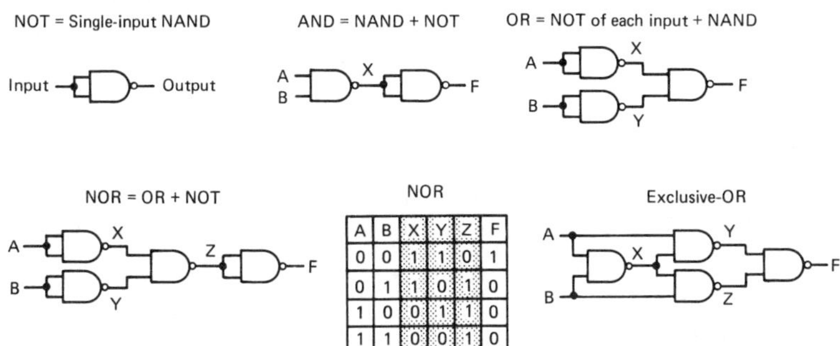

A	B	X	Y	Z	F
0	0	1	1	0	1
0	1	1	0	1	0
1	0	0	1	1	0
1	1	0	0	1	0

Fig. 15.01

stage-by-stage truth table, as has been done for the NOR gate made from four NAND gates. Note:

(i) NOT = single-input NAND (or NOR), made by joining *all* the inputs together;
(ii) AND = NAND followed by NOT;
(iii) OR = NOT of each input followed by NAND;
(iv) NOR = OR followed by NOT.

These facts can often be used to minimize the number of ICs required for a system.

(d) Astable and monostable multivibrators from NAND gates

The *astable* circuit in Fig. 15.02(*a*) uses two CMOS inverters A and B (i.e. two single-input NANDs). Output B is the input to A and, if at switch-on it is 'high', output A, and so also the astable output V_o, will be 'low'. Since input B must be 'low' (if its output is 'high'), C will start to *charge* up through R. When the right side of C is positive enough, i.e. 'high', output B begins to go 'low', driving input A 'low' and making V_o 'high'. As a result the voltage on the right side (as well as the left) of C rises suddenly (due to the charge on it not having time to change). The resulting *positive feedback* reinforces the direction input B was going and confirms the 'high' state of V_o.

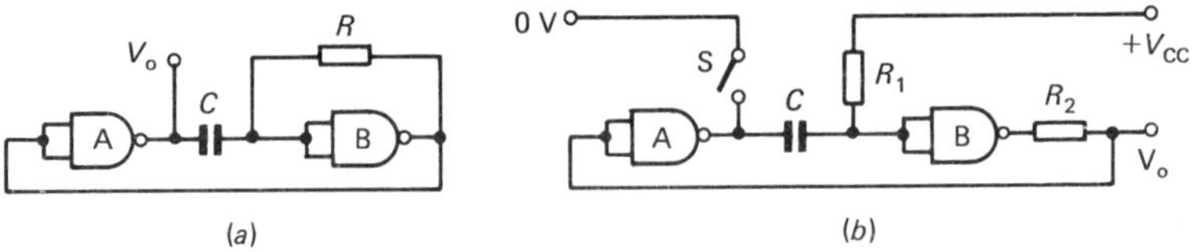

(a) (b)

Fig. 15.02

Because the input to B is now 'high' and its output 'low', C starts to *discharge* through R and when the input voltage to B becomes 'low', B switches back to its first state with a 'high' output. This in turn makes V_o 'low'. The process is repeated to give a square wave output V_o, at a frequency determined by the values of the capacitance C and resistance R.

The *monostable* circuit of Fig. 15.02(*b*) is in its stable state with output B (V_o) 'low' because its input is 'high' via R_1. Closing S briefly takes output A from 'high' (due to its 'low' input from output B) to low and so forces input B 'low' and V_o 'high'. C then charges through R_1 until input B is high enough to switch its output back to the stable state with V_o 'low'. The length of the pulse so produced depends on the values of C and R_1.

15.5 Codes

Normally we count on the scale of ten or *decimal system* using the ten digits 0 to 9. When the count exceeds 9, we place a 1 in a second column to the left of the units column to represent tens. A third column to the left of the tens column gives hundreds and so on. The values of successive columns starting from the right are 1, 10, 100, etc., or in powers of ten 10^0, 10^1, 10^2, etc.

(a) Binary code

Counting in electronic systems is done by digital (i.e. two-state) circuits on the scale of two or *binary system* using the digits (bits) 0 and 1 which are usually represented by 'low' and 'high' voltage levels respectively. Many more columns are required since the number after 1 in binary is 10, i.e. 2 in decimal, but the digit 2 is not used in the binary system.

Successive columns in this case represent, from the right, powers of two, i.e. 2^0, 2^1, 2^2, 2^3, etc., or in decimal 1, 2, 4, 8, etc. Table 15.3 shows how the decimal numbers from 0 to 15 are coded in the binary system; a four-bit code is required. Note that the *least significant bit* (l.s.b.) is on the extreme right in each number and the *most significant bit* (m.s.b.) on the extreme left. The hexadecimal column will be explained in section (b) opposite.

For larger numbers, higher powers of two have to be used. For example:

$$\text{decimal } 29 = 1 \times 16 + 1 \times 8 + 1 \times 4 + 0 \times 2 + 1 \times 1$$

$$= 1 \times 2^4 + 1 \times 2^3 + 1 \times 2^2 + 0 \times 2^1 + 1 \times 2^0$$

$$= 11101 \text{ in binary (a five-bit code).}$$

(b) Hexadecimal code

One of the obvious disadvantages of binary code is the representation of even quite small numbers by a long string of bits. The hexadecimal (hex) code is more compact and therefore less liable to error by a programmer (see section 20.7(*a*)). In it, the range of decimal digits, i.e. 0 to 9, is extended by adding the letters A to F for the numbers 10 to 15 respectively, as shown in Table 15.3.

Table 15.3 Decimal, binary and hexadecimal codes

Decimal		**Binary**				**Hexadecimal**
10^1 (10)	10^0 (1)	2^3 (8)	2^2 (4)	2^1 (2)	2^0 (1)	16^0 (1)
0	0	0	0	0	0	0
0	1	0	0	0	1	1
0	2	0	0	1	0	2
0	3	0	0	1	1	3
0	4	0	1	0	0	4
0	5	0	1	0	1	5
0	6	0	1	1	0	6
0	7	0	1	1	1	7
0	8	1	0	0	0	8
0	9	1	0	0	1	9
1	0	1	0	1	0	A
1	1	1	0	1	1	B
1	2	1	1	0	0	C
1	3	1	1	0	1	D
1	4	1	1	1	0	E
1	5	1	1	1	1	F

For larger numbers, successive columns from the right represent powers of sixteen (hence hexadecimal), i.e. $16^0(1)$, $16^1(16)$, $16^2(256)$, etc. For example:

$$\text{decimal } 16 = 1 \times 16^1 + 0 \times 16^0 = 10 \text{ in hex}$$

$$\text{decimal } 29 = 1 \times 16^1 + 13 \times 16^0 = 1\text{D in hex}$$

$$\text{decimal } 407 = 1 \times 16^2 + 9 \times 16^1 + 7 \times 16^0 = 197 \text{ in hex}$$

$$\text{decimal } 940 = 3 \times 16^2 + 10 \times 16^1 + 12 \times 16^0 = 3\text{AC in hex}$$

(c) Binary coded decimal (BCD)

This is a popular, slightly less compact variation of binary but in practice it allows easy conversion back to decimal for display purposes. Each *digit* of a decimal number is coded in binary instead of coding the whole number. For example:

$$\text{decimal } 29 = 0010 \quad 1001 \quad \text{in BCD}$$

$$\text{decimal } 407 = 0100 \quad 0000 \quad 0111 \quad \text{in BCD}$$

In binary, decimal 407 requires only nine bits and is 110010111:

$$\begin{aligned}\text{decimal } 407 &= 1 \times 256 + 1 \times 128 + 0 \times 64 + 0 \times 32 + 1 \times 16 + 0 \times 8 \\ &\quad + 1 \times 4 + 1 \times 2 + 1 \times 1 \\ &= 1 \times 2^8 + 1 \times 2^7 + 0 \times 2^6 + 0 \times 2^5 + 1 \times 2^4 + 0 \times 2^3 \\ &\quad + 1 \times 2^2 + 1 \times 2^1 + 1 \times 2^0\end{aligned}$$

It is obviously essential to know which particular binary code is being used.

15.6 Encoders

While computers and other digital electronic systems work in some form of binary code, humans prefer the decimal code. Code converters are therefore necessary to convert one code into another. An *encoder* (usually at the input of a system) makes conversions into binary, often from decimal.

(a) Encoding

Encoding circuits take many forms. The basic principles of one type of decimal to BCD encoder is shown in Fig. 15.03. The array, called a *rectangular diode matrix*, contains switches numbered 0 to 9 (which may all be on the same keyboard) representing the corresponding decimal numbers.

If, for example, switch 3 is pressed, diodes D_4 and D_3 are connected to supply + and conduct. Current flows through R_A and R_B, creating voltages across them which make outputs A and B go 'high'. The other two outputs, C and D stay 'low' since they are connected to ground via R_C and R_D respectively. The BCD output is then 0011 (3 in decimal).

Check for yourself that the correct 'bit' patterns are obtained from DCBA when each of the other switches is closed.

(b) Hexadecimal keyboard encoder IC

This type of encoder converts a hexadecimal input into a four-bit binary output. In Fig. 15.04 the CMOS 74C922 encoder IC is shown with eight inputs supplied from sixteen switches, wired on a keyboard in a matrix of four rows and four columns.

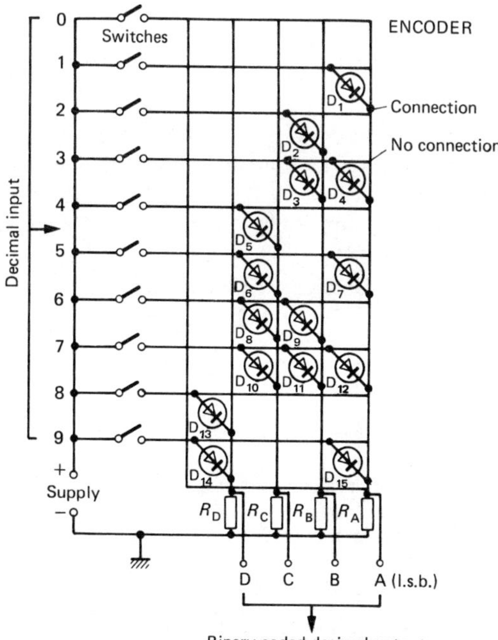

Fig. 15.03

C_1 is an external capacitor whose value (e.g. 0.01 μF) decides the rate of scanning (strobing) of the switches by an internal oscillator to detect the one that has been pressed. The other external capacitor C_2 (e.g. 0.1 μF) eliminates contact bounce on the switches. This occurs when two contacts are pushed together. They do not stay in contact at first but make a series of rapid, imperfect contacts producing unwanted pulses. The IC contains a circuit which, together with C_2, stops this. The line above 'output enable' in the encoder pin connection diagram indicates that pin 13 must be connected to 0 V for the encoder to work. This is the accepted way of showing that an input is 'active' when 'low' and is used in other pin connection diagrams.

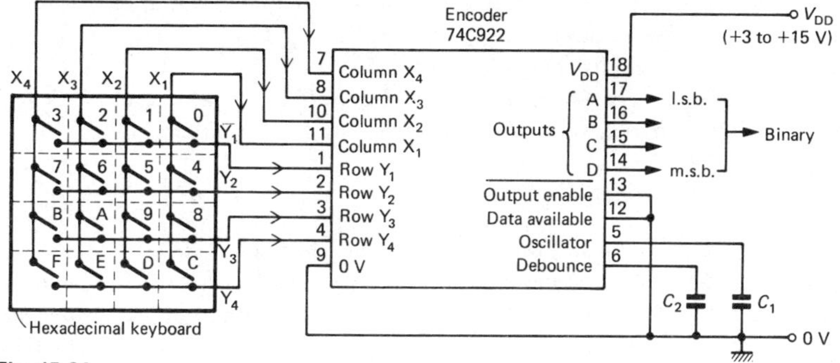

Fig. 15.04

15.7 Decoders

A *decoder* (usually at the output end) converts from binary into decimal.

(a) Decoding

A circuit for a BCD-to-decimal decoder is shown in Fig. 15.05. It uses ten four-input AND gates and as well as the four binary coded decimal inputs A, B, C and D it also requires their complements $\overline{A}$, $\overline{B}$, $\overline{C}$ and $\overline{D}$ (pronounced not A, not B, not C and not D). The latter are supplied by four inverters (NOT gates) in the inputs.

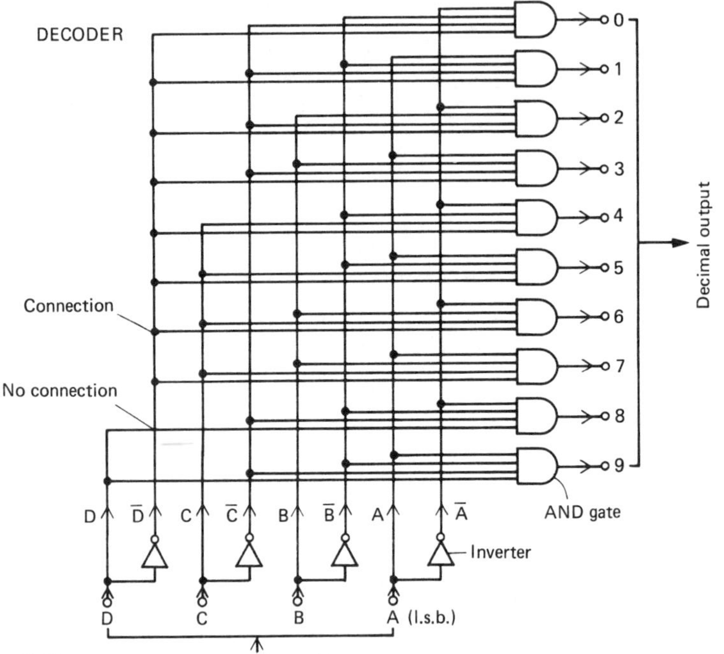

Fig. 15.05

Remembering that an AND gate gives a 'high' (1) output only when all its inputs are 'high', you should check that, for example, an input of 0110 (i.e. D = 0, C = 1, B = 1, A = 0) and its complement 1001 (i.e .$\overline{D}$ = 1, $\overline{C}$ = 0, $\overline{B}$ = 0, $\overline{A}$ = 1) is decoded to give a 'high' output for decimal 6 only. If, for instance, each of the ten outputs were used to drive one LED, then that for digit 6 would light up.

(b) BCD decoder-driver ICs

A BCD (i.e. four-bit binary) number is applied to the four inputs D (m.s.b.), C, B, A (l.s.b.) and is converted into seven outputs, a, b, c, d, e, f, g, each capable of driving directly one segment of a seven-segment LED decimal display (with suitable current-limiting resistors). Fig. 15.06(*a*) gives the connections for the CMOS 4511B decoder-driver IC supplying a common-cathode LED display.

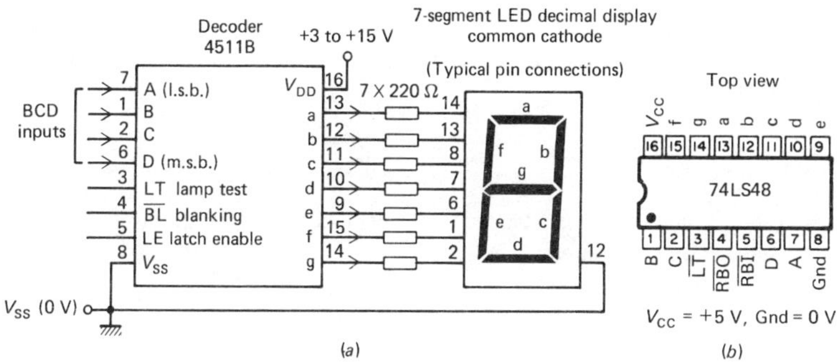

Fig. 15.06

As an example of the decoding action, suppose D = 0, C = 0, B = 1 and A = 1, the BCD input is thus 0011 (3 in decimal) and the five outputs a, b, c, d, g, needed to light the five LED segments making a '3', all go 'high'. If the BCD number exceeds 9, all outputs go 'low' and the display is blank.

Normally $\overline{\text{LT}}$ (lamp test) is kept 'high'; if it is 'low', it is 'active' and all outputs go 'high' and test the display by lighting every segment, no matter what the inputs are. $\overline{\text{BL}}$ (blanking) is also usually held 'high', but if it goes 'low' (and $\overline{\text{LT}}$ is 'high'), all outputs go 'low' and blank out the display. When LE (latch enable) is 'low', the seven outputs follow the input changes; if LE is 'high', the last input is stored or 'latched' and held on display. Table 15.4 sums up the behaviour of the IC.

Table 15.4 Truth table for BCD decoder-driver IC

$\overline{\text{LT}}$ $\overline{\text{BL}}$ LE	**Inputs** D C B A	**Outputs** a b c d e f g	**Display**
0 X X	X X X X	1 1 1 1 1 1 1	8
1 0 X	X X X X	0 0 0 0 0 0 0	blank
1 1 0	0 0 1 1	1 1 1 1 0 0 1	3
1 1 0	1 0 1 0	0 0 0 0 0 0 0	blank
1 1 1	X X X X	*	*

X= does not matter if 1 or 0

* = stores input present LE was 0, i.e. latches

The equivalent low power Schottky TTL common-cathode decoder is the 74LS48; its pin connections are similar to those of the 4511B and are shown in Fig. 15.06(*b*). $\overline{\text{RBO}}$ (ripple blanking output) and $\overline{\text{RBI}}$ (ripple blanking input) are used for multidigit displays.

15.8 Adders

One kind of circuit required for electronic arithmetic is the adder—it performs binary addition.

(a) Half-adder

A half-adder adds *two bits at a time* and has to deal with four cases (essentially three since two are the same). They are:

0	0	1	1
+0	+1	+0	+1
0	1	1	10

In the fourth case, 1 plus 1 equals 2, which in binary is 10, i.e. the right-hand column is 0 and 1 is carried to the next column on the left. The circuit for a half-adder must therefore have two inputs, i.e. one for each bit to be added, and two outputs, i.e. one for the *sum* and one for the *carry*.

One of the many ways of building a half-adder from logic gates uses an exclusive-OR gate and an AND gate. From the truth tables in Figs. 15.07(*a*) and (*b*), you can see that the output of the exclusive-OR gate is always the *sum* of the addition of two bits (being 0 for 1 + 1), while the output of the AND gate equals the *carry* of the two-bit addition (being 1 only for 1 + 1). Therefore, if both bits are applied to the inputs of both gates at the same time, binary addition occurs. The circuit is shown with its two inputs A and B in Fig. 15.07(*c*), along with the half-adder truth table.

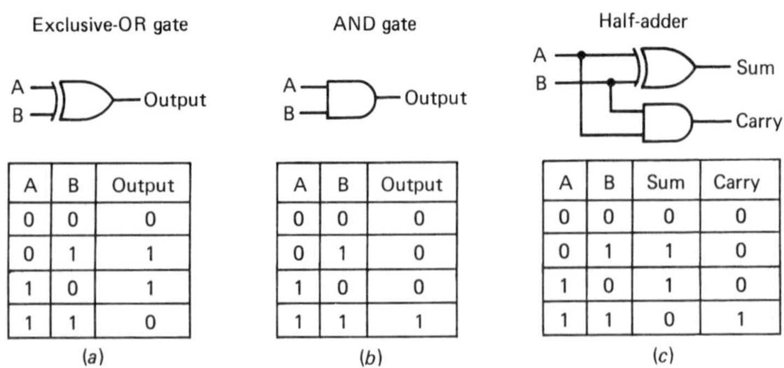

A	B	Output
0	0	0
0	1	1
1	0	1
1	1	0

(*a*)

A	B	Output
0	0	0
0	1	0
1	0	0
1	1	1

(*b*)

A	B	Sum	Carry
0	0	0	0
0	1	1	0
1	0	1	0
1	1	0	1

(*c*)

Fig. 15.07

(b) Full-adder

This adds *three bits at a time*, a necessary operation when two multi-bit numbers are added. For example, to add 3 (11 in binary) to 3 we write:

$$\begin{array}{r} 11 \\ +11 \\ \hline 110 \\ \hline \end{array}$$

The answer 110 (6 in decimal) is obtained as follows. In the least significant (right-hand) column we have:

$$1 + 1 = \text{sum } 0 + \text{carry } 1$$

In the next column three bits have to be added because of the carry from the first column, giving:

$$1 + 1 + 1 = \text{sum } 1 + \text{carry } 1$$

A full-adder circuit therefore needs three inputs, A, B, and C, and two outputs (one for the sum and the other for the carry). It is realized by connecting two half-adders (HA) and an OR gate as in Fig. 15.08(*a*). We can check that its produces the correct answer by putting A = 1, B = 1 and C = 1, as done in Fig. 15.08(*b*). The first half-adder HA_1 has both inputs 1 and so gives a sum of 0 and a carry of 1. HA_2 has inputs of 1 and 0 and gives a sum of 1 (i.e. the sum output of the full-adder) and a carry of 0. The inputs to the OR gate are 1 and 0 and, since one of the inputs is 1, the output (i.e. the carry output of the full-adder) is 1. The addition of 1 + 1 + 1 is therefore sum 1 and carry 1. The truth table with the other three-bit input combinations is given in Fig. 15.08(*c*). You may like to check that the circuit does produce the correct outputs for them.

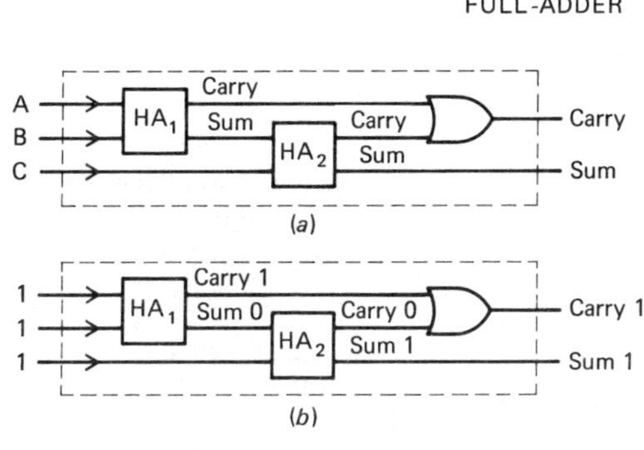

(*a*)

(*b*)

Inputs			Outputs	
A	B	C	Sum	Carry
0	0	0	0	0
0	0	1	1	0
0	1	0	1	0
1	0	0	1	0
0	1	1	0	1
1	0	1	0	1
1	1	0	0	1
1	1	1	1	1

(*c*)

Fig. 15.08

(c) Multi-bit adder

Two multi-bit binary *numbers* are added by connecting adders in parallel. For example, to add two four-bit numbers, four adders are needed, as shown in Fig. 15.09(*a*) for the addition of 1110 (decimal 14) and 0111 (decimal 7). Follow it through to see that the sum is 10101 (decimal 21). (Strictly speaking the full-adder (FA_1) need only be a half-adder since it only handles two bits.)

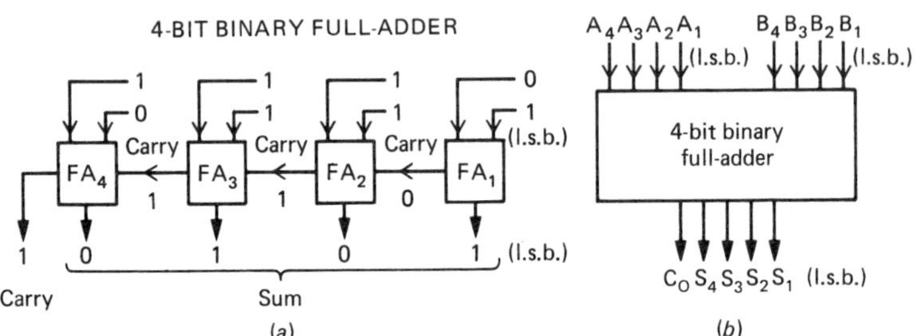

Fig. 15.09

The largest binary numbers that can be added by a four-bit adder are 1111 and 1111, i.e. 15 + 15 = 30. By connecting more full-adders to the left end of the system, the capacity increases.

In the block diagram for a four-bit parallel adder, shown in Fig. 15.09(*b*), the four-bit number $A_4A_3A_2A_1$ is added to the $B_4B_3B_2B_1$, A_1 and B_1 being the least significant bits (l.s.b.). S_4, S_3, S_2 and S_1 represent the sums and C_O, the most signficant bit (m.s.b.) of the answer, is the carry of the output.

A disadvantage of the adder in Fig. 15.09(*a*) is that each stage must wait for the carry from the preceding one before it can decide its own sum and carry, i.e. the carry ripples down from right to left. This is called 'ripple carry addition' and it can be speeded up by including in the adder extra circuits which predict immediately all the carries. In such a 'look-ahead carry adder', all bits are added at the same time giving much faster addition.

(d) Four-bit full-adder ICs

The pin connections for the CMOS 4008B IC and the low power Schottky TTL version, the 74LS283, are shown in Fig. 15.10. For the addition of larger-bit numbers, the 'carry out' pin (C_O) on one IC is connected to 'carry in' (C_I) on another IC which handles the next four most significant bits.

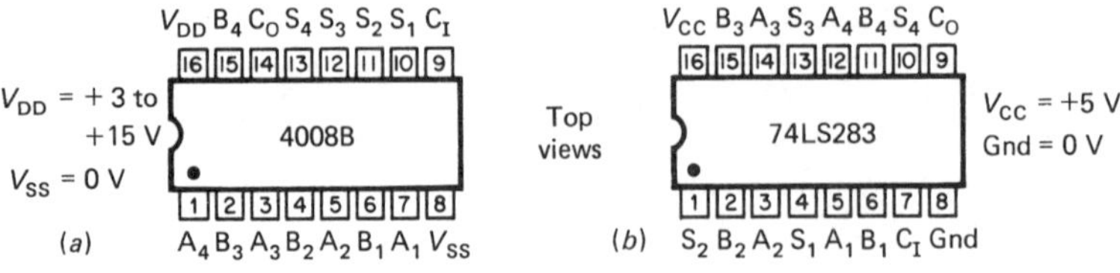

Fig. 15.10

15.9 Magnitude comparators

Sometimes two binary numbers A and B have to be compared. Often we need to decide whether A = B, for instance, when it is essential to know whether a certain total has been reached in a counting operation. Occasionally there may also be a need to check whether A > B or A < B. Such comparisons are made by a magnitude comparator.

(a) Comparing two one-bit numbers

One possible circuit and its truth table are given in Fig. 15.11. Consider the first line of the table. One bit is A = 1 and the other is B = 0, (i.e. A > B) so $\overline{A}$ = 0 and $\overline{B}$ = 1. The inputs A and $\overline{B}$ to the upper AND gate are therefore both 'high', making its output C 'high'. On the other hand, the inputs $\overline{A}$ and B to the lower AND gate are both 'low', making its output D 'low'. Since one of the inputs to the NOR gate is 'high' (i.e. C), its output E is 'low'. The only 'high' output is C so this is the output that shows when A > B. By working through the circuit in the same way for the other three input conditions, you will find that only output E is 'high' when A = B (i.e. A = 1 and B = 1 or A = 0 and B = 0) and only output D is 'high' when A < B (i.e. A = 0 and B = 1).

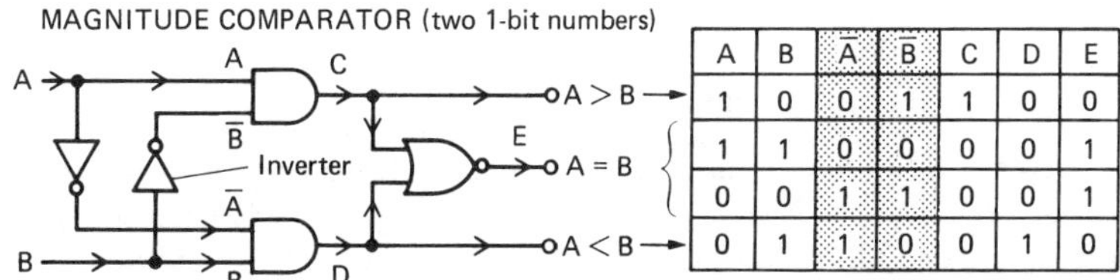

A	B	$\overline{A}$	$\overline{B}$	C	D	E
1	0	0	1	1	0	0
1	1	0	0	0	0	1
0	0	1	1	0	0	1
0	1	1	0	0	1	0

Fig. 15.11

(b) Comparing two two-bit numbers

In the circuit of Fig. 15.12 the two two-bit numbers A (= A_2A_1) and B (= B_2B_1) are compared a bit at a time, starting with the most significant bits, i.e. A_2 and B_2. MC_1 and MC_2 are magnitude comparators, like the one in Fig. 15.11, which compare two one-bit numbers; MC_1 compares A_1 and B_1, while MC_2 compares A_2 and B_2. The E_1 input to the upper two AND gates is kept 'high' so 'enabling' these gates to give a 'high' output when A_2 and B_2 are 'high'.

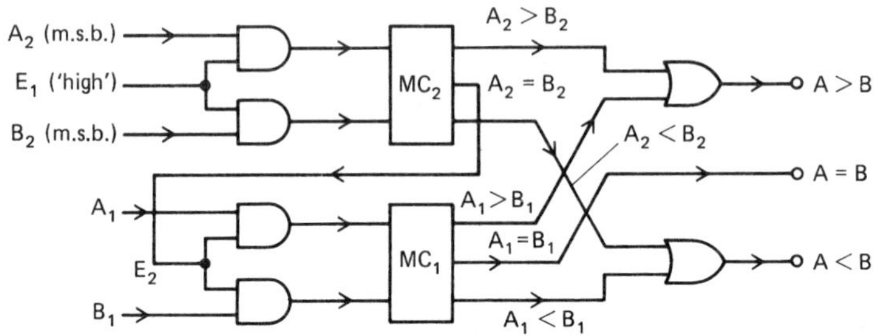

Fig. 15.12

First, suppose A > B, say A is 11 and B is 10, then $A_2 = 1$ and $B_2 = 1$. Both inputs to MC_2 are therefore 1s and so the only MC_2 output that goes 'high' is $A_2 = B_2$. This makes the E_2 input to the two lower AND gates 'high' and 'enables' them, allowing MC_1 to compare A_1 and B_1. Since $A_1 = 1$ and $B_1 = 0$, only the $A_1 > B_1$ output from MC_1 goes 'high' and thereby causes the output from the upper OR gate to go 'high', i.e. the A > B output.

Second, suppose A < B, say A is 01 and B is 10, then $A_2 = 0$ and $B_2 = 1$. In that case the $A_2 = B_2$ output from MC_2 is 'low' and this 'inhibits' (i.e. prevents) the operation of MC_1. But the $A_2 < B_2$ output from MC_2 goes 'high', so giving a 'high' output from the lower OR gate, i.e. the A < B output.

Work out for yourself what happens if A = B, say both are 10.

(c) Four-bit magnitude comparator ICs

These compare two four-bit numbers $A_4A_3A_2A_1$ and $B_4B_3B_2B_1$, A_1 and B_1 being the least significant bits. The pin connections for the CMOS 4585B IC and the low power Schottky TTL equivalent, the 74LS85, are given in Fig. 15.13. Inputs A > B, A = B and A < B are used to follow expansion when two or more comparators are connected.

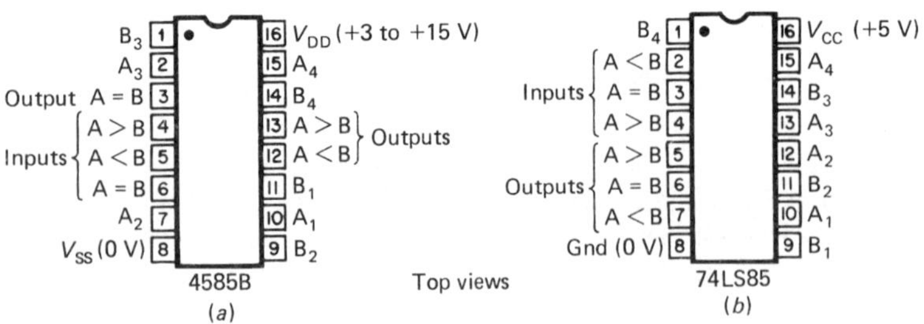

Fig. 15.13

15.10 Multiplexers

Multiplexers are another type of decision-making digital IC and are important in electronics. Their use in the transmission of signals will be considered in section 18.15.

(a) Multiplexing

A multiplexer or data selector selects one of its *several inputs* and transmits the data (a 1 or a 0) at that input to its *one output*. A 4-to-1 line multiplexer behaves like a four-position single-pole rotary switch, Fig. 15.14(*a*). The input selected by the multiplexer is decided by the 'command' signals at its *select* input.

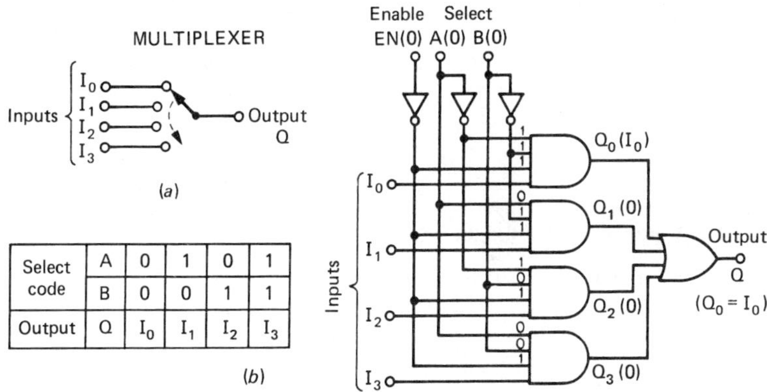

Select code	A	0	1	0	1
	B	0	0	1	1
Output	Q	I_0	I_1	I_2	I_3

Fig. 15.14

A circuit for a 4-to-1 multiplexer is shown in Fig. 15.14(*b*) with a truth table. The select code is applied in binary to A and B, A being the l.s.b. For example, if the select code is 00, input I_0 appears at the output Q, so long as the circuit is enabled, i.e. EN is 'low'. You can check this by working out the inputs to the four AND gates (they are as marked on the diagram). They show that outputs Q_1, Q_2 and Q_3 from the bottom three gates are all 'low' (whatever inputs I_1, I_2 and I_3 may be), while the output Q_0 from the top AND gate is 'high' if $I_0 = 1$ and 'low' if $I_0 = 0$, i.e. $Q_0 = I_0$. If $Q_0 = 1$, output Q from the OR gate is 1 but if $Q_0 = 0$, it is 0, that is $Q = Q_0 = I_0$. The other three select codes can be checked similarly.

(b) Multiplexer ICs

The pin connections for the CMOS 4539B 4-to-1 multiplexer and its low power Schottky TTL equivalent the 74LS153, are shown in Fig. 15.15. Each IC has two multiplexers with its own enable input ($\overline{EN}$: normally kept 'low') but both halves share the same two select inputs (A and B).

Other multiplexers available include: 8-to-1 (4512B and 74LS151) and 2-to-1 (4019B and 74LS157).

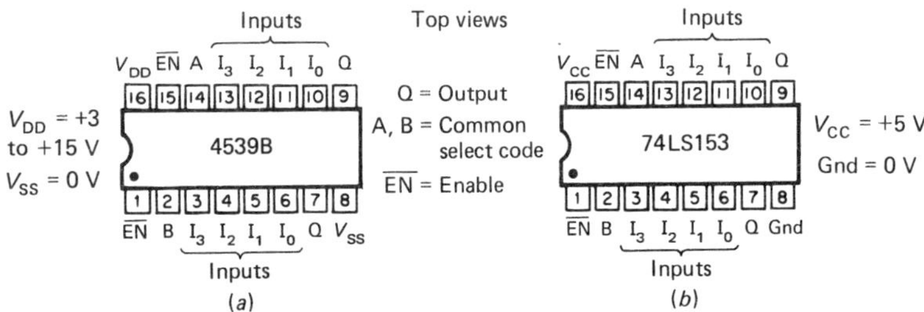

Fig. 15.15

(c) Multiplexed digital display

This is an example of one of the many uses of multiplexing. The obvious way to display a decimal number with several digits on, for example, a watch or calculator, is to use a BCD decoder-driver and seven-segment LED display for each decimal digit. All digits are then lit at the same time.

However, multiplexing provides a method in which all digits share one BCD decoder-driver and, therefore, circuit connections and power consumed are much reduced. The idea is to scan the digits rapidly in turn (at a frequency in the range 500 Hz to 1 kHz) so that they all appear to be on due to the persistence of vision (i.e. the eye sees a light for a fraction of a second after it has been switched off). During the short time each digit is alight, the input to the decoder-driver is connected to the appropriate counter for that digit.

The block diagram for a four-digit display is shown in Fig. 15.16. The upper multiplexer, which takes the outputs from the BCD counters and supplies them as inputs to the BCD decoder-driver, is a four-pole-four-throw (4P4T) type. The lower multiplexer is a single-pole-four-throw (SP4T) type and if the display is a common-cathode one (as here), it energizes each digit by grounding its cathode in turn at the appropriate time. (Using a common-anode display and decoder-driver, it would go to the positive of a supply rather than ground.) The digits are run at higher than normal current to make them brighter when they are on during the scanning cycle. The scanning oscillator synchronizes the various parts of the system which must also be matched to work together.

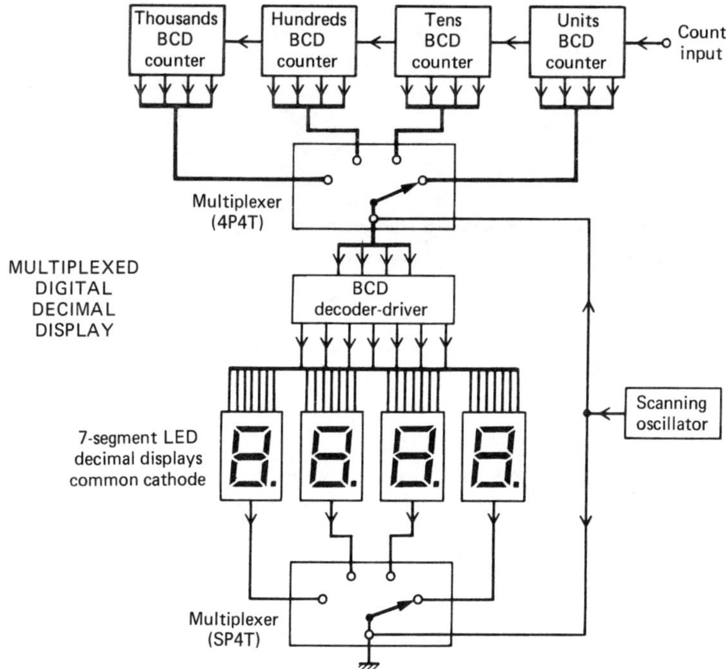

Fig. 15.16

15.11 Arithmetic and logic unit

An arithmetic and logic unit (ALU) is a more complex IC than any discussed so far. As we will see in section 20.1, it is at the heart of a computer. There are two alternative ways or *modes* of operation.

(a) Arithmetic mode

In this case a variety of arithmetic operations are performed in binary, including addition, subtraction, multiplication and division. Addition and adders have already been described. As a further illustration of binary computation and its implementation electronically, we now consider subtraction.

One method, called the *one's complement* method, is best explained by an example. Suppose we have to subtract 0110 (6) from 1010 (10). We proceed as follows:

(i) Form the one's complement of 0110; this is done by changing 1s to 0s and 0s to 1s giving 1001.

(ii) Add the one's complement from (i) to the number from which the subtraction is to be made, i.e. add 1001 and 1010:

```
  1010
 +1001
 -----
 10011
```

(iii) If there is a 1 carry in the most significant position of the total in (ii), remove it and add it to the remaining four bits to get the answer:

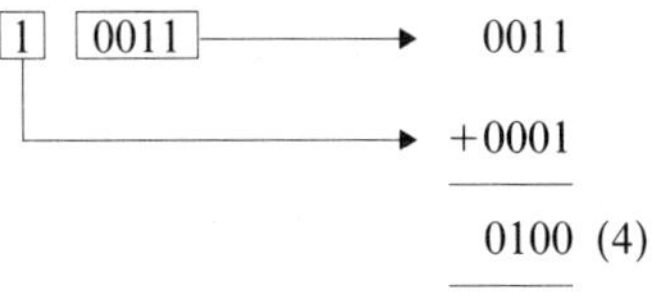

The 1 carry that is added is called the 'end-around carry' (EAC). When there is an EAC, the answer to the subtraction is positive (as above); if there is no EAC, the answer is negative and is in one's complement form. For example, subtracting 0101 (5) from 0011 (3), we get:

```
  0011
 +1010  (one's complement of 0101)
 -----
  1101  → −0010
```

The one's complement of 0101 (i.e. 1010) has been added to 0011 to give 1101. There is no EAC and so to get the final answer we take the one's complement of 1101, i.e. 0010 and put a negative sign in front to give −0010 (−2).

The method, which obviously works and can be justified theoretically, is much used in ALUs because it means that binary subtraction, as well as addition, can be done by an adder since the one's complement of a number is easily obtained by an inverter. A simplified circuit of a four-bit subtractor is shown in Fig. 15.17, where the four-bit number B is subtracted from the four-bit number A. There is an EAC which is applied to the carry-in input of the adder and the answer is S.

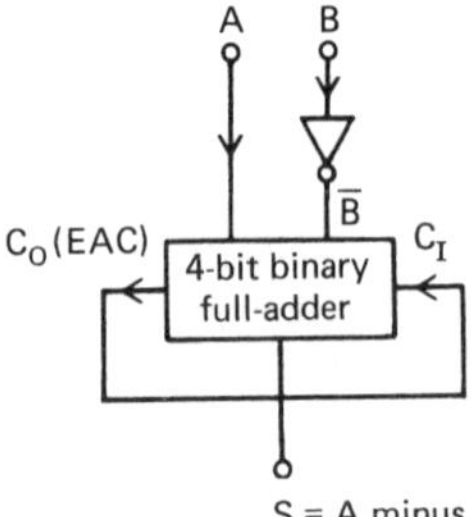

Fig. 15.17

Multiplication (see section 20.7(b)) and division can be performed by repeated addition and subtraction respectively.

(b) Logic mode

Logical operations such as those done by logic gates are performed in this mode. As an example, suppose the four-bit 'word', A, is dcba (a 'word' in computer jargon is a number of bits) and that we need to know whether bit c is a 1 or a 0. The important technique of *bit masking* is used and requires the ALU to act as an AND gate.

To carry out the operation we AND A with another word, B, where B is 0100. The ALU operates on each pair of corresponding bits in the two words individually and since it is only opposite bit c in word A that a 1 appears in word B, the output (result) of the AND operation can only be 0100 if c is a 1, otherwise it will be 0000 (i.e. c is a 0).

(c) Four-bit ALU ICs

The pin connections for two four-bit ALUs are shown in Fig. 15.18—the CMOS 4581B and the low power Schottky TTL 74LS181. Both have 24 pins and a width of 0.6 inch (compared with 0.3 inch for most ICs) and are LSI types.

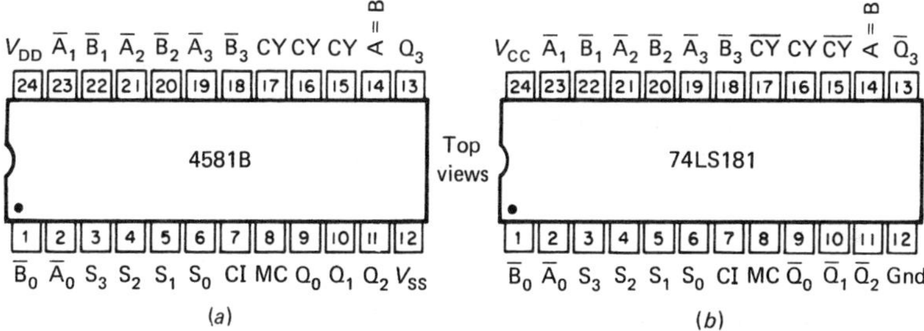

Fig. 15.18

Pin 8 (mode control: MC) is made 'low' for the arithmetic mode and 'high' for the logic mode. Two four-bit words $A_3A_2A_1A_0$ and $B_3B_2B_1B_0$ are operated on, the result appearing at the four outputs $Q_3Q_2Q_1Q_0$. In each mode 16 different operations can be performed, selected by a four-bit binary code at pins 3, 4, 5, 6 (select: S_3, S_2, S_1, S_0). In arithmetic operations pin 7 (CI) accepts the carry-in from another IC. Pins 15, 16 and 17 (CY) are concerned with carry-out and allow high speed counting with other ALUs using words of greater length. Pin 14 (A = B) enables the ALU to be used as a four-bit magnitude comparator; it goes 'high' when each bit of word A is the same as the corresponding bit of word B, provided MC is 'low', CI is 'high' and the binary code on S_3, S_2, S_1, S_0 is 0110.

Manufacturers publish 'function tables' showing the operations performed (in both arithmetic and logic modes) when each of the 16 possible combinations are set up on S_3, S_2, S_1, S_0.

15.12 Revision questions and Problems

1. Explain the term 'combinational logic'. Name a type of digital circuit which uses that logic.

2. What do these abbreviations stand for: SSI, MSI, LSI, VLSI, TTL, CMOS?

3. For TTL and CMOS, compare their
   a) supply voltage requirements,
   b) power consumption,
   c) switching speed,
   d) input impedance.

4. State the precautions to be observed with (a) CMOS ICs, (b) TTL ICs.

5. How many input leads are needed for an IC having
   a) quad two-input NAND gates,
   b) a hex inverter,
   c) dual four-input NOR gates?

6. a) Draw the symbols for each of the following gates:
   (i) inverter, (ii) AND, (iii) NAND, (iv) NOR, (v) OR, (vi) exclusive-OR.
   b) State what the input conditions must be for each of the gates in a) to give a 'high' output.
   c) Draw five diagrams to show how the other gates listed in a) can be made from NAND gates.

7. Make the following code conversions.
   a) Decimal numbers 4, 13, 21, 38, 64 to binary.
   b) Binary numbers 111, 11001, 101010, 110010 to decimal.
   c) Decimal numbers 6, 11, 15, 31, 83, 300 to hexadecimal.
   d) Hexadecimal numbers 3, A, D, 1E, 1A5 to decimal.
   e) Decimal numbers 9, 17, 28, 370, 645 to BCD.

8. What is the job of (a) an encoder, (b) a decoder?

9. Add the following pairs of binary numbers and check your answers by converting the binary numbers to decimal:
   (a) 10 + 01; (b) 11 + 10; (c) 101 + 011; (d) 1011 + 0111.

10. What is the function of (a) a half-adder, (b) a full-adder?
    Write a truth table for each.

11. What is a magnitude comparator?

12. a) Explain the term multiplexer (data selector).
    b) State three advantages of multiplexing a multi-bit digital decimal display.

13. Subtract the following pairs of binary numbers by the one's complement method and check your answers by converting the binary numbers to decimal:
    (a) 1001 − 0101; (b) 11001 − 00111; (c) 0011 − 0110; (d) 01100 − 10101.

14. What does ALU stand for? What does it do?

UNIT 16

Memory-type circuits

16.1 Introduction

(a) Sequential logic

The digital ICs to be considered in this Unit are 'memory-type' circuits having flip-flops (bistable multivibrators) as their main components. They use *sequential* logic, i.e. their outputs depend not only on their *present* inputs (as was the case with the combinational logic circuits of Unit 15) but on *past* ones as well. The order or sequence in which inputs are applied is important.

This chief requirement of a sequential logic circuit is that it should have some kind of memory to 'remember' the earlier inputs. The memory is usually obtained by feedback connections which ensure that, when the input changes, the effect of the previous input is not lost. For handling information (data) in binary code, flip-flops, with their two stable states representing 1s and 0s, can do the job and are readily switched from one state to the other by appropriate inputs.

(b) Clocked logic

Sequential logic circuits and systems are either clocked (synchronous) or asynchronous. Today, most belong to the first group, in which the output does not respond immediately to an input change but waits until it receives a *clock* or *enabling* pulse.

Clocking is important in large digital systems containing hundreds of interconnected flip-flops. If the same clock pulse is applied to them all simultaneously, they change together, i.e. are synchronized. This reduces the risk of timing sequences being upset by one part racing ahead (called the 'race' condition). Changes can also progress in an orderly way, a step at a time, as commanded by the clock pulses. The technique makes more elaborate circuits feasible. Clocked logic circuits cause fewer operating problems than asynchronous circuits. In the latter there is no common control and any change ripples through the system, a change in one part producing a change in other parts and so on.

A 'clock' in digital electronics is a *pulse generator*, usually crystal controlled (see section 12.6) with a very steady repetition frequency, 100 MHz or higher being typical. There are two main types of clocking or triggering—*level* and *edge*. In the first, changes occur when the *level* of the clock pulse is 1 or 0—see section 14.7; in the second, they happen either during the *rise* of the clock pulse from 0 to 1 (called rising- or positive-edge triggering) or during the *fall* from 1 to 0 (called falling- or negative-edge triggering—see section 14.10). Most modern clocked logic is edge triggered, generally on the rising edge, but examples of both will be given—see sections 16.4 and 17.1.

Clock pulses should have fast rise and fall times (less than 5 μs) and inputs coming from switches, keyboards, etc., should be 'debounced'. Section 17.6 (Schmitt triggers) explains how both conditions are met in practice.

16.2 SR flip-flop

(a) Basic SR flip-flop

The discrete component version of the SR (set-reset) flip-flop was considered in section 14.7. It can be built in other ways—one is shown in Fig. 16.01(*a*) using NAND gates. There are two inputs S and R and two outputs Q and $\overline{Q}$. N_1 and N_2 are inverters (e.g. one-input NAND gates) and the feedback required in a sequential logic circuit is obtained by connecting each output as one of the inputs to the gate which controls the other output. The outputs therefore depend on outputs as well as inputs.

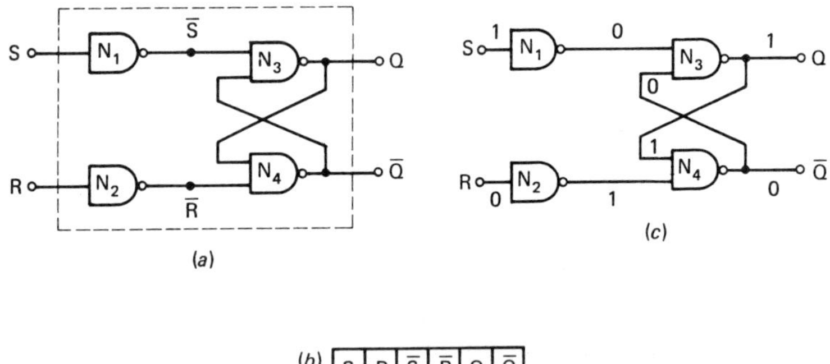

(*b*) SR FLIP-FLOP

S	R	$\overline{S}$	$\overline{R}$	Q	$\overline{Q}$
1	0	0	1	1	0
0	0	1	1	1	0
0	1	1	0	0	1
0	0	1	1	0	1
1	1	0	0	1	1

Fig. 16.01

The most important property of a flip-flop is that it can exist in one of two stable states, either Q = 1 (and $\overline{Q}$ = 0), which is the 1 ('high') state or Q = 0 (and $\overline{Q}$ = 1) called the 0 ('low') state. Each state in effect stores one bit of 'information', i.e. a 1 when Q = 1 or a 0 when Q = 0, and the flip-flop acts as a *one-bit memory*.

The complete behaviour of the circuit can be worked out by going through the sequence of input conditions shown in the truth table in Fig. 16.01(*b*), remembering that the output from a NAND gate is always 1 *unless* both inputs are 1s. For example if S = 1 and R = 0, as in Fig. 16.1(*c*), then $\overline{S}$ = 0 and $\overline{R}$ = 1. Q must be a 1 since one of its inputs (that from N_1) is a 0. Both inputs to N_4 are therefore 1s and so $\overline{Q}$ = 0.

You should note the following points.

(i) When S = 1 and R = 0, then Q = 1 (and $\overline{Q}$ = 0); this is the '1' stable state and the flip-flop is 'set'.

(ii) When S = 0 and R = 1, then Q = 0 (and $\overline{Q}$ = 1); this is the '0' stable state and the flip-flop is 'reset'.

(iii) When S = 0 and R = 0, then Q (and $\overline{Q}$) can be either 1 or 0, depending on the state before this input condition existed. The previous output state is retained by the circuit, i.e. no change occurs as shown by the second and fourth rows of the truth table. To see this, suppose in Fig. 16.01(*c*) that S becomes 0, R being 0. Even though one input to N_3 will now be 1, Q stays 1 because the other input to N_3 is still 0. Similarly, starting from S = 0 and R = 1 (third row in truth table), if R becomes 0, the same reasoning shows that Q and $\overline{Q}$ do not change.

(iv) When S = 1 and R = 1, then Q = 1 (and $\overline{Q}$ = 1). Unlike the other three cases, here Q and $\overline{Q}$ are the same. The circuit is stable while S = R = 1 but if S and R change simultaneously from 1 to 0, an undesirable 'race condition' develops. This is due to one of the output gates (N_3 or N_4) unavoidably switching slightly faster than the other (because of the near impossibility of making two absolutely identical gates) and we cannot predict whether Q or $\overline{Q}$ will be 1. The circuit action is said to be *indeterminate* and in circuits containing SR flip-flops the condition S = R = 1 should not be allowed to arise.

(b) Clocked SR flip-flop

The circuit and symbol of Figs. 16.02(*a*) and (*b*) are for a basic SR flip-flop with a third input, called clock CK or enable EN, added for use in synchronous systems.

Before clocking, outputs A and B from NAND gates N_1 and N_2 respectively are both 1s, whatever the inputs to S and R. And since A and B correspond to $\overline{S}$ and $\overline{R}$ in Fig. 16.01(*a*), the second and fourth input rows in the truth table in Fig. 16.01(*b*) for the basic SR flip-flop, show that outputs Q and $\overline{Q}$ will retain their present values, irrespective of any changes in S and R. The flip-flop is said to be *inhibited* or *disabled*.

When clocked, the flip-flop is enabled and A and B (and so also Q and $\overline{Q}$) depend on the S and R inputs. The behaviour of the circuit is shown in the truth table in Fig. 16.02(*c*), which you should check through row by row using the circuit. The input condition S = R = 1 is not used for the reason given in (a) (iv) above.

SR flip-flop ICs are not common, since, as we will see in the following sections, other more versatile types of flip-flop can do the same job.

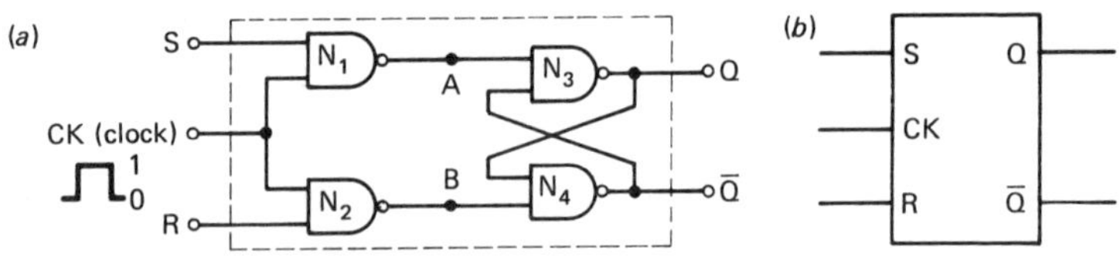

CLOCKED SR FLIP-FLOP

(c)

Inputs		During clock pulse		Outputs before clock pulse		Outputs after clock pulse		Comments
S	R	A	B	Q	$\overline{Q}$	Q	$\overline{Q}$	
0	0	1	1	1	0	1	0	No change in outputs
0	0	1	1	0	1	0	1	
1	0	0	1	1	0	1	0	Flip-flop sets with Q = 1 and $\overline{Q}$ = 0
1	0	0	1	0		1	0	
0	1	1	0	1	0	0	1	Flip-flop resets with Q = 0 and $\overline{Q}$ = 1
0	1	1	0	0	1	0	1	
1	1	0	0	1	0	1	1	This input not allowed
1	1	0	0	0	1	1	1	

Fig. 16.02

16.3 T-type flip-flop

The discrete component version of this type, also called the *triggered* or *toggling flip-flop*, was considered in section 14.10. The circuit in Fig.16.03(*a*) shows how a clocked SR flip-flop built from four NAND gates is modified to become a T-type flip-flop. The Q output from N_3 is fed back to the R input on N_2 and the $\overline{Q}$ output from N_4 is returned to the S input on N_1. The symbol for a T-type flip-flop is given in Fig. 16.03(*b*).

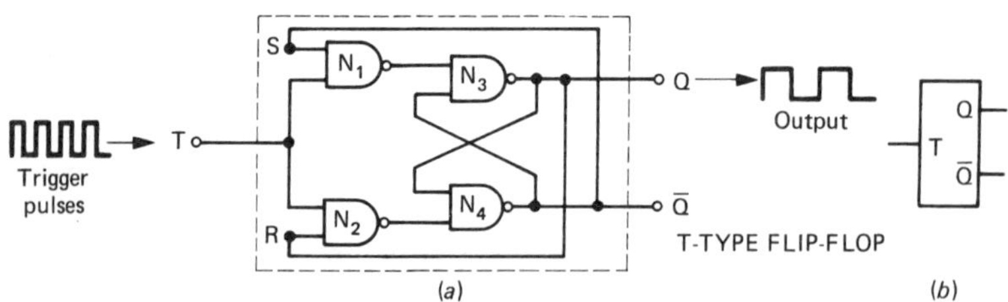

Fig. 16.03

When a succession of *very brief* trigger pulses is applied to the clock input, renamed the T input, each pulse causes the flip-flop outputs to change state from 1 to 0 or vice versa, i.e. to toggle.

Suppose Q = 0 and $\overline{Q}$ = 1 just before the first trigger pulse arrives. The feedback connections will therefore make S = 1 and R = 0. When the pulse arrives making T = 1, both inputs to N_1 will be 1 and its output goes to 0 (from the NAND rule: output 'high' unless all inputs 'high'). As a result, Q is set to 1 because the input from N_1 to N_3 is now 0. The flip-flop has changed state and in so doing, because of the feedback, S is now 0 and R is 1. The next trigger pulse makes the output from N_2 a 0 (since both its inputs are 1) and the flip-flop resets with $\overline{Q}$ = 1 and Q = 0.

In effect each incoming trigger pulse is steered alternately to the S and R inputs, causing the flip-flop to set and reset repeatedly, i.e. to toggle. Note that Q = 1 only once every two trigger pulses, that is, the output has half the frequency of the input. T-type flip-flops divide the input frequency by two and are the basic components of binary counters. For reliable operation the trigger pulse must not last longer than the time it takes the gates to switch (typically less than 15 nanoseconds = 0.015 μs). Otherwise, during one long trigger pulse, the flip-flop could repeatedly reset and set again as a result of Q and $\overline{Q}$ changing and being fed back to S and R. It would be no use as a binary counter because the frequency of the Q output would not be half that of the trigger pulses since Q is a 1 more than once per trigger pulse. We will see in section 16.5 how the risk of this problem arising is eliminated by the use of 'master-slave' flip-flops.

T-type flip-flop ICs are not available separately since other types of flip-flop can be made to toggle.

16.4 D-type flip-flop (or data latch)

(a) Action

The D-type flip-flop is also a modified SR type but it has only one input (D) and so the unpredictable state of S = R = 1 cannot arise. The circuit and symbol are shown in Figs. 16.04(*a*) and (*b*). There is an inverter N_5 (e.g. a one-input NAND gate) connected to the input so that the R input is always the complement of the S (and D) input. The action is simpler than the clocked SR flip-flop.

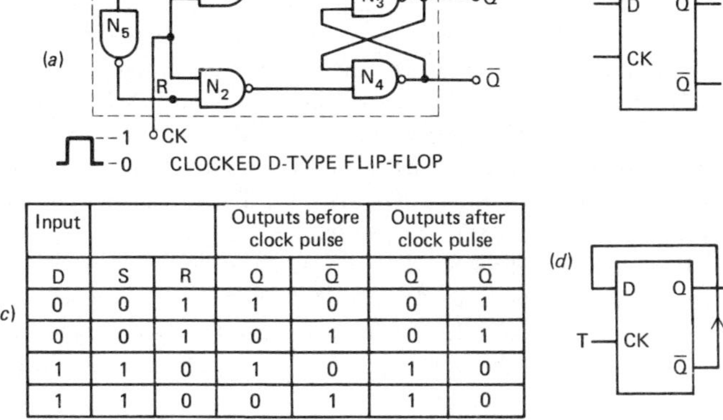

Input			Outputs before clock pulse		Outputs after clock pulse	
D	S	R	Q	$\overline{Q}$	Q	$\overline{Q}$
0	0	1	1	0	0	1
0	0	1	0	1	0	1
1	1	0	1	0	1	0
1	1	0	0	1	1	0

Fig. 16.04

After the clock input CK has gone from 0 to 1, the bit of data (1 or 0) at the D input is transferred to the Q output. If the D input changes while CK = 1, Q follows it, i.e. Q and D stay the same. At the end of the clock pulse when CK has fallen from 1 to 0, Q retains the last value of D when CK was 1. While CK = 0, Q does not change (even if D does) until the next clock pulse arrives at CK. The circuit is also called a *data latch* because when CK = 1, Q latches on to the bit of data at D and stores it temporarily. The truth table in Fig. 16.04(*c*) shows that the output equals the input one clock pulse earlier.

The D in D-type stands for *data* or *delay*; the latter term refers to the fact that a bit at the data input is held back and does not reach the Q output until CK = 1.

(b) Use as T-type flip-flop

A D-type flip-flop can be converted to a T-type by connecting the $\overline{\mathrm{Q}}$ output to the D input as in Fig. 16.04(*d*). Successive clock pulses, that are sufficiently brief, then cause toggling as follows.

Suppose the first clock pulse leaves Q = 1 and $\overline{\mathrm{Q}}$ = 0 because D is 1. Feedback then makes D = 0 and so after the second clock pulse, Q = 0 and $\overline{\mathrm{Q}}$ = 1. D now becomes 1 and the third clock pulse makes Q = 1 and $\overline{\mathrm{Q}}$ = 0 again and so on.

(c) Quad data latch ICs

These contain four identical D-type flip-flops. The two ICs to be considered are both level triggered.

In the low power Schottky TTL 74LS75 IC, Fig. 16.05(*a*), each flip-flop has its own D input and Q and $\overline{\mathrm{Q}}$ outputs; they are controlled in pairs by two enable inputs (EN: pin 13 for D_0 and D_1, pin 4 for D_2 and D_3). While EN is 'high', the D inputs are transferred to the corresponding Q outputs and their complements appear at the $\overline{\mathrm{Q}}$ outputs, i.e. if D = 1 then Q = 1 and $\overline{\mathrm{Q}}$ = 0. When EN goes 'low', the bits present at the D inputs when EN was 'high' are stored at the Q outputs until EN goes 'high' again. To store four bits simultaneously the two enables are connected in parallel.

In the CMOS 4042B IC, Fig. 16.05(*b*), all four flip-flops are controlled by two 'command' controls, namely clock CK and polarity POL. If these are in the same state, i.e. both 'high' or both 'low', then as soon as a D input is applied, it is transferred to the corresponding Q output (and its complement to $\overline{\mathrm{Q}}$). This behaviour is shown by the first four rows of the truth table, Fig. 16.05(*c*). If the state of CK is then changed so that it is different from that of POL, the Q (and $\overline{\mathrm{Q}}$) values are stored (latched). A subsequent change at any of the D inputs does not affect the Q (or $\overline{\mathrm{Q}}$) values—so long as CK and POL are still different. The fifth and sixth rows in the truth table summarize the two possibilities in this case.

Quad latches are useful for obtaining a reading on a rapidly changing numerical display that is counting pulses (which would otherwise be seen as a blur) coming at a high rate from a counter. The latch is connected between the counter and the decoder (e.g. binary to decimal) feeding the display, as in Fig. 16.06. When the latch is enabled by receiving an appropriate signal, it stores the count (as a four-bit binary number) and holds it for display until the next enabling signal arrives. In the meantime, the counter can carry on counting.

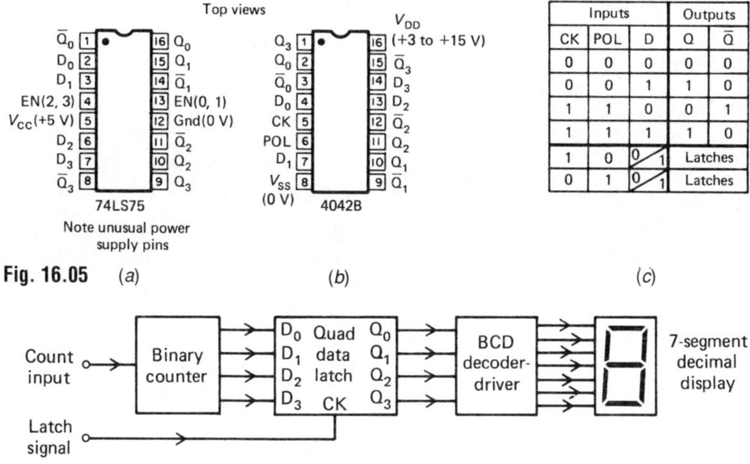

Inputs			Outputs	
CK	POL	D	Q	$\overline{Q}$
0	0	0	0	0
0	0	1	1	0
1	1	0	0	1
1	1	1	1	0
1	0	0/1	Latches	
0	1	0/1	Latches	

Fig. 16.05 (*a*) (*b*) (*c*)

Fig. 16.06

(d) Dual D-type flip-flop ICs with set and reset

The low power Schottky TTL 74LS74 and the CMOS 4013B ICs consist of two independent D-type flip-flops; both are rising-edge triggered. They have direct set and reset inputs (despite the fact that previously steps were taken to design clocked logic circuits that prevent unwanted inputs and 'race conditions') which allow the flip-flop to be operated independently of the clocked input. The direct inputs are used, for example, to reset a group of counting flip-flops to zero or to preset ('jam') certain data into a store. Set is also known as *preset* (Pr) and reset as *clear* (Cr). Pin connections are given in Figs. 16.07(*a*) and (*b*). In the direct mode they behave as SR flip-flops: the 74LS74 sets (i.e. $Q = 1$ and $\overline{Q} = 0$) when $\overline{S} = 0$ and $\overline{R} = 1$ and resets (i.e. $Q = 0$ and $\overline{Q} = 1$) when $\overline{S} = 1$ and $\overline{R} = 0$; the 4013B sets when $S = 1$ and $R = 0$ and resets when $S = 0$ and $R = 1$. This happens regardless of other inputs.

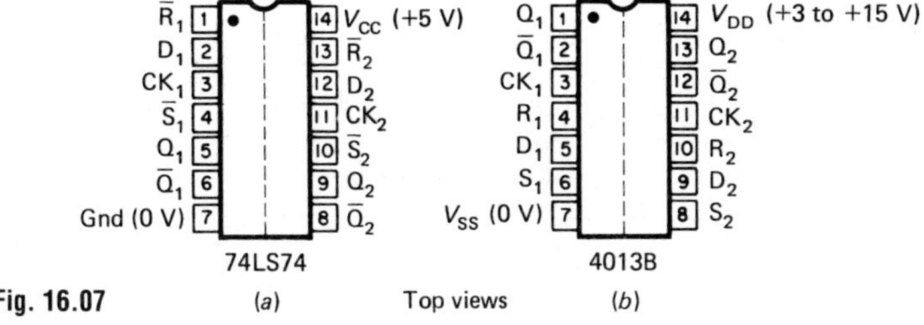

Fig. 16.07 (*a*) Top views (*b*)

In the clocked mode, the bit of data (1 or 0) at the D input is transferred to the Q output during the rising edge (⎍) of a clock pulse. Q stores this value when the clock pulse falls (⎍) and does not change until the next rising edge arrives. For clocking, the 74LS74 requires $\overline{S} = \overline{R} = 1$ and the 4013B requires $S = R = 0$.

Two further points should be noted. First, the set and reset inputs override the clock input. Second, toggling occurs if D and $\overline{Q}$ are connected, as explained before.

16.5 JK flip-flop

(a) Action

The clocked JK flip-flop is about as versatile as a flip-flop can be. It does not have a disallowed input condition: all four input combinations do something meaningful, thereby making it useful in a wide range of general and special applications.

A basic circuit and the symbol are shown in Figs. 16.08(*a*) and (*b*). There are two inputs, called J and K (for no obvious reason) and two sets of feedback connections. In actual ICs provision is also usually made for direct setting and resetting. The operation of the circuit when clocked is fully described by the truth table in Fig. 16.08(*c*). Note that J = K = 1 (equivalent to S = R = 1 in an SR flip-flop) is allowed and causes toggling. Its action can be summed up by saying that:

(i) if J = K = 1, it is a T-type flip-flop;
(ii) if J and K are different, it is a D-type flip-flop;
(iii) if J = K = 0, it appears to do nothing but in fact the same state is clocked back into it.

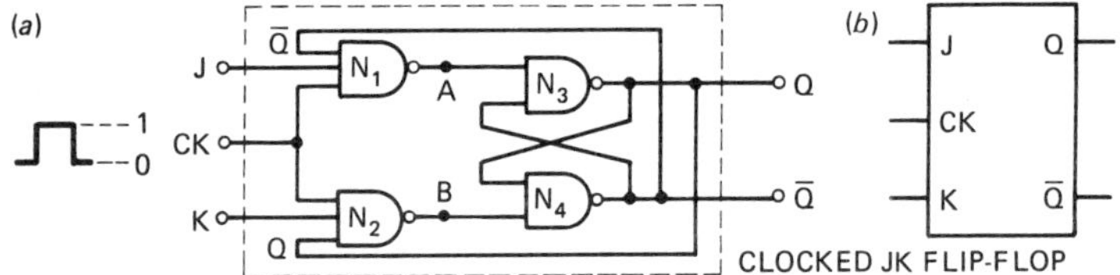

(*c*)

Inputs		Outputs before clock pulse		During clock pulse		Outputs after clock pulse		Comments
J	K	Q	$\bar{Q}$	A	B	Q	$\bar{Q}$	
0	0	1	0	1	1	1	0	No change in outputs
0	0	0	1	1	1	0	1	
1	0	1	0	1	1	1	0	Stays at or sets to Q = 1 and $\bar{Q}$ = 0
1	0	0	1	0	1	1	0	
0	1	1	0	1	0	0	1	Stays at or resets to Q = 0 and $\bar{Q}$ = 1
0	1	0	1	1	1	0	1	
1	1	1	0	1	0	0	1	Toggles
1	1	0	1	0	1	1	0	

Fig. 16.08

(b) Dual JK flip-flop ICs

Pin connections for the low power Schottky TTL 74LS76 IC, which is level triggered, and the CMOS 4027B IC, which is rising-edge triggered, are given in Figs. 16.09(*a*) and (*b*). Both contain two independent JK flip-flops with direct set ($\bar{S}$,S) and reset ($\bar{R}$,R) inputs. They operate in either the clocked or the direct mode.

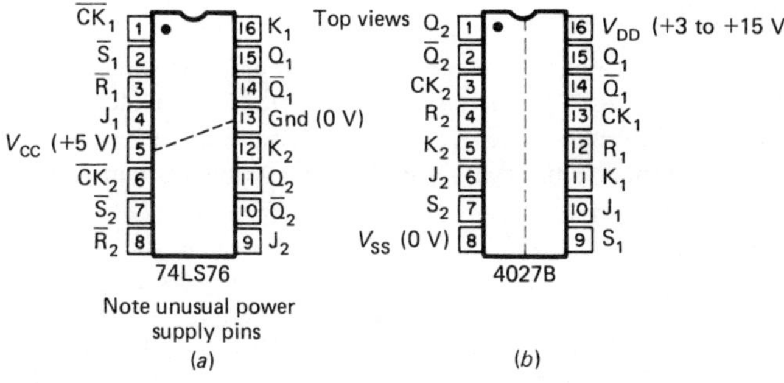

Fig. 16.09

In the direct mode they behave as SR flip-flops; the 74LS76 sets (i.e. Q = 1 and $\overline{Q}$ = 0) when $\overline{S}$ = 0 and $\overline{R}$ = 1 and resets (i.e. Q = 0 and $\overline{Q}$ = 1) when $\overline{S}$ = 1 and $\overline{R}$ = 0; the 4027B sets when S = 1 and R = 0 and resets when S = 0 and R = 1. All this happens regardless of other inputs.

In the clocked mode changes occur in the 74LS76 when the clock pulse level falls from 1 to 0, and in the 4027B on the rising edge of a clock pulse. For clocking, the 74LS76 requires $\overline{S}$ = $\overline{R}$ = 1 and the 4027B requires S = R = 0. The direct inputs ($\overline{S}$, S and $\overline{R}$,R) override the clock inputs in both cases.

(c) Master-slave flip-flops

If the duration of a clock pulse is greater than the switching time of a flip-flop, outputs can 'race' round to inputs (because of the feedback connections) causing uncontrolled switching and unreliable operation. This problem was mentioned in section 16.3 when a brief reference was made to the use of T-type flip-flops as counters.

To prevent such undesirable effects, circuits have been developed using two flip-flops, one called the *master* and the other the *slave*. The arrangement for a JK flip-flop is shown in the block diagram of Fig. 16.10. The action occurs in two stages, controlled by the clock pulse. During the first stage, while the master is enabled and stores the data which is present at its J,K inputs, the slave is disabled (by the inverter at its clock input) and thereby isolated from the master. In the second stage, the master is disabled and isolated from the inputs while the slave, now enabled, accepts the data from the master and transfers it to the output. The risk is thus removed of output changes from the slave being fed back to the input of the master before the clock pulse has ended. This solves the 'race' problem.

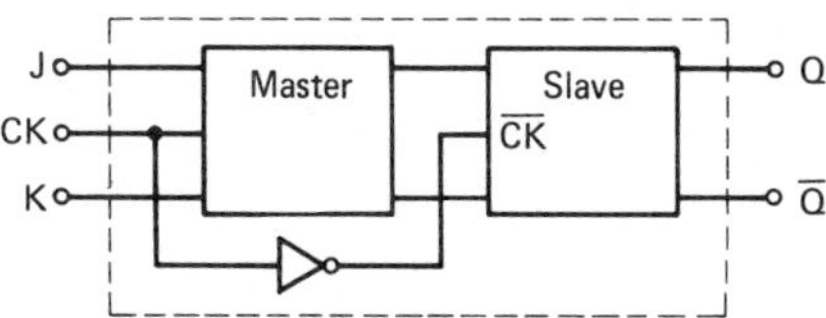

Fig. 16.10

Most modern flip-flops use the master-slave principle.

16.6 Shift registers

A shift register stores a binary number and shifts it out when required. It consists of several D-type or JK flip-flops connected in series, one for each bit (0 or 1) in the number. The bits may be fed in and out serially, i.e. one after the other, or in parallel, i.e. all together. Shift registers are used in, for example, calculators to store two binary numbers before they are added.

(a) Types

In the four-bit *serial-input-serial-output* (SISO) type shown in Fig. 16.11(*a*), clocked D-type flip-flops are used, the Q output of each one being applied to the D input of the next. The bits are loaded one at a time, usually from the left, and move one flip-flop to the right every clock pulse. Four pulses are needed to enter a four-bit number such as 0101 (5) and another four to move it out serially. This is the simplest type.

(*a*) Serial-input-serial-output shift register

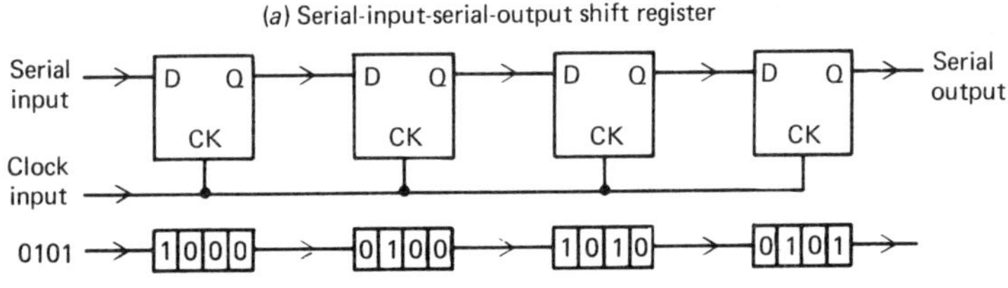

(*b*) Parallel-input-parallel-output shift register

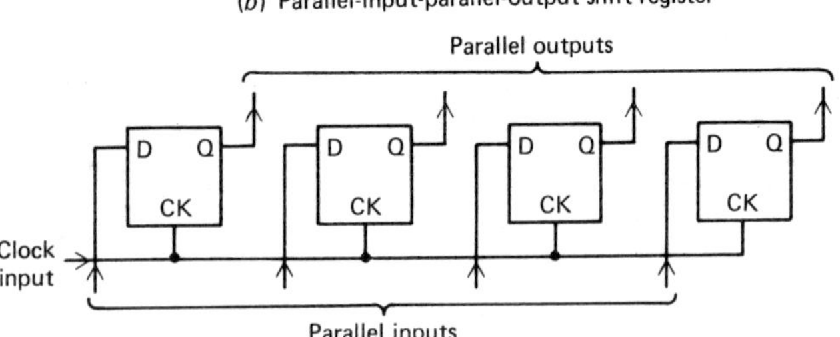

Fig. 16.11

In the *parallel-input-parallel-output* (PIPO) type shown in Fig. 16.11(*b*), all bits enter their D inputs simultaneously and are transferred together to their respective Q outputs (where they are stored) by the same clock pulse. They can then be shifted out in parallel.

Two other arrangements are serial-input-parallel-output (SIPO) and parallel-input-serial-output (PISO).

(b) Four-bit universal shift register ICs

The pin connections for the low power Schottky TTL 74LS95B and the CMOS 4035B are given in Figs. 16.12(*a*) and (*b*). Both can be used in any of the four ways mentioned above, hence the term 'universal'.

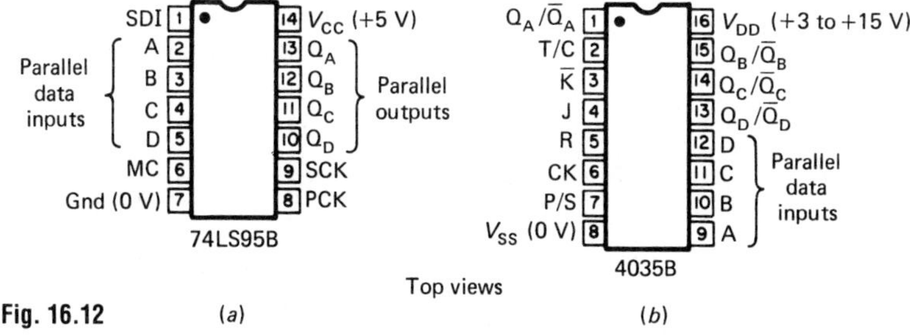

Fig. 16.12 (*a*) (*b*)

The 74LS95B is serial loaded to the right by having mode control MC 'low', presenting data one bit at a time to serial data in SDI and applying clocking pulses to serial clock SCK. On the falling edge of each pulse the bit of data at SDI is shifted to output Q_A, that at Q_A goes to Q_B, that at Q_B goes to Q_C, that at Q_C goes to Q_D and that at Q_D moves out of the register. Parallel loading is achieved by making MC 'high' and when a clock pulse is applied to parallel clock PCK, the data at inputs A, B, C and D are all transferred at the same time to Q_A, Q_B, Q_C and Q_D respectively where they are stored.

The 4035B is serial loaded by holding parallel/series $\overline{\text{P}}$/S and reset R 'low' and true/complement T/C 'high'. The J and $\bar{K}$ inputs, which belong to a JK flip-flop controlling the input to the four D-types that follow, are connected to form the serial input for data. Data stored at the outputs Q_A, Q_B, Q_C and Q_D are shifted one stage to the right on the rising edge of a clock pulse applied to clock CK. If T/C is made 'low', all outputs give the complements, $\bar{Q}$ of the data in the register, i.e. 0 for 1 and vice versa. For parallel loading, R is held 'low' while T/C and P/S are made 'high'. The data are applied to inputs A, B, C and D and are stored at Q_A, Q_B, Q_C and Q_D respectively when a clock pulse is applied. If R is made 'high', all outputs go 'low' in both serial and parallel modes.

16.7 Counters

Counters are widely used in electronic systems to determine, for example, the number of objects passing on a conveyor belt or the number of operations performed in a digital computer. They consist of flip-flops (master-slave) connected so that they toggle when the pulses to be counted are applied to the clock input. Counting is done in binary code, the 'high' and 'low' states represent the bits 1 and 0 respectively.

There are two main types of counter—ripple and synchronous.

(a) Ripple counter

In Fig. 16.13(*a*) a three-bit binary ripple counter is shown using three JK flip-flops having J = K = 1 (for toggling) and the Q output of each flip-flop feeding the clock input CK of the next. To explain the action, suppose that the flip-flops are triggered on the falling edge of a pulse and that initially their outputs Q_0, Q_1 and Q_2 have all been reset to zero.

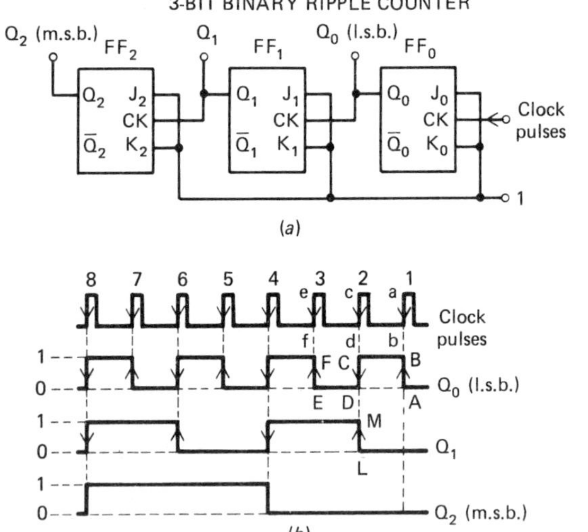

Number of clock pulse	Outputs		
	Q_2	Q_1	Q_0
0	0	0	0
1	0	0	1
2	0	1	0
3	0	1	1
4	1	0	0
5	1	0	1
6	1	1	0
7	1	1	1
8	0	0	0

(*c*)

Fig. 16.13

On the falling edge ab of the first clock pulse, shown in Fig. 16.13(*b*), Q_0 switches from 0 to 1. The resulting rising edge AB of Q_0 is applied to CK of FF_1, which does not change state because AB is not a falling edge. Hence $Q_2 = 0$, $Q_1 = 0$ and $Q_0 = 1$, giving a binary count of 001.

The falling edge cd of the second clock pulse makes FF_0 change state again and Q_0 goes from 1 to 0. The falling edge CD of Q_0 switches FF_1 this time, making $Q_1 = 1$. The rising edge LM of Q_1 leaves FF_2 unchanged. The count is now $Q_2 = 0$, $Q_1 = 1$ and $Q_0 = 0$, i.e. 010.

The falling edge ef of the third clock pulse to FF_0 changes Q_0 from 0 to 1 again but the rising edge EF does not switch FF_1, leaving $Q_1 = 1$, $Q_2 = 0$ and the count 011. The action thus ripples along the flip-flops, each one waiting for the previous flip-flop to supply a falling edge at its clock input before changing state.

Fig. 16.13(*c*) shows that the states of Q_0, Q_1 and Q_2 give the count in binary of the clock pulses, Q_0 being the l.s.b.

(b) Synchronous counter

In this type all flip-flops are clocked simultaneously and the *propagation delay time* (i.e. the time lapse between the clock pulse being applied and the output of the counter changing) is much less than for a ripple counter with a large number of flip-flops. Synchronous counters are therefore used for high-speed counting but their circuits are more complex than those of ripple types and will not be considered.

(c) Modulo

The modulo of a counter is the number of output states it goes through before resetting to zero. A counter with three flip-flops counts from 0 to $(2^3 - 1) = 8 - 1 = 7$; it has eight different output states representing the decimal numbers 0 to 7 and is a modulo-8 counter. A counter with *n* flip-flops counts from 0 to $2^n - 1$ and has 2^n states, i.e. it is a modulo-2^n counter.

(d) Up- and down-counters

In an up-counter like that in Fig. 16.13(*a*), the binary number represented by the output increases with each clock pulse. In a down-counter the binary number decreases by one for each clock pulse. To convert the up-counter in Fig. 16.13(*a*) into a down-counter, the $\overline{Q}$ output (instead of the Q) of each flip-flop is coupled to the CK input of the next, the count still being given by the Q outputs.

(e) Dividers

In a modulo-8 counter one output pulse appears at Q_2 for every eight clock pulses. That is, if *f* is the frequency of a regular train of clock pulses, that of the pulses from Q_2 is *f*/8. A modulo-8 counter is therefore a 'divide-by-8' circuit as well. Counters are used as dividers in digital watches.

(f) Four-bit binary counter ICs

The low power Schottky TTL 74LS93, Fig. 16.14(*a*), is a ripple up-counter, triggered on the falling edge of a clock pulse. FF_0 is a separate flip-flop with its own clock input $\overline{CK_0}$ and output Q_0; it is a modulo-2, divide-by-2 counter. FF_1, FF_2 and FF_3 form a divide-by-8, modulo-8 counter with clock input $\overline{CK_1}$ and three outputs Q_1, Q_2 and Q_3 (m.s.b.). (The lines above the clock inputs indicate that the IC is falling edge triggered.)

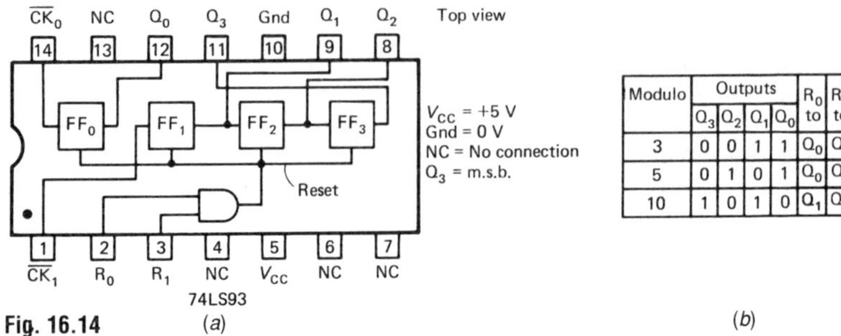

Modulo	Outputs				R_0 to	R_1 to
	Q_3	Q_2	Q_1	Q_0		
3	0	0	1	1	Q_0	Q_1
5	0	1	0	1	Q_0	Q_2
10	1	0	1	0	Q_1	Q_3

Fig. 16.14 (*a*) (*b*)

Both counters can be used together as a four-bit, modulo-16 counter if Q_0 is joined to $\overline{CK}_1$ and clock pulses are entered at $\overline{CK}_0$, at least one of reset inputs R_0 and R_1 being 'low'. Resetting to zero occurs when both R_0 and R_1 are 'high'. Other modulo counters are obtained if R_0 and R_1 are connected to different outputs, as shown in the table of Fig. 16.14(*b*). For example, linking R_0 to Q_1 and R_1 to Q_3 gives a modulo-10 counter since, when the count is 1010 (decimal 10), $Q_3 = 1$, $Q_2 = 0$, $Q_1 = 1$ and $Q_0 = 0$ and the output from the AND gate in the IC (fed by R_0 and R_1) goes 'high', so resetting the counter to zero.

The CMOS 4516B, Fig. 16.15(*a*), is a synchronous, up/down-counter, triggered on the rising edge of a clock pulse CK. The count increases by 1 if up/down (U/$\overline{D}$) is 'high' and decreases by 1 if it is 'low'. Normally reset R, carry in $\overline{CI}$ and preset enable PE are 'low'. If R = 1, the counter resets to zero. If $\overline{CI}$ = 1, the counter stops and this input can be used to inhibit or enable counting. If PE = 1, the counter reads the total count on the four preset inputs, P_0, P_1, P_2 and P_3 (m.s.b.). Fig. 16.15(*b*) summarizes the behaviour.

For counts above fifteen, two or more counters can be joined in ripple mode if carry out $\overline{CO}$ of the first feeds CK of the second and so on. Alternatively if $\overline{CO}$ of one is taken to $\overline{CI}$ of the next they are in synchronous mode.

The CMOS 4017B, Fig. 16.15(*c*), is a *decade* counter with ten decoded outputs (Q_0 to Q_9) and each goes 'high' in turn on the rising edge of successive clock pulses provided reset R and clock enable $\overline{CE}$ are both 'low'. When R = 1, the counter resets to zero and then Q_0 = 1 and all other outputs are 'low'. R must be returned to zero for counting to start again. Counting stops if $\overline{CE}$ = 1.

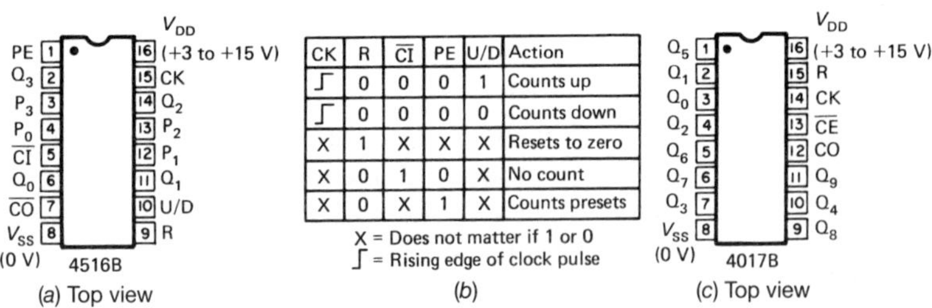

CK	R	$\overline{CI}$	PE	U/$\overline{D}$	Action
⎍	0	0	0	1	Counts up
⎍	0	0	0	0	Counts down
X	1	X	X	X	Resets to zero
X	0	1	0	X	No count
X	0	X	1	X	Counts presets

X = Does not matter if 1 or 0
⎍ = Rising edge of clock pulse

(*a*) Top view (*b*) (*c*) Top view

Fig. 16.15

16.8 Semiconductor memories

(a) Organization

The memory of an electronic system stores *data*, i.e. the information to be processed and the program of instructions to be carried out. An LSI semiconductor memory consists of an array of memory cells, each one storing one bit (0 or 1) of data. The array is organized so that the bits are in groups or 'words' of typically 1, 4, 8 or 16 bits.

Every word has its own location or *address* in the memory which is identified by a certain binary number. The first word is at address zero, the second at one, the third at two and so on. The table in Fig. 16.16 shows part of the contents of a sixteen-word four-bit memory, i.e. it has sixteen addresses each storing a four-bit word. For example in the location with address 1110 (decimal 14), the data stored is the four-bit word 0011 (decimal 3).

Address					Data				
Decimal	Binary m.s.b.			l.s.b.	Binary m.s.b.			l.s.b.	Decimal
0	0	0	0	0	0	1	0	1	5
1	0	0	0	1	1	1	0	0	12
2	0	0	1	0	0	1	1	0	6
3	0	0	1	1	1	0	0	1	9
⋮	⋮	⋮	⋮	⋮	⋮	⋮	⋮	⋮	⋮
14	1	1	1	0	0	0	1	1	3
15	1	1	1	1	0	1	1	1	7

Fig. 16.16

In a *random access memory* all words can be located equally quickly, i.e. access is random and it is not necessary to start at address zero. In a *sequential memory*, however, it is necessary to start at address zero and work through each address in turn until the desired one is reached.

(b) Types

There are two main types—read only memories, i.e. ROMs, and read and write memories which are confusingly called RAMs (because they allow random access, as ROMs also do). A RAM not only lets the data at any address be 'read' but it can also have new data 'written' into any address. Whereas a RAM normally loses the data stored almost as soon as the power to it is switched off, i.e. it is a *volatile* memory, a ROM does not, i.e. it is a *non-volatile* memory. ROMs are used for permanent storage of fixed data such as the program in a calculator or a computer.

RAMs are of two types—*static* and *dynamic*. In a static RAM (SRAM) each memory cell consists basically of a bistable whose contents are fixed until the cell is written into or the power is switched off. The memory cells in a dynamic RAM (DRAM) are tiny capacitors which become charged but due to leakage currents the charged cells have to be 'refreshed' regularly, e.g. every millisecond, from within the chip itself. A charged capacitor represents a 1 and an uncharged one a 0. Non-volatile DRAM ICs are now available with inbuilt low voltage backup batteries.

For high capacity memories, DRAMs cost less than SRAMs on account of their smaller size due to the use of MOSFETs and the need for only one transistor per memory cell compared with two for a SRAM. They also consume less power. Today single DRAM chips with capacities of 64 megabytes (see below) can be bought and within a few years this figure is likely to be in the gigabyte range.

(c) Structure

The structure of a sixteen-word four-bit RAM is shown by the block diagram of Fig. 16.17; that for a ROM is similar but it has no 'write' provision.

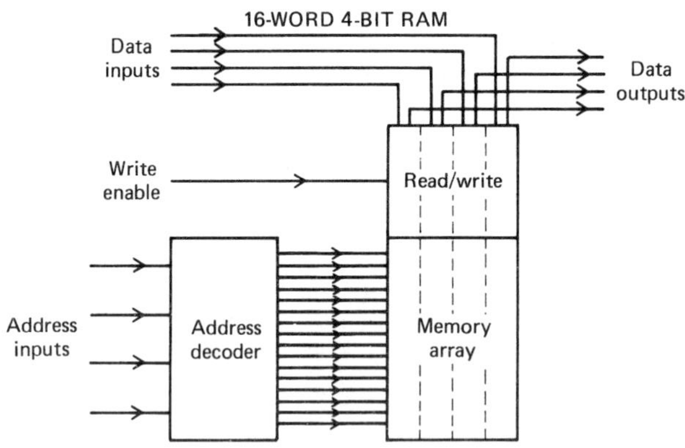

Fig. 16.17

To write a word into a particular address, the four-bit binary number of the address is applied to the address inputs and the word (also in binary) to the data inputs. When write enable is at the appropriate logic level (say 'high'), the word is stored at the correct address in the memory array, as located by the address decoder.

To read a word stored at a certain address, the address code is applied as before and the word appears at the data outputs if write enable is at the required logic level (say 'low').

(d) Storage capacity

A sixteen-word four-bit memory has a storage capacity of 16 × 4 or 64 bits (it has 64 memory cells) and is limited to four-bit words (i.e. the decimal numbers 0 to 15). To store the letters of the alphabet (in a binary code) as well as numbers, a memory for eight-bit words is required.

An eight-bit word is called a *byte*. In computer language the symbol K (capital K) is used to represent 1 024 (2^{10}). For example, a memory with a capacity of 4K bits stores 4 × 1 024 = 4 096 bits, i.e. 512 words if it is organized in bytes or 1 024 words for a four-bit organization. Do not confuse K with k (small k) standing for kilo-, i.e. 1 000.

A memory with four address inputs has 2^4 or 16 locations (since four bits give any of the sixteen decimal numbers from 0 to 15). One with eight address inputs has 2^8 or 256 locations.

(e) Sixteen-word four-bit RAM ICs

The pin connection for the low power Schottky TTL 74LS89 and the CMOS 40114B are given in Figs. 16.18(*a*) and (*b*); they are identical. The l.s.b. of the address goes to address input A and of the data to data input DI_0. $\overline{DO}_0$ gives the l.s.b. of the *complement* of the data output; if the *true* data output is required, the data input must be inverted by an inverter in the data input line before it is written into the memory.

To write data in, memory enable, $\overline{ME}$, and write enable, $\overline{WE}$, must both be 'low'. To read data, $\overline{ME}$ must be 'low' and $\overline{WE}$ 'high'.

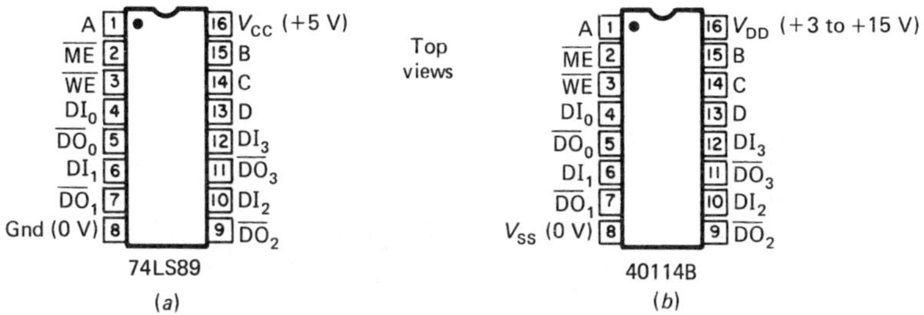

Fig. 16.18

(f) PROMs and EPROMs

The programmable ROM or PROM lets the user 'burn' the pattern of bits, i.e. the program, into a ROM by applying a high voltage which fuses a link in the circuit. The disadvantage of this type is that it does not allow changes or corrections to be made later. When it is necessary to make alterations, for instance during the development of a program, an erasable PROM, or EPROM, is used in which the program is stored electrically and is erased by exposure to ultraviolet radiation before reprogramming. An EEPROM is an electrically erasable PROM.

16.9 Memory circuits

The operation of the address decoder and of one of the one-bit read/write memory cells in a sixteen-word four-bit RAM will now be outlined.

(a) Address decoder

The job of the address decoder is to ensure that a word is written into or read out of the correct address. If does this by converting the binary number of each address into a different decimal number.

The circuit is the same as that described in Fig. 15.5, and part of it is shown again in Fig. 16.19 connected to a simplified (and incomplete) block diagram of the memory array of sixteen 'rows' and four 'columns' of one-bit memory cells MC (of which there should be sixty-four). Each decimal output from the decoder goes to the address line of one row of four memory cells and when that decimal output goes 'high', it enables all the cells in that row to read or write.

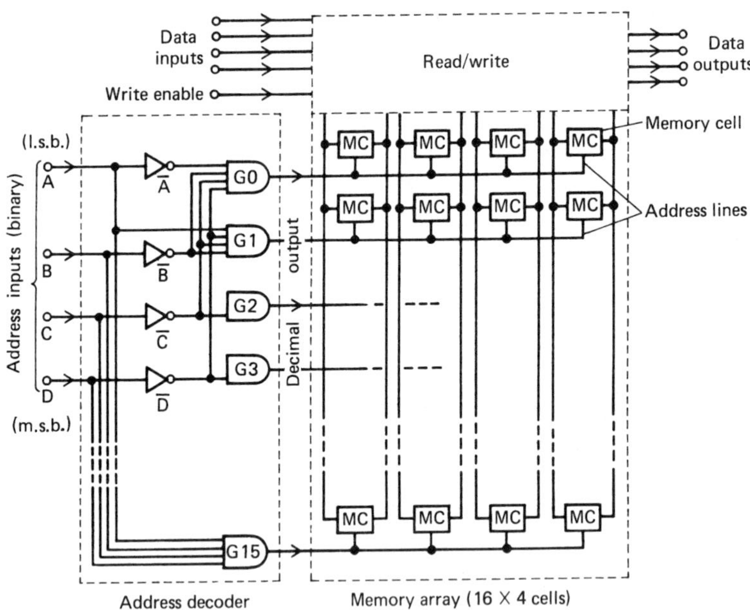

Fig. 16.19

Suppose we wish to locate address zero. The decoder has to produce a 1 at the output of AND gate G0 (and no other) when the binary code for address zero, i.e. 0000, is applied to the address inputs A, B, C and D. G0 only does this if all its inputs are 'high' and so we have to invert the four address inputs to obtain $\overline{A}$, $\overline{B}$, $\overline{C}$ and $\overline{D}$, i.e. 1111, and apply these to G0.

For address fifteen, the binary code is 1111 and in this case A, B, C and D can be connected directly to AND gate G15, as shown in Fig. 16.19. Other gates must receive the correct combination of A or $\overline{A}$, B or $\overline{B}$, C or $\overline{C}$, D or $\overline{D}$ depending on the address code.

In practice, the setting up of an address in, for example, a computer, is done automatically by appropriate circuits when a switch on a keyboard is pressed.

(b) One-bit read/write bistable memory cell

Its action can be explained using the circuit of Fig. 16.20 which contains an SR flip-flop. To write in or read out data, the address input (from the decoder) must be 'high'. To write, write enable must also be 'high'; if the data input is a 1 or a 0, then S = 1 or 0 (and R = 0 or 1). Therefore Q = 1 or 0 and the data output is 1 or 0, i.e. what is written in. If write enable is 'low', S = R = 0, Q (and $\overline{Q}$) are unaffected by the data input and the memory reads, i.e. data output is 1 or 0 if Q = 1 or 0.

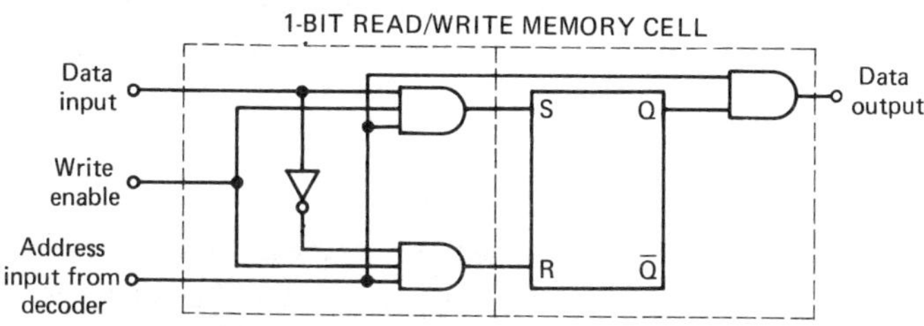

Fig. 16.20

In a sixteen-word four-bit memory, depending on whether we wish to write or read, the data input or output for say the m.s.b., is connected to the data input or output of all the cells in the column that handles the m.s.b. (Fig. 16.19). However, only the cell in the row whose address line is 'high' responds, as do the other three cells in the same row dealing with the other three bits of the four-bit word.

16.10 Revision questions and Problems

1. Explain the terms
   a) sequential logic,
   b) clocked logic,
   c) level triggering,
   d) edge triggering,
   e) synchronous system.

2. Draw a circuit for a clocked SR flip-flop. State what it does when clocked if
   a) S = R = 0;
   b) S = 1, R = 0;
   c) S = 0, R = 1;
   d) S = R = 1.
   Why is d) not allowed?

3. A 100 kHz square wave drives one T-type flip-flop. What is the frequency of the output?

4. a) Draw a circuit for a clocked D-type flip-flop. What does it do? Why is it also called a data latch?
   b) How is a D-type flip-flop converted to a T-type flip-flop?
   c) How can a numerical display be prevented from flickering when it is changing rapidly?

5. Draw a circuit for a clocked JK flip-flop. State what it does when clocked if
a) J = K = 0; c) J = 0, K = 1;
b) J = 1, K = 0; d) J = K = 1.

6. What is the 'race' problem? Explain how it is solved in a master-slave flip-flop.

7. a) What does a shift register do?
b) Draw block diagrams for a four-bit shift register which is
(i) serial loaded,
(ii) parallel loaded.
c) How many clock pulses are needed to shift an eight-bit binary number into an eight flip-flop serial shift register?

8. Draw the block diagram for a binary ripple up-counter containing four D-type flip-flops. How would you modify it to be a down-counter?

9. What advantage does a synchronous counter have over a ripple counter?

10. a) How many output states are there in a counter which has
(i) 1, (ii) 2, (iii) 3, (iv) 4, (v) 5 flip-flops?
b) What is the highest decimal number each counter in a) can count before resetting?

11. a) What is meant by the modulo of a counter?
b) How many flip-flops are needed to build counters of modulo-
(i) 2, (ii) 5, (iii) 7, (iv) 10, (v) 30?
c) If f is the frequency of the clock pulses applied to a modulo-10 counter, what is the frequency of the output pulses from the last flip-flop?

12. The block diagram of a modulo-16 binary up-counter is shown in Fig. 16.21; it has a reset input R which resets the counter to zero (i.e. $Q_0 = Q_1 = Q_2 = Q_3 = 0$) when it goes 'high'. Copy and modify the circuit by adding an AND gate so that it becomes a modulo-10 counter. (*Hint*—see Fig. 16.14(*a*).)

13. a) Explain the following terms as they are used in connection with a memory: data, bit, word, address, write, read, random access, volatile, byte.
b) Distinguish between
(i) a ROM and a RAM,
(ii) a PROM and an EPROM.
c) What is the storage capacity of a 64-word eight-bit memory in
(i) bits,
(ii) bytes?
d) How many locations does a memory with five address inputs have?

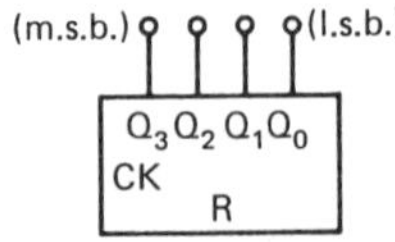

Fig. 16.21

14. Draw the block diagram for a sixteen-word four-bit RAM. State what each 'block' does.

UNIT 17

More digital circuits

The first three sections in this Unit are devoted to digital ICs in which op amps, operating in one of the ways described in Unit 13, are essential parts of the more complex circuits of a timer, a digital-to-analogue converter and an analogue-to-digital converter. Section 17.4 considers the use of the op amp in a digital role as a multivibrator and section 17.5 outlines its action as a waveform generator. Schmitt trigger ICs are treated in section 17.6.

The Unit concludes the subject of digital ICs with sections on some further properties of TTL and CMOS, followed by a short treatment of the basic gate circuits used to build them.

17.1 Timers

Astable and monostable multivibrators are used for timing operations in many electronic systems. The popular eight-pin 555 timer IC can act either as a 'square' wave oscillator (i.e. an astable) or as a single pulse generator (i.e. a monostable). It works on any d.c. supply from 3 to 15 V and can source or sink up to 200 mA at its output. Basically it consists of an SR flip-flop (see section 16.2) whose set and reset are each controlled by an op amp voltage comparator (see section 13.8).

(a) Astable operation

The block diagram and pin connections are shown in Fig. 17.01; R_1, R_2, C_1 and C_2 are external components. (Note that the circle is omitted from the transistor symbol in an IC.) Threshold (pin 6) is joined to trigger (pin 2). Initially C_1 charges up through R_1 and R_2 and, when the voltage across it just exceeds $\frac{2}{3}V_{CC}$, the output from the threshold comparator (with a reference voltage of $\frac{2}{3}V_{CC}$ on its other input from the voltage divider chain formed by the three equal resistors R in series across V_{CC}) goes 'high' and resets the flip-flop, i.e. $\overline{Q}$ goes 'high'. This has two results. First, the output from the IC (pin 3) goes 'low' (due to the inverting buffer output stage) and second, Tr_1 switches on (since its base is now positive), allowing C_1 to discharge through it and R_2.

When the voltage across C_1 has fallen to just below $\frac{1}{3}V_{CC}$, the output from the trigger comparator (with a reference voltage of $\frac{1}{3}V_{CC}$ at its other input from the three-resistor chain) goes 'high' and sets the flip-flop. $\overline{Q}$ therefore goes 'low', with two results. First, the output from the IC goes 'high' and second, Tr_1 turns off (since its base is no longer positive) so letting C_1 charge up to $\frac{2}{3}V_{CC}$ again through

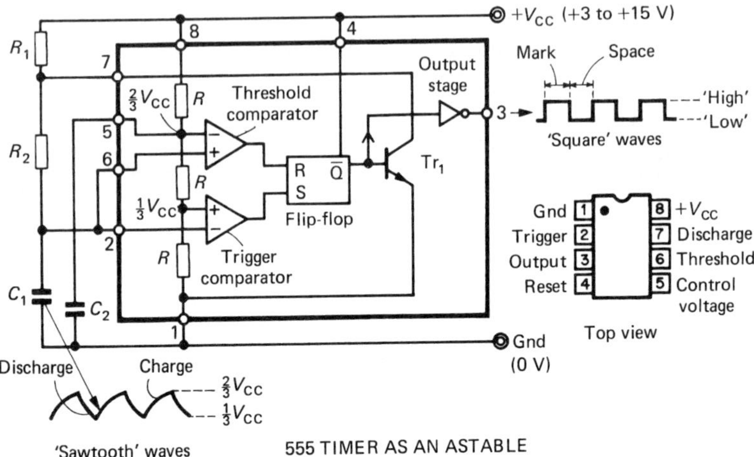

Fig. 17.01

R_1 and R_2, as it did at the start. This cycle is repeated continuously giving an oscillatory output with a rectangular waveform which is 'high' while C_1 is charging and 'low' while it discharges.

(b) Astable properties

The frequency f of the oscillations can be shown mathematically to be given by:

$$f = \frac{1.44}{(R_1 + 2R_2)C_1} \text{ Hz}$$

R_1 and R_2 are in Ω and C_1 is in F. Alternatively it may be more convenient to use MΩ and μF. The maximum operating frequency is about 1 MHz, the minimum being limited by the value and leakage of C_1. The graphs in Fig. 17.02 show how R_1, R_2 and C_1 affect f: R_1 and R_2 should not be less than 1 kΩ.

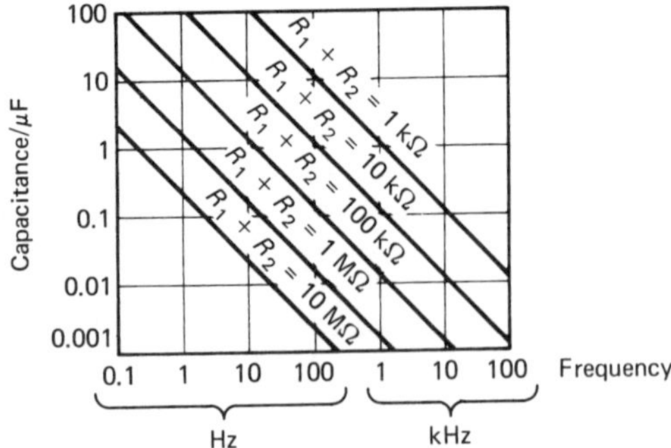

Fig. 17.02

The mark–space ratio of the output waveform is one, i.e. the waveform is square, if R_2 is large compared with R_1. A range of ratios is possible by suitably selecting R_1/R_2 since in general the output is 'high' for longer than it is 'low', owing to C_1 charging through both R_1 and R_2 but discharging only through R_2.

Control voltage (pin 5) normally goes to 0 V via a 0.01 μF capacitor (C_2) but if a voltage of between 45% and 90% of V_{CC} is applied to it, the value of f can be varied independently of R_1, R_2 and C_1, i.e. frequency modulation occurs. For example, suppose the control voltage is $\frac{1}{2}V_{CC}$. The reference voltage on the threshold comparator will now be $\frac{1}{2}V_{CC}$ (instead of $\frac{2}{3}V_{CC}$) and so C_1 need only charge up to $\frac{1}{2}V_{CC}$ from $\frac{1}{3}V_{CC}$. Similarly when C_1 discharges through R_2, the voltage across it has only to fall from $\frac{1}{2}V_{CC}$ to $\frac{1}{3}V_{CC}$. Both these actions take less time and so f increases.

Reset (pin 4) is the 'direct reset' for the flip-flop and should be connected to $+V_{CC}$; if its voltage falls below about 0.7 V, the astable stops.

(c) Monostable operation

The connections are given in Fig. 17.03. In this case only one external resistor, R_1, is required and threshold (pin 6) is joined to discharge (pin 7). One rectangular output pulse is produced when the circuit is triggered by the falling (negative-going) edge of an external pulse applied to trigger (pin 2). It then returns to its one stable state ('low' output) to await the next trigger pulse. The trigger pulse can be obtained for example, by switching S_1 from X (i.e. $+V_{CC}$) to Y (i.e. 0 V) and back to X again, the operation being completed in a time which must be less than the output pulse time.

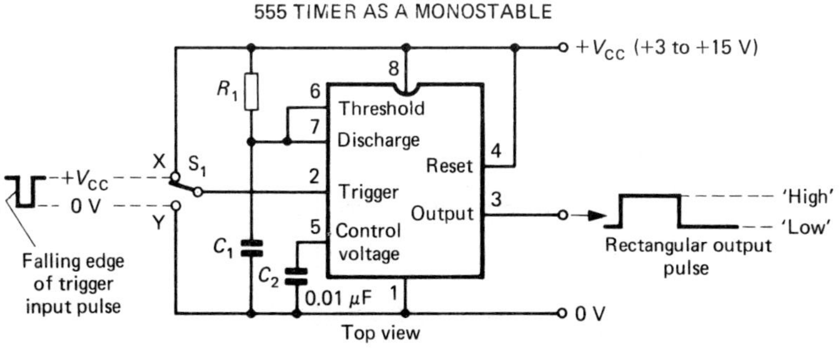

Fig. 17.03

Try to work out how the circuit works by drawing a diagram like that in Fig. 17.01 (but with the monostable connections) and start from the fact that when the trigger voltage drops to $\frac{1}{3}V_{CC}$ (on its fall from V_{CC}), the output of the trigger comparator goes 'high'.

The time T of the output pulse can be shown to be given by:

$$T = 1.1R_1 \times C_1$$

For example, if $R_1 = 3.3\ \text{M}\Omega$ and $C_1 = 1\,000\ \mu\text{F}$, then $T = 3\,630\ \text{s} \approx 1$ hour, which is about the maximum reliable pulse time.

17.2 Digital-to-analogue converter

There are many occasions when digital signals have to be converted to analogue ones. For example, a digital computer is often required to produce a graphical display on the screen of a cathode ray tube (CRT). This involves using a digital-to-analogue (D/A) converter to change the two-level digital output (the 1s and the 0s) from the computer, into a continuously varying analogue voltage for the input to the CRT, so that it can deflect the electron beam and make it 'draw pictures' at high speed.

(a) D/A conversion

The principle of the *binary weighted resistor* method is shown in Fig. 17.04 for a four-bit input. It is so called because the values of the resistors, R, $2R$, $4R$ and $8R$ in this case, increase according to the binary scale, the l.s.b. of the digital input having the largest value resistor. The circuit uses an op amp as a *summing amplifier* (see section 13.5) with a feedback resistor R_f. S_1, S_2, S_3 and S_4 are digitally controlled electronic SPDT switches.

Each switch connects the resistor in series with it to a fixed reference voltage V_{REF} when the input bit controlling it is a 1, and to ground (0 V) when it is a 0. The input voltages V_1, V_2, V_3 and V_4 applied to the op amp by the four-bit input (via the resistors) therefore have one of two values , either V_{REF} or 0 V.

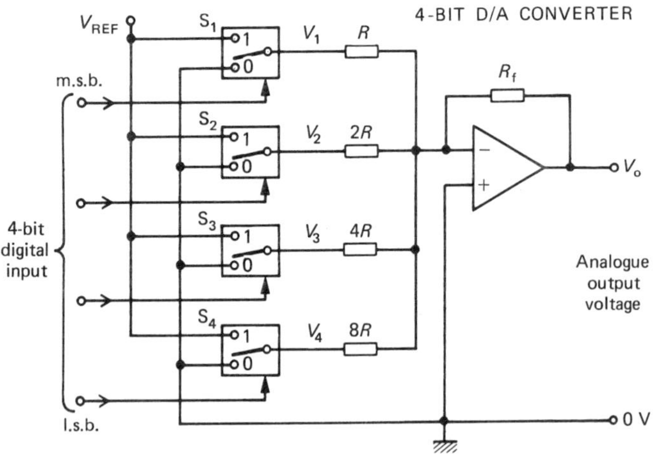

Fig. 17.04

Using the formula derived in section 13.5, the analogue output voltage V_o from the op amp is:

$$V_o = -\left(\frac{R_f}{R} \times V_1 + \frac{R_f}{2R} \times V_2 + \frac{R_f}{4R} \times V_3 + \frac{R_f}{8R} \times V_4\right)$$

If $R_f = 1\ \text{k}\Omega = R$, then: $V_o = -(V_1 + \frac{1}{2}V_2 + \frac{1}{4}V_3 + \frac{1}{8}V_4)$

With a four-bit input of 0001 (decimal 1), S_4 connects 8R to V_{REF}, making $V_4 = V_{REF}$; S_1, S_2 and S_3 connect $2R$, $4R$ and $8R$ respectively to 0 V, making $V_1 = V_2 = V_3 = 0$. And so, if $V_{REF} = -8$ V, we get:

$$V_o = -\left(0 + 0 + 0 + \frac{(-8)}{8}\right) = -\frac{(-8)}{8} = +1\text{ V}$$

With a four-bit input of 0110 (decimal 6), S_2 and S_3 connect $2R$ and $4R$ to V_{REF}, making $V_2 = V_3 = V_{REF} = -8$ V; S_1 and S_4 connect R and $8R$ to 0 V, making $V_1 = V_4 = 0$. Therefore:

$$V_o = -\left(0 + \frac{(-8)}{2} + \frac{(-8)}{4} + 0\right) = -(-4 - 2) = +6\text{ V}$$

From these two examples we see that the analogue output voltage V_o is directly proportional to the digital input. V_o has a 'stepped' waveform with a 'shape' that depends on the binary input pattern, Fig. 17.05. The waveform is not 'smooth' but it does vary.

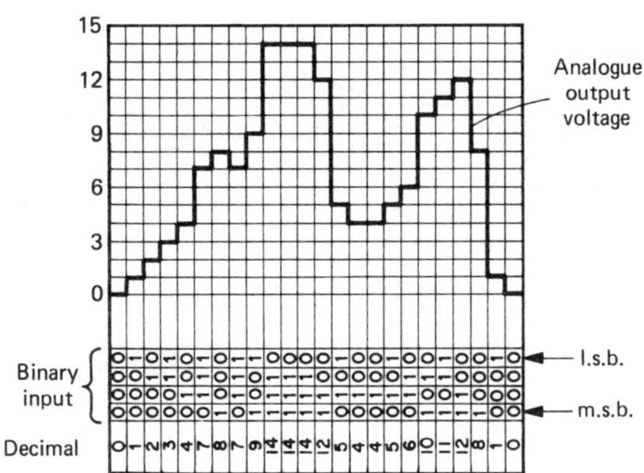

Fig. 17.05

(b) Eight-bit D/A converter IC

The pin connections for the ZN425E are shown in Fig. 17.06. It operates on a voltage ($+V_{CC}$) in the range +4.5 V to +5.5 V and produces a reference voltage of +2.55 V, available at pin 16 (V_{REF} out), which it uses itself if pin 16 is connected to pin 15 (V_{REF} in). To stabilize the reference voltage, a capacitor (e.g. 0.22 μF) is connected between pins 16 and 1. Logic select (pin 2) is held 'low' when the bit inputs (pins 5, 6, 7, 9, 10, 11, 12 and 13) are driven externally. The bit inputs (5.5 V max) can be supplied from either TTL or CMOS.

For maximum accuracy the analogue output voltage (pin 14) should be fed out via a buffer amplifier.

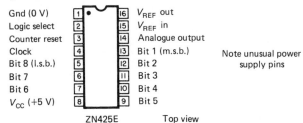

Function	Pin	Pin	Function
Gnd (0 V)	1	16	V_{REF} out
Logic select	2	15	V_{REF} in
Counter reset	3	14	Analogue output
Clock	4	13	Bit 1 (m.s.b.)
Bit 8 (l.s.b.)	5	12	Bit 2
Bit 7	6	11	Bit 3
Bit 6	7	10	Bit 4
V_{CC} (+5 V)	8	9	Bit 5

ZN425E   Top view

Note unusual power supply pins

Fig. 17.06

17.3 Analogue-to-digital converter

In many modern measuring instruments the reading is frequently displayed digitally, but the input, from, for example, a transducer measuring temperature, is in analogue form. An analogue-to-digital (A/D) converter is then needed.

(a) A/D conversion

The block diagram for a four-bit 'counter' type A/D conversion circuit is shown in Fig. 17.07(*a*), with waveforms to help you to follow the action. An op amp is again used but in this case as a voltage comparator. The analogue input voltage V_2 (shown here as a steady d.c. voltage) is applied to the non-inverting (+) input; the inverting (−) input is supplied by a ramp generator with a repeating sawtooth waveform voltage V_1, Fig. 17.07(*b*).

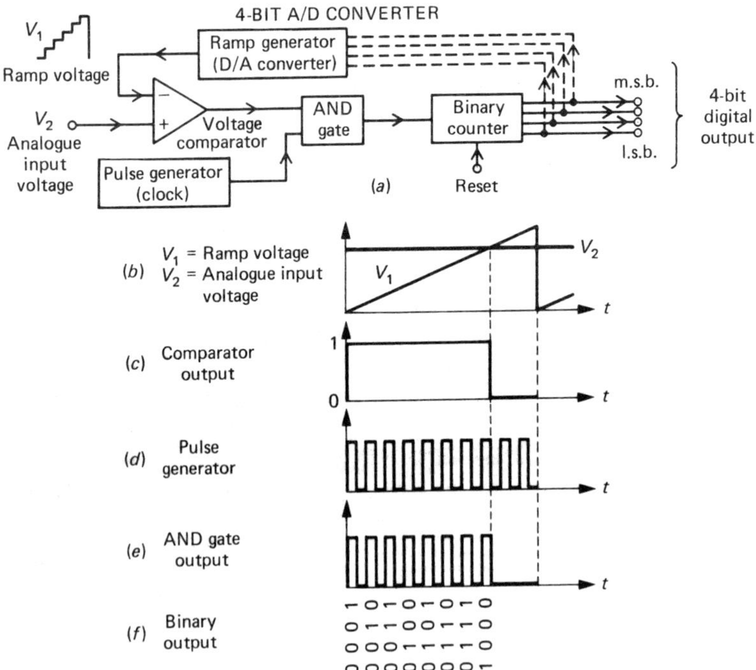

Fig. 17.07

The output from the comparator is applied to one input of an AND gate and is 'high' (a 1) until V_1 equals (or exceeds) V_2, when it goes 'low' (a 0), as in Fig. 17.07(*c*). The other input of the AND gate is fed by a steady train of pulses from a pulse generator (a 'clock'), as shown in Fig. 17.07(*d*). When both these inputs are 'high', the gate opens and gives a 'high' output, i.e. a pulse. From Fig. 17.07(*e*), you can see that the number of pulses so obtained from the AND gate depends on the 'length' of the comparator output pulse, i.e. on the time taken by V_1 to reach V_2. This time is proportional to the analogue voltage, if the ramp is linear. The output pulses from the AND gate are recorded by a binary counter and, as shown in Fig. 17.07(*f*), are the digital equivalent of the analogue input voltage V_2.

In practice the ramp generator is a D/A converter which takes its digital input from the binary counter, shown by the dashed lines in Fig. 17.07(*a*). As the counter advances through its normal binary sequence, a staircase waveform with equal steps (i.e. a ramp) is built up at the output of the D/A converter, like that shown by the first four 'steps' in Fig. 17.05. Similar conversion techniques are used in digital voltmeters—see section 21.7.

(b) Eight-bit A/D converter IC

The ZN425E (Fig. 17.06) may also be used for analogue-to-digital conversion by adding an external voltage comparator and a gate control for the clock input (pin 4) and holding logic select (pin 2) 'high'. The analogue output V_1 from the ZN425E (pin 14) is applied to the inverting input of the comparator, Fig. 17.08, and the analogue input voltage V_2 to be converted is applied to the non-inverting input. V_1 is a rising staircase voltage (due to the action of the internal binary counter) and the 'ramping' continues until V_1 equals V_2. At this point any further clock pulses to the ZN425E are stopped by the gate control and the eight-bit output gives the converted digital equivalent of V_2.

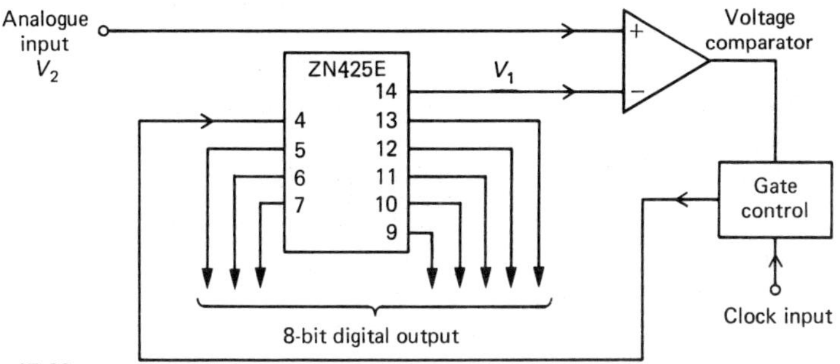

Fig. 17.08

17.4 Op amp as a multivibrator

An op amp voltage comparator in its saturated condition can operate as any of the three types of multivibrator if suitable external components are connected and positive feedback used.

(a) Bistable operation

The circuit is shown in Fig. 17.09(*a*). The inverting input V_1 is the voltage across R_1. A certain fraction, β, of the output V_o is fed back to the non-inverting input (i.e. the feedback is positive) and is given by $V_2 = +\beta V_o$.

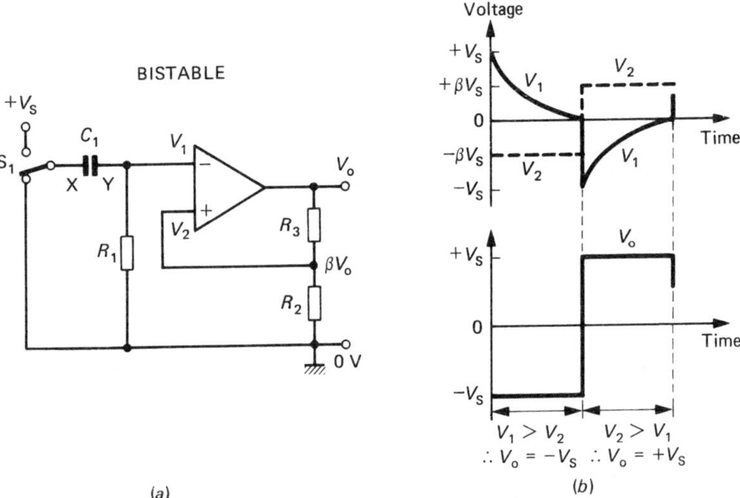

Fig. 17.09

Suppose S_1 is switched to $+V_S$. V_1 rises at once to $+V_S$ (since C_1 has not had time to charge and no voltage can exist across an uncharged capacitor) and so V_o goes negative. V_2 also goes negative but not to the same extent because $V_2 = +\beta V_o$, where $\beta = R_2/(R_2 + R_3)$ which is less than 1. V_1 is therefore greater than V_2 and the difference is enough to drive the op amp rapidly into negative saturation, i.e. V_o is almost equal to $-V_S$, since $V_o = A_o(V_2 - V_1)$ as we saw in section 13.8. This stable state is maintained even though V_1 falls exponentially to zero (see section 5.6) as C_1 charges up through R_1 until the voltage across it is $+V_S$. (V_1, now 0 V, is still greater than V_2, which is negative.)

To switch to the other stable state, S_1 is connected to 0 V, causing V_1 to fall suddenly to $-V_S$. (This happens because C_1 was fully charged with voltage V_S across its plates, therefore if plate X is connected to 0 V, Y must drop to $-V_S$ since C_1 has not had time to discharge.) Hence V_o goes positive (owing to V_1 going negative), as does $V_2(= +\beta V_o)$ and so V_2 is now greater than V_1 and the op amp goes into positive saturation with V_o almost equal to $+V_S$.

The graphs in Fig. 17.09(*b*) show how V_1, V_2 and V_o vary during one cycle, i.e. when S_1 is switched from 0 V to $+V_S$ to 0 V.

(b) Astable operation

In this case, switching occurs automatically. The circuit is shown in Fig. 17.10(*a*) and various voltage waveforms are shown in Fig. 17.10(*b*).

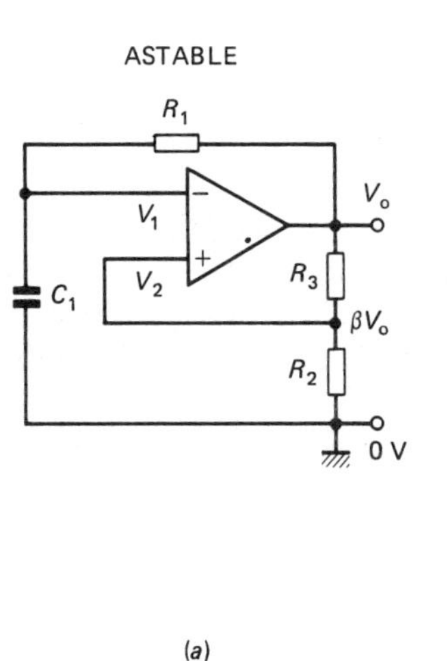

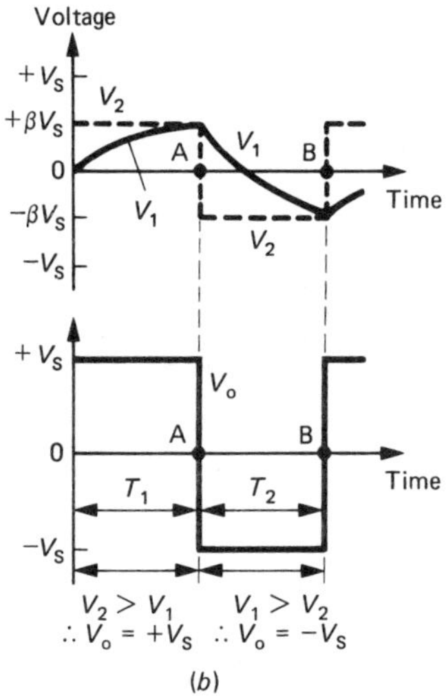

Fig. 17.10

Suppose the output voltage V_o is positive at a particular time. As in the bistable circuit, a certain fraction, β, of V_o, is fed back as the non-inverting input voltage V_2 which equals βV_o where $\beta = R_2/(R_2 + R_3)$. V_o is also fed back via R_1 to the inverting terminal and V_1 rises (exponentially) as C_1 is charged. After a time which depends on the time constant $C_1 \times R_1$, V_1 exceeds V_2 sufficiently for the op amp to switch into negative saturation, i.e. $V_o = -V_S$ (point A on graphs). Also, the positive feedback makes V_2 go negative (where $V_2 = -\beta V_o = -\beta V_S$).

C_1 now starts to charge up in the opposite direction making V_1 fall rapidly (during time AB on graphs) and eventually become more negative than V_2. The op amp therefore switches again to its positive saturated state with $V_o = +V_S$ (point B on graphs). This action continues indefinitely, the periodic time T given by:

$$T = T_1 + T_2 = 2.3C_1R_1 \log_{10}(1 + 2R_2/R_3)$$

(c) Monostable operation

In the circuit of Fig. 17.11(*a*), when S_1 is pressed and released, one positive pulse is produced whose duration depends on the time constant $C_1 \times R_1$.

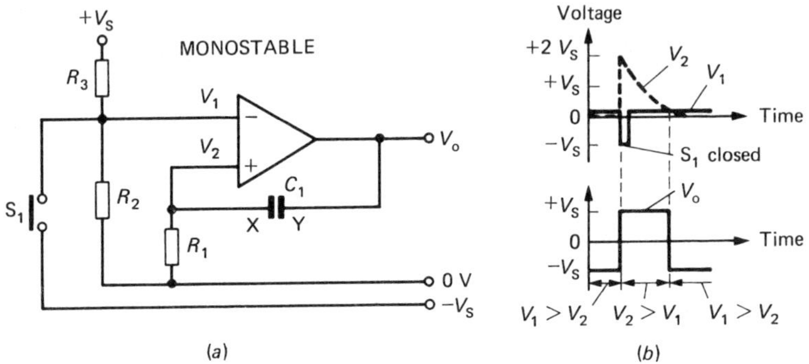

Fig. 17.11

With S_1 open, the + input is grounded via R_1, i.e. $V_2 = 0$ V and the − input is positive owing to the small voltage at the junction of the voltage divider formed by R_2 and R_3. Therefore V_1 is greater than V_2 and the op amp is negatively saturated, i.e. $V_o = -V_S$. Also C_1 is charged with plate X positive relative to plate Y. (More exactly, X is at 0 V and Y at $-V_S$ approx.)

When S_1 is closed briefly, the − input is connected momentarily to $-V_S$, i.e. $V_1 = -V_S$ and so V_2 is greater than V_1. The op amp becomes positively saturated, i.e. $V_o = +V_S$. This makes Y rise to $+V_S$ and X to near $+2V_S$, i.e. $V_2 \approx +2V_S$, thereby holding the op amp in positive saturation even when S_1 opens again. C_1 starts to discharge through R_1, V_2 gradually falls to zero and when it is less than V_1, the op amp reverts to negative saturation again. Fig. 17.11(*b*) shows waveforms for V_1, V_2 and V_o.

17.5 Waveform generators

Waveform generators produce square, triangular, sine, sawtooth and pulse waveforms of high accuracy. The frequency, which is stable over a wide range of temperature and supply voltage, can be set externally in the range 0.001 Hz to 1 MHz.

(a) Principles of operation

Square and triangular waveforms are obtained by using an op amp as an astable, as explained in section 17.4(b). Referring to the circuit and graphs in Fig. 17.10, the output voltage V_o provides square waves and the voltage V_1 across C_1 is the source of triangular waves. The latter have 'exponential' sides, but they are 'straightened' by extra circuitry which ensures that C_1 is charged by a constant current, rather than by the exponential one supplied through R_1.

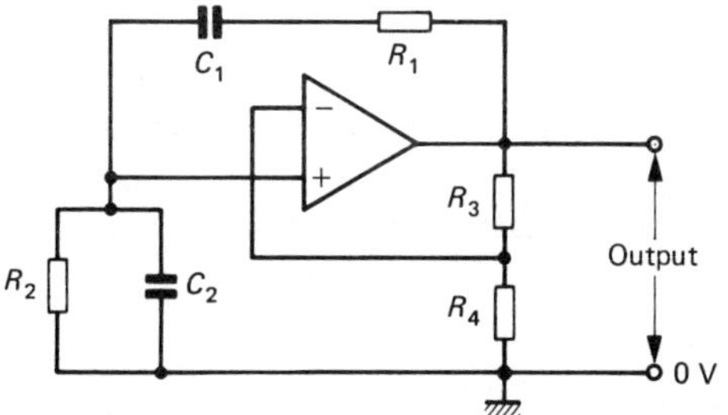

Fig. 17.12

Sine waves can be generated by an op amp using a Wien network circuit similar to that in Fig. 12.17 (see section 12.7) for a discrete-component *RC* oscillator. An op amp circuit is given in Fig. 17.12. The frequency-selective network R_1C_1, R_2C_2, applies positive feedback to the non-inverting (+) input. Negative feedback is supplied via R_3 and R_4 to the inverting (−) input. As before, the frequency f is given by $f = 1/(2\pi RC)$ where $R = R_1 = R_2$ and $C = C_1 = C_2$.

(b) Waveform generator IC

The pin connections for the 8038 are given in Fig. 17.13(*a*) and a circuit in which it is used as an a.f. oscillator is shown in Fig. 17.13(*b*). The mark–space ratio may be changed by adjusting the 'duty cycle' variable resistor, thus enabling sawtooth and pulse outputs to be obtained. (The *duty cycle* equals the ratio of the time the mark lasts to the total time the mark and space last.) To prevent overloading, the outputs should not be connected to a load less than 10 kΩ without the use of a buffer.

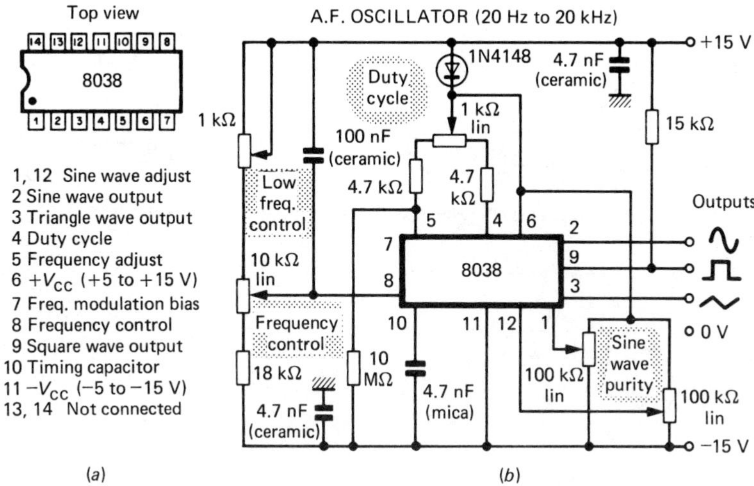

Fig. 17.13

17.6 Schmitt triggers

Reliable operation of digital circuits requires input waveforms to rise and fall rapidly. This ensures that transistors spend only a very short time in the linear region between saturation and cut-off, where they are high-gain amplifiers and liable to pick up unwanted signals. The Schmitt trigger, considered in section 14.11 in discrete-component form, is useful in cases where trouble could arise, as we will see in (b) below.

(a) Schmitt trigger ICs

Two popular IC versions are the low power Schottky TTL 74LS132 and the CMOS 4093B. Both are quad two-input NAND gates, as shown in Fig. 17.14(*a*). (Note the Schmitt trigger sign.) These can be used as normal logic gates but they have very rapid switching actions. As with any NAND gate, if one or all inputs are 'low', the output is 'high' and if all inputs are 'high' the output is 'low'. Their Schmitt trigger properties for a 5 V supply are given in Fig. 17.14(*b*).

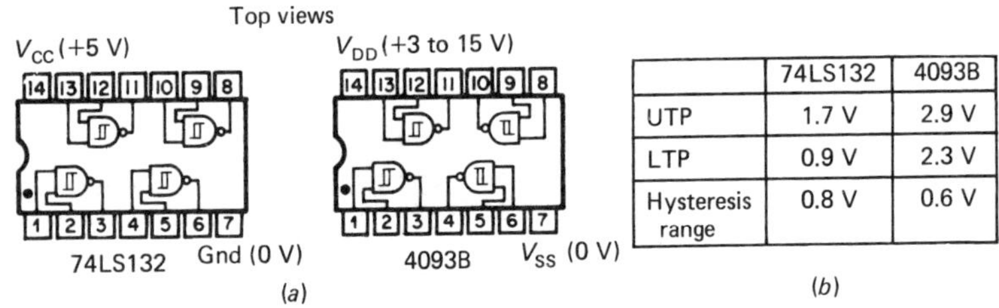

	74LS132	4093B
UTP	1.7 V	2.9 V
LTP	0.9 V	2.3 V
Hysteresis range	0.8 V	0.6 V

(*b*)

Fig. 17.14

(b) Uses

Some uses of Schmitt trigger ICs are shown in Fig. 17.15 where, in all cases, both inputs are joined and the gate behaves as an inverter.

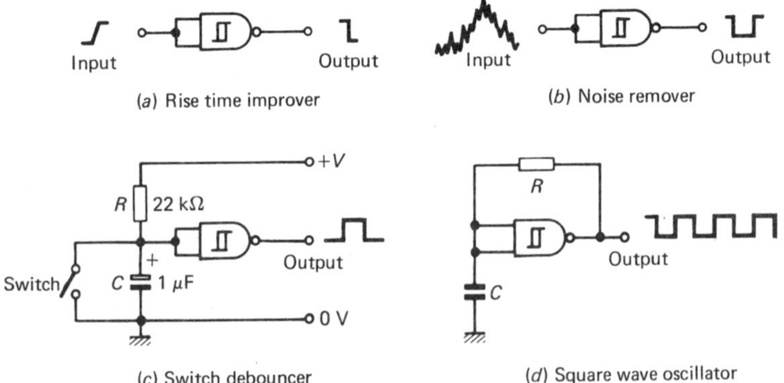

Fig. 17.15

In Fig. 17.15(*a*) the trigger is a *rise time improver* which converts a slowly changing input into one with a fast rise time. It would also turn a sine wave into a square wave.

In Fig. 17.15(*b*) we have a '*noise*' *remover*. Unwanted, high frequency spikes on the input signal (due, for example, to sparking at the contacts of an electrical machine in a factory) are eliminated.

In Fig. 17.15(*c*) it is a *switch debouncer* which, due to the discharging action of *C*, prevents contact bounce at the switch—see section 15.6.

In Fig. 17.15(*d*) it is a *square wave oscillator*, i.e. an astable producing a continuous train of pulses with fast rise and fall times suitable for use as a clock in a digital circuit. Oscillations occur because *C* starts to charge up at switch-on and when its voltage reaches the upper trip point (UTP), the output of the Schmitt trigger rapidly goes 'low' (since its input is 'high'). *C* then starts to discharge through *R* and when its voltage drops to the lower trip point (LTP), the Schmitt output quickly switches to 'high', allowing *C* to recharge to the UTP and so on. The process is repeated again and again and 'good' square waves are produced. If $R = 390\ \Omega$, the output frequency f is given by:

$$f \approx \frac{2\,000}{C}\ \text{Hz}$$

C is in μF.

17.7 Noise immunity and interfacing

(a) Noise immunity

This refers to the maximum unwanted stray voltage an IC can tolerate without a false change of output state. It depends on the difference between the 'high' and 'low' logic levels. The greater the difference the better the noise immunity.

Suppose that the 'low' output of an IC is 0.4 V and that 0.5 V of 'noise' is picked up (due to stray electric or magnetic fields) on the connection between this IC and another it is driving. The input to the second IC, instead of being 0.4 V will be 0.9 V. If the system is TTL, you can see from the logic level diagram of Fig. 17.16 that, being greater than 0.8 V, this would not qualify as a 'low' input. The second IC could therefore be on the point of wrongly changing its output state. This would not happen with CMOS ICs since, on a 5.0 V supply, switching does not occur until 1.5 V.

False triggering may occur similarly in the 'high' state, again more so with TTL than with CMOS. In fact CMOS can tolerate voltage fluctuations ('noise') up to about 40% of the supply voltage; its noise immunity is very good, much better than that of TTL. The fact that CMOS ICs can be operated from unstabilized power supplies is a direct result of this.

(b) Interfacing

The connection of ICs of one logic family to those of another (to obtain the advantages of both) or to an input/output transducer is known as interfacing.

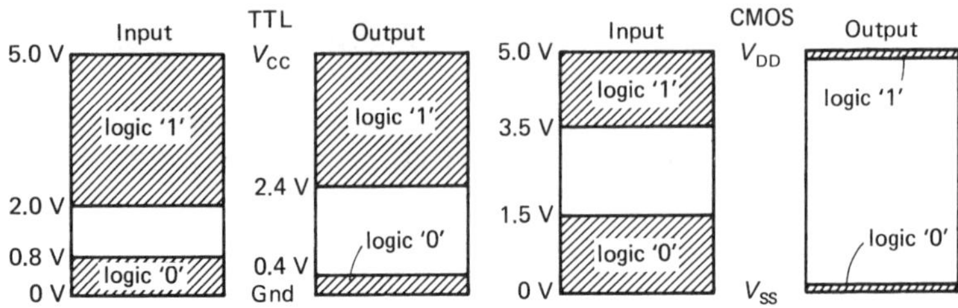

Fig. 17.16

(i) TTL to CMOS When operated from a common 5.0 V supply, Fig. 17.16 shows that the logic levels for TTL and CMOS are different, i.e. the two families are not compatible even though a TTL IC has the current capability to drive any number of CMOS ICs. For example, the worst-case value of a TTL 'high' output (2.4 V) is lower than the minimum input required by CMOS (3.5 V) to ensure switching. An external 'pull-up' resistor *R* (2.2 kΩ) connected as in Fig. 17.17(*a*) overcomes the problem.

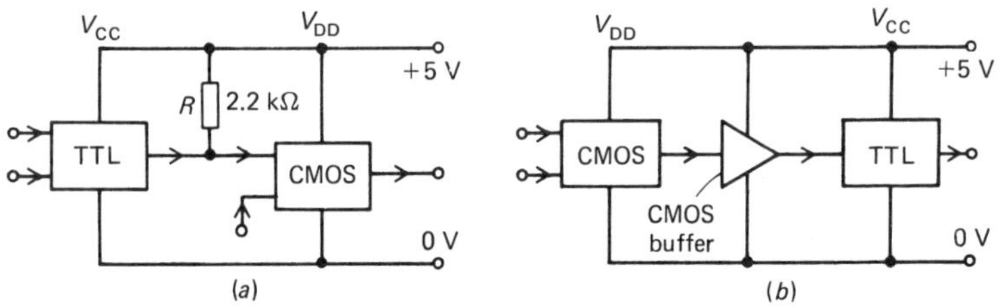

Fig. 17.17

(ii) CMOS to TTL A CMOS IC can drive one low power Schottky TTL IC but not a standard one unless a buffer is interposed, as in Fig. 17.17(*b*).

(iii) TTL and CMOS to LEDs and lamps LEDs and lamps are often used to indicate the logic levels of ICs. Two LED interface circuits are given in Figs. 17.18(*a*) and (*b*). In Fig. 17.18(*a*) the LED lights for a 'high' output from the IC which acts as a source of the LED current. In Fig. 17.18(*b*) the LED lights for a 'low' output from the IC which must be able to sink the LED current.

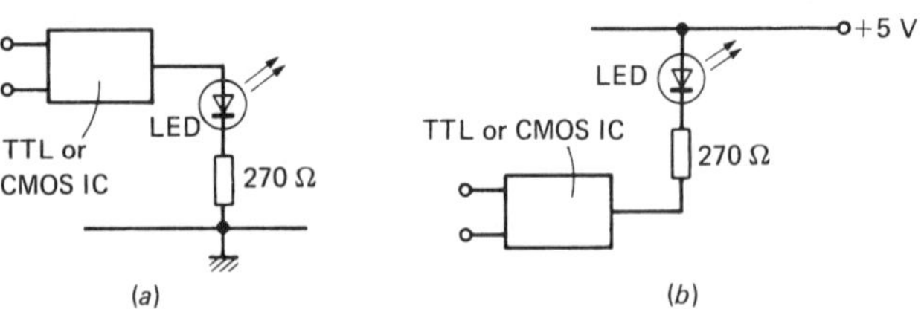

Fig. 17.18

Filament lamps, being higher current loads, are best driven via interface transistors. Two arrangements are shown, using one transistor in Fig. 17.19(*a*) and using a current-amplifying Darlington pair (see section 11.6) in Fig. 17.19(*b*). In both cases the lamp lights when the output from the IC is 'high'.

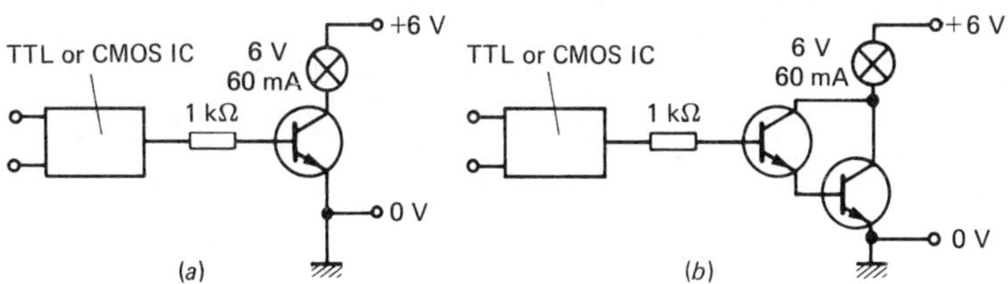

Fig. 17.19

17.8 TTL circuitry

To illustrate the more complex kind of circuitry used in digital TTL ICs, a NAND gate will be considered. It is typical because, as we have seen, all digital circuits (combinational and sequential) can be built from NAND gates.

(a) TTL NAND gate

This is a development of an earlier type of NAND gate in which diodes at the inputs are replaced by a special multiple-emitter n-p-n transistor in the IC (not normally made as a discrete component) which gives rise to the term transistor-transistor logic. (Strictly speaking it should be multiple-emitter transistor logic, i.e. METL.)

A simplified circuit of a two-input NAND gate is shown in Fig. 17.20(*a*). The input transistor Tr_1 operates essentially as three diodes, D_1, D_2 and D_3, Fig. 17.20(*b*). With the base bias resistor R_1, Tr_1 acts as an AND gate whose output provides the input to the inverting (output) transistor Tr_2. The whole circuit acts as a NAND gate, as we will now see.

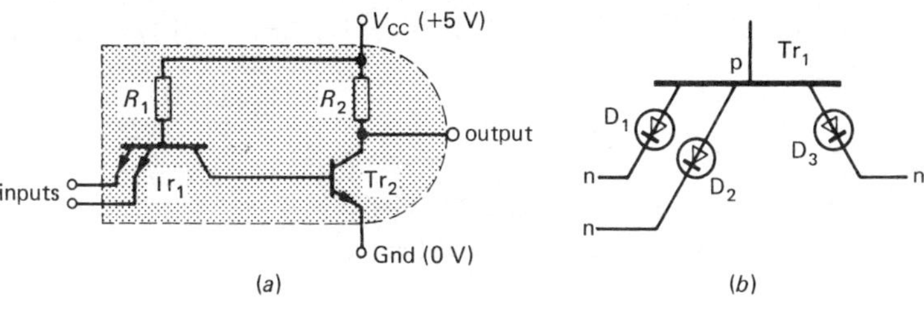

Fig. 17.20

If both inputs to Tr_1 are 'high', e.g. near +5 V (or are unconnected), no current flows from the base to either emitter since they are not forward biased, i.e. D_1 and D_2 do not conduct. However, current can flow through R_1 and the base-collector junction of Tr_1, i.e D_3 conducts, and then to ground via the base-emitter junction of Tr_2. This current is large enough to saturate Tr_2, making the resistance of Tr_2 very small compared with that of its load R_2. The output voltage is thus 'low' (0 to 0.4 V).

If either or both inputs to Tr_1 are joined to ground (0 V) and thereby go 'low', sufficient base current flows through R_1 and the base-emitter junction(s) of Tr_1 to cause transistor action, i.e. D_1 and/or D_2 conduct. As a result, Tr_1 passes a large momentary collector current to ground in the form of the positive charges stored in the p-type base of Tr_2, which goes negative. Tr_2 switches off rapidly, i.e. its resistance is very high and a 'high' output voltage (2.4 to 5.0 V, typically 3.3 V) is obtained.

Summing up, the circuit acts as a NAND gate since the output is 'high' unless both inputs are 'high'.

(b) Further features of TTL ICs

(i) Power supply specification As mentioned earlier a stabilized supply of 5.0 V ± 0.25 V must be used. Higher voltages cause too large currents through the base-emitter junction of Tr_1. Lower voltages reduce the difference between the 'high' and 'low' levels and could result in faulty switching.

(ii) Fan-out Each input in the 'high' state sources a current of 40 μA from the previous circuit or device and in the 'low' state the input sinks 1.6 mA. Each output in the 'high' state can source 400μA and in the 'low' state it can sink 16 mA. TTL ICs can therefore accept (sink) much more current than they can provide (source). Each gate is able to 'feed' ten other TTL gates, i.e. their fan-out is ten.

(iii) Schottky TTL In standard TTL the bipolar transistors switch between cut-off and saturation. In Schottky TTL ICs, a special diode, called a Schottky diode, which has an aluminium and n-type silicon junction, is integrated with a bipolar transistor as in Fig. 17.21. It turns on at 0.4 V (compared with 0.6 V for a p-n silicon junction) and stops the transistor saturating by allowing current to flow through the diode rather than through the collector-base junction of the transistor. More rapid switching occurs. Schottky TTL uses unsaturated logic.

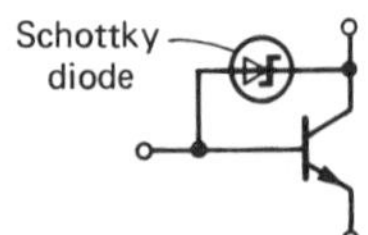

Fig. 17.21

17.9 CMOS circuitry

A NOT gate will be considered as a simple, typical example of CMOS circuitry.

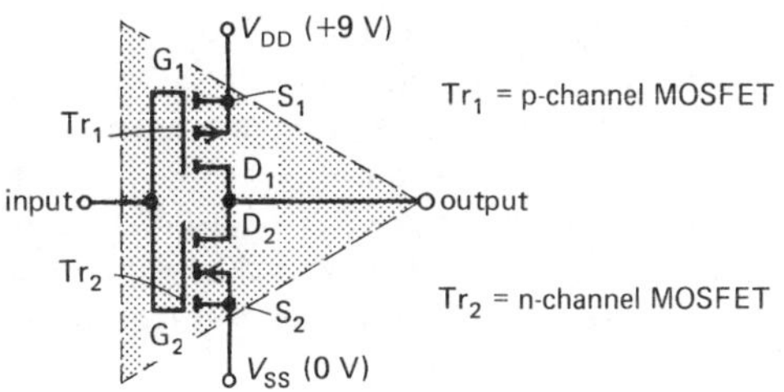

Fig. 17.22

(a) CMOS NOT gate

The basic circuit is shown in Fig. 17.22. It consists of one p-channel enhancement MOSFET Tr_1 (see section 8.8) and one n-channel type Tr_2 joined in series in the same chip to form a complementary pair. The source S_1 and substrate of Tr_1 go to supply positive ($V_{DD} = +9$ V); the source S_2 and substrate of Tr_2 go to ground ($V_{SS} = 0$ V). The gates G_1 and G_2 are connected together at the input terminal and the output is taken from the junction of the two drains D_1 and D_2.

From the $I_D - V_{GS}$ characteristic of an n-channel enhancement MOSFET (Fig. 8.15(*b*), p. 102), you can see that it allows current to pass from drain to source only when V_{GS} is positive, i.e. when the gate is positive with respect to the source. A p-channel type conducts when its gate is negative with respect to the source.

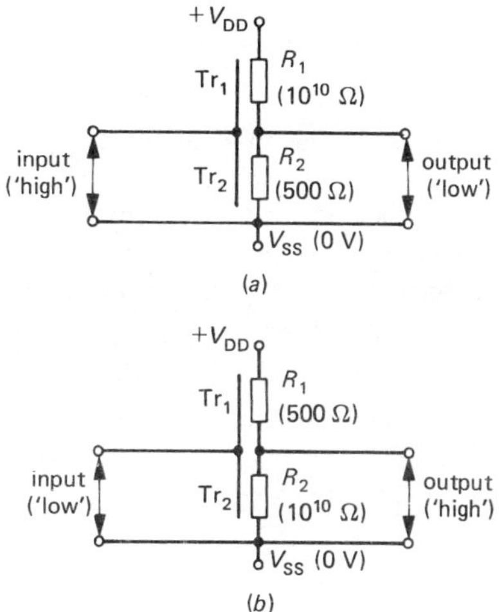

Fig. 17.23

If the input is 'high', e.g. +9 V, the n-channel transistor conducts, behaving as a small resistor R_2 (of about 500 Ω). The p-channel transistor is cut off and acts as a very large resistor R_1 (of about 10^{10} Ω). The circuit is then equivalent to a two-resistor voltage divider in which the output is 'low' (0 V), since it is taken from across the small resistor R_2, as shown in Fig. 17.23(*a*).

If the input is 'low', e.g. 0 V, the gate G_1 of Tr_1 is negative with respect to its source S_1 (at say +9 V) and so it switches on. In this case it acts as a very low resistance R_1 (500 Ω). The equivalent circuit is shown in Fig. 17.23(*b*) and the output, being taken across a very large resistance R_2 (about 10^{10} Ω) is 'high'.

The circuit acts as a NOT gate or inverter. Normally protection diodes are included at the input to reduce the chance of electrostatic breakdown.

(b) Further features of CMOS ICs

(i) Operating current For both 'high' and 'low' inputs the supply (quiescent) current through the two MOSFETs in Fig. 17.22 is tiny (of the order of nano-amperes, i.e. 10^{-9} A) because one of them is always cut off and has a resistance of about 10^{10} Ω. However both MOSFETs are partly on during every change of onput state and a pulse of current is drawn from the supply. The more often this happens, the greater is the average current taken.

In general, while CMOS ICs working at a switching frequency of 5 kHz draw only one-thousandth of the current of a similar standard TTL IC, at 5 MHz they require about the same current as TTL.

(ii) Power supply The minimum voltage required depends on the threshold voltage of a MOSFET, i.e. the voltage to turn it on and this is just over 1 V upwards. The maximum is limited by the breakdown characteristics of the IC. Any supply in the range of +3 to +15 V is suitable.

(iii) Input impedance This is very high (about 10^{12} Ω) because the gates of the MOSFETs form one plate of a small capacitor, the other plate being the semi-conductor material on the other side of the very thin layer of insulating silicon oxide which acts as the dielectric. Input current is therefore needed only briefly to charge or discharge the gate capacitor. In effect the input behaves as if it were on open circuit, i.e. it has infinite impedance.

(iv) Fan-out The near-zero input current of a CMOS IC accounts for its high fan-out of about 50. The output of a typical CMOS gate can source or sink about 5 mA on a 5 V supply and about 10 mA on a 10 V supply.

(v) Output swing The unloaded output voltage can swing from zero to the full positive supply voltage (V_{DD}) owing to the absence of saturation voltages and forward biasing junction voltages such as occur with TTL.

17.10 Revision questions and Problems

1. What happens in a 555 timer working as an astable if:
   a) a voltage is applied to the control voltage pin of say 4.5 V (on a 9 V supply),
   b) the reset pin voltage falls below about 0.7 V?

2. The circuit of Fig. 17.24 is for a simple two-tone door bell using a 555 timer. When S_1 is on, a note is produced in the speaker and C_3 charges up. When S_1 is switched off, a note of higher pitch comes from the speaker for a short time and then stops.
   a) Which components decide the pitch of the note when S_1 is
      (i) on,
      (ii) off?
   b) Why does the second note only last for a short time? (*Hint*—your answer to Question 1b) gives a clue.)
   c) What would be the effect of omitting C_3 from the circuit?

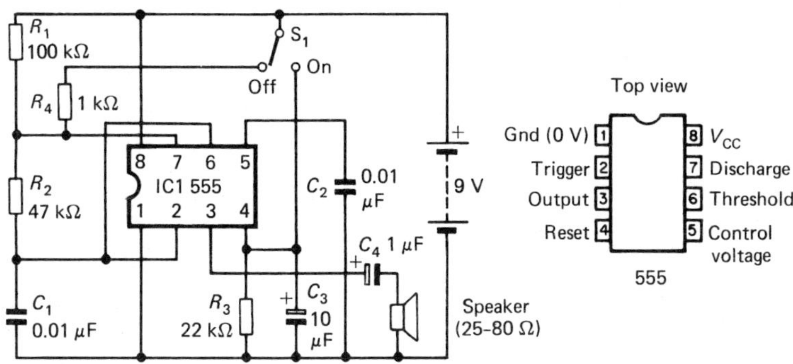

Fig. 17.24

3. What is a D/A converter? What does it do? What part does an op amp play in its action?

4. What is an A/D converter? What does it do? What part does an op amp play in its action?

5. Draw the circuit and explain how it works, for an op amp used as
   a) a bistable,
   b) an astable,
   c) a monostable.

6. State four uses of a Schmitt trigger.

7. What is meant by 'noise immunity'? Why is it better for CMOS than TTL? What effect does it have on power supply requirements?

8. a) Explain the term 'interfacing'.
b) Why are TTL and CMOS not compatible (even if the latter uses a 5 V power supply)?
c) Draw three circuits that will enable
(i) a TTL IC to drive a CMOS IC,
(ii) a CMOS IC to drive a standard TTL IC,
(iii) a TTL or CMOS IC to drive a low voltage filament lamp.

9. Explain the term 'fan-out'. What is its value for
a) CMOS,
b) TTL?

10. a) Possible circuits for a logic level indicator are given in Fig. 17.25(*i*), (*ii*) and (*iii*). In which circuit(s) will a logic level 1 (i.e. +5 V) applied to the input light up the LED?

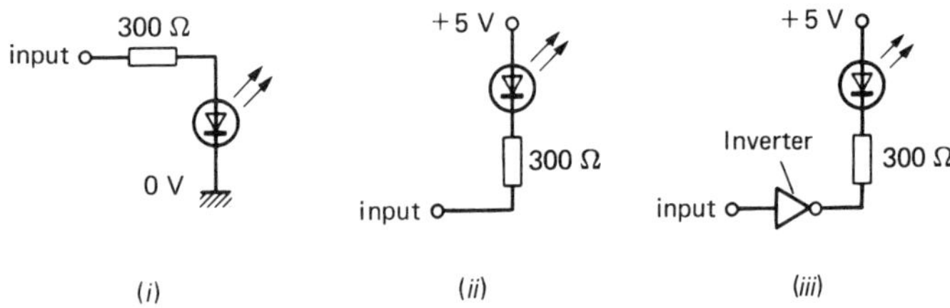

Fig. 17.25

b) The table for typical TTL input and output sourcing and sinking currents on a 5 V supply, shows that a TTL gate can sink much more current than it can source (e.g. for an output, 16 mA compared with 400 μA).

TTL	Logic 1 (source)	Logic 0 (sink)
Output	400 μA	16 mA
Input	40 μA	1.6 mA

If when checking TTL logic levels we wish to make a logic 1 light an LED (needing typically 10 mA) *and* to avoid overloading a logic 1 output connected to the indicator input (so preventing the logic 1 output voltage from falling into the indeterminate region between 2.4 V and 0.4 V in Fig. 17.16 and producing faulty switching) which circuit should be used for both conditions to be satisfied?

c) From the table above work out how many other TTL gate inputs can be driven by one TTL gate output.
d) The output of a CMOS device can sink or source about 10 mA on a 9 V supply. Which circuit(s) in Fig. 17.25 would satisfy the two requirements in b) if a 9 V supply is used and a 680 Ω resistor replaced the 300 Ω one?

PART FIVE

Electronic systems

UNIT 18

Radio, television, audio and communication systems

18.1 Radio waves

The radio waves from a transmitting aerial can reach a receiving aerial in one or more of three different ways, as shown in Fig. 18.01, depending on the frequency of the waves, the aerial and the power of the transmitter.

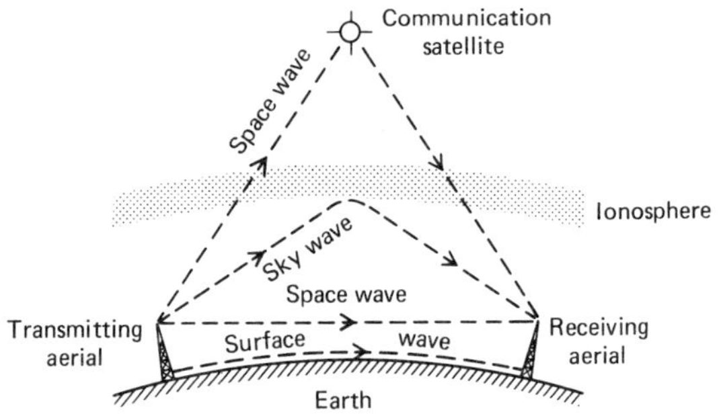

Fig. 18.01

(a) Surface or ground wave

This travels along the surface of the ground, following the curvature of the earth. Its range is limited mainly by the extent to which energy is absorbed from it by the ground. Poor conductors such as sand absorb more strongly than water and the higher the frequency the greater the absorption. The range may be about 1 500 km at low frequencies (long waves) but only a few kilometres for v.h.f. waves.

(b) Sky wave

This travels skywards and, if it is below a certain *critical frequency* (typically 30 MHz), is returned to earth by the *ionosphere*. This consists of layers of air molecules, stretching from about 60 km above the earth to 500 km, which have become positively charged through the removal of electrons by the sun's ultraviolet radiation. On striking the earth the sky wave bounces back to the ionosphere where it is once again 'reflected' down to earth and so on until it is completely attenuated.

The critical frequency varies with the time of day, the seasons and the eleven-year sun-spot cycle. Sky waves of low, medium and high frequencies have a range of several thousand kilometres but at v.h.f. and above (and sometimes at h.f.) they usually pass through the ionosphere into outer space.

If both the surface wave and the sky wave from a particular transmitter are received at the same place, interference can occur if the two waves are out of phase. When the phase difference varies, the signal 'fades', i.e. goes weaker and stronger.

(c) Space wave

For v.h.f., u.h.f. and s.h.f. signals, only the space wave, which travels in a more or less straight line from transmitter to receiver, is effective. When the transmitting aerial is at the top of a tall mast standing on high ground, a range of up to about 150 km is possible on earth if obstacles such as hills, buildings or trees do not block the path to the receiving aerial.

Space waves are also used in satellite communication systems to send signals many thousands of kilometres. Frequencies in the s.h.f. (microwave) band are employed since they can penetrate the ionosphere. Signals from the ground transmitting station are received and amplified by the satellite, then retransmitted to the ground receiving station. The world-wide system of communication satellites is managed by INTELSAT (International Telecommunication Satellite Organization) which has over 20 satellites, all orbiting the earth at a height of about 36 000 km and spaced so that global communication is possible. Since each satellite takes 24 hours to orbit the earth once, it appears to be at rest from a particular place on earth and, for this reason, it is called a synchronous geostationary satellite.

Table 18.1 (overleaf) shows some applications of the various frequency bands.

18.2 Aerials

(a) Transmitting aerials

When a.c. flows into a transmitting aerial, radio waves of the same frequency, f, as the a.c. are radiated if the length of the aerial is comparable with the wavelength, λ (lambda), of the waves. λ can be calculated from the following equation which is true for any wave motion:

$$v = f \times \lambda$$

v is the speed of radio waves, which is the speed of light, namely 300 million metres per second (3×10^8 m s^{-1}). The greater f is, the smaller λ will be, since v is fixed. For example, if $f = 1$ kHz $= 10^3$ Hz, then $\lambda = v/f = 3 \times 10^8/10^3 = 3 \times 10^5 = 30\,000$ m, but if $f = 100$ MHz $= 10^8$ Hz, then $\lambda = 3$ m. Therefore, if aerials are not to be too large, they must be supplied with r.f. currents (>20 kHz) from the transmitter.

Table 18.1 Applications of frequency bands

Frequency	Surface wave	Sky wave	Space wave
Low (l.f.) (long waves) 30 kHz–300 kHz	Medium range communication	Long range communication	
Medium (m.f.) (medium waves) 300 kHz–3 MHz	Local sound broadcasts	Distant sound broadcasts	
High (h.f.) (short waves) 3 MHz–30 MHz		Distant sound broadcasts; communication	
Very high (v.h.f.) 30 MHz–300 MHz			FM sound broadcasts; TV; mobile systems
Ultra high (u.h.f.) 300 MHz–3 GHz			TV
Super high (s.h.f.) (microwaves) above 3 GHz			Radar; communication via satellites; telephone links

A common type is the *dipole* aerial consisting of two vertical or horizontal conducting rods (or wires), each having a length of one-quarter of the required wavelength (i.e. $\lambda/4$) and centre fed, as shown in Fig. 18.02(*a*). It behaves as a series LC circuit whose resonant frequency depends on its length, since this determines its inductance (which any conductor has) and its capacitance, arising from the capacitor formed by each conductor and the air between them. The radiation pattern of Fig. 18.02(*b*) shows that a vertical dipole emits radio waves equally in all horizontal directions.

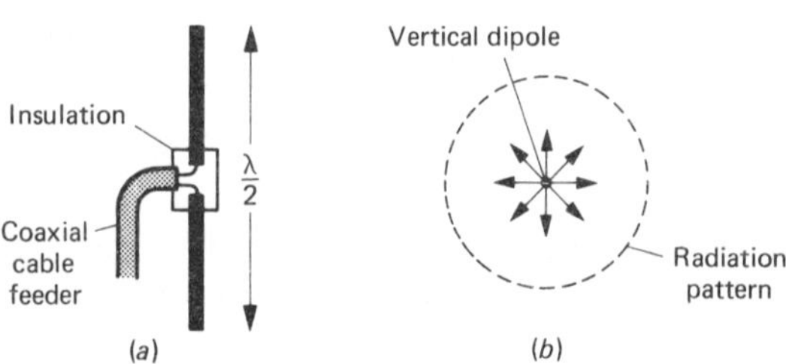

Fig. 18.02

Concentration of much of the radiation in one direction can be achieved by placing a *reflector* behind the dipole and a *director* in front of it, Fig. 18.03(*a*). The reflector is about 5% longer than $\lambda/2$ and the director is about 5% shorter than $\lambda/2$. Both are conductors and, if the distance of each from the dipole is correctly chosen (about $\lambda/4$ or less), the radiation they emit (arising from the voltage and current induced in them by the radiation from the dipole) adds to that from the dipole in the wanted direction and substracts from it in the opposite direction. Fig. 18.03(*b*) shows the kind of radiation pattern obtained with this aerial, called a *Yagi* array.

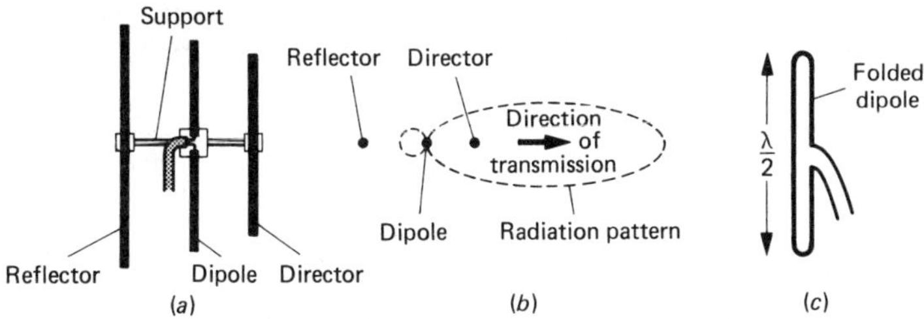

Fig. 18.03

A radio communication system based on, for example, h.f. sky wave transmission, uses different frequencies to ensure reliability, since the critical frequency changes. Economy also demands the use of the same aerial. In such cases, resonant aerials like the dipole are unsuitable and more complex types such as the *rhombic* are employed.

In the s.h.f. (microwave) band *dish* aerials in the shape of huge curved metal bowls are used. The radio waves fall on them from a small dipole at their focus, which is fed from the transmitter, and are reflected as a narrow, highly directed beam, as shown in Fig. 18.04(*a*).

For maximum transfer of the r.f. power produced by a transmitter to the aerial, the impedance of the aerial must be matched to that of the cable connecting it to the transmitter—called the 'feeder'. At resonance, a dipole, for example, has a purely resistive input impedance of 73 Ω and matching is achieved by a coaxial cable (Fig. 18.04(*b*)) feeder, as in Fig. 18.02(*a*).

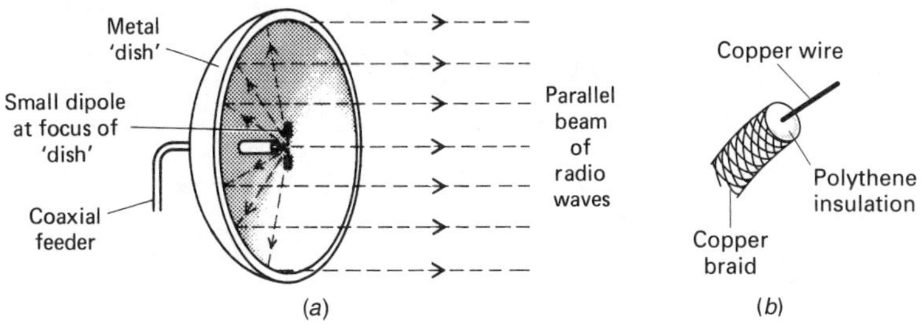

Fig. 18.04

(b) Receiving aerials

Whereas transmitting aerials may handle many kilowatts of power, receiving aerials, though basically similar, deal with only a few picawatts due to the voltages (of a few microvolts) and currents induced in them by passing radio waves. Common types are the dipole, the vertical whip and the ferrite rod.

Dipoles with reflectors and directors are a common sight for television (where the reflector is often a metal plate with slots and the dipole is folded, as in Fig. 18.03(*c*), to restore the impedance of 73 Ω since this is reduced by the extra elements) and FM radio reception. They give maximum output when (i) their length is correct ($\lambda/2$), (ii) they are lined up on the transmitter, and (iii) they are vertical or horizontal if the transmitting aerial is vertical or horizontal, i.e. if the transmission is vertically or horizontally polarized.

A vertical whip aerial is a tapering metal rod (often telescopic), used for car radios. In most portable medium/long wave radio receivers the aerial consists of a coil of wire wound on a ferrite rod, as in Fig. 6.08(*b*) of section 6.4. Ferrites are non-conducting magnetic substances which 'draw in' nearby radio waves. The coil and rod together act as the inductor L in an LC circuit tuned by a capacitor C. For maximum gain, the aerial, being directional, should be lined up so as to be end-on to the transmitter.

Small dish aerials for the reception of satellite TV (microwave) signals are now commonplace.

18.3 Amplitude modulated (AM) radio

(a) Amplitude modulation

An aerial requires r.f. voltages in order to emit radio waves but speech and music produce a.f. voltages. The transmission of sound by radio therefore involves combining a.f. and r.f. in some way. It is done in a transmitter by a process called *modulation*. Amplitude modulation is common, being used in medium, long and short wave broadcasting.

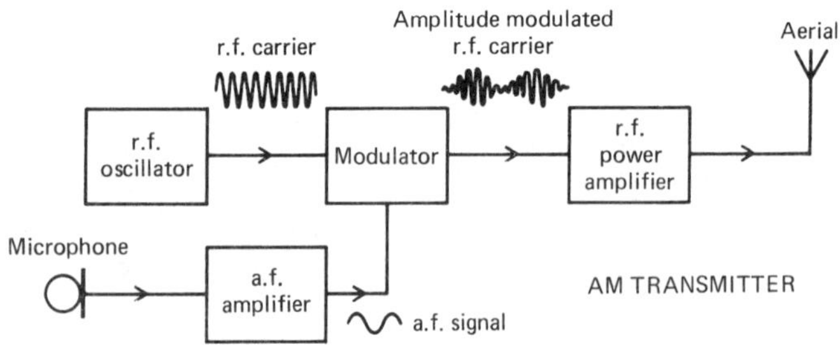

Fig. 18.05

A block diagram for an AM transmitter is shown in Fig. 18.05. In the modulator the amplitude of the r.f. carrier from the r.f. oscillator (usually crystal controlled) is varied at the frequency of the a.f. signal from the microphone. Fig. 18.06 shows the effect of combining the signal wave and the carrier wave. The *modulation depth*, *m*, is defined by

$$m = \frac{\text{signal peak}}{\text{carrier peak}} \times 100\%$$

For a signal peak of 1 V and a carrier peak of 2 V, $m = \frac{1}{2} \times 100 = 50\%$. If m exceeds 100%, distortion occurs, but, if it is too low, the quality of the sound at the receiver is poor; a value of 80% is satisfactory.

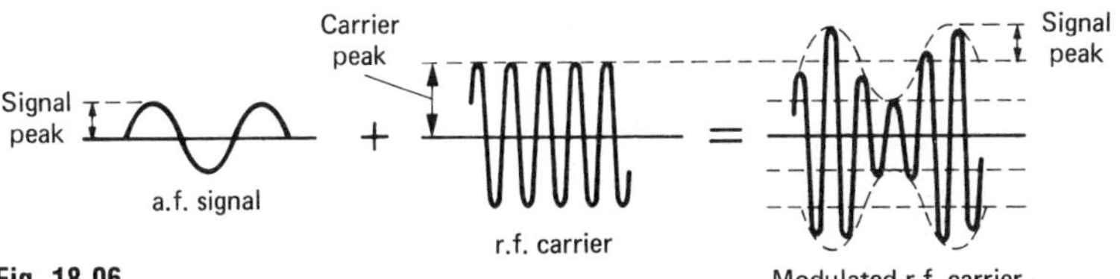

Fig. 18.06

A wave with a sine waveform consists of just one frequency and is the simplest possible. Since an AM r.f. carrier is non-sinusoidal (i.e. its amplitude changes), it must contain other frequencies—see section 3.3. It can be shown that, if a sinusoidal r.f. carrier of frequency f_c is amplitude modulated by a sinusoidal a.f. signal of frequency f_s, three different sinusoidal, constant amplitude r.f. waves are present in the AM carrier. Their frequencies are f_c, $f_c + f_s$ (the upper side frequency) and $f_c - f_s$ (the lower side frequency), as shown in Fig. 18.07(*a*). All can be detected separately.

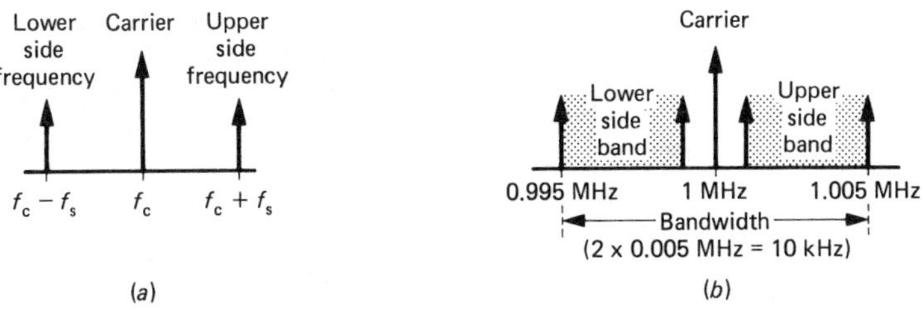

Fig. 18.07

In practice, the carrier is modulated by a range of a.fs, and each produces a pair of frequencies. The result is a band of frequencies, called the upper and lower sidebands, on either side of the carrier. For example, if f_c = 1 MHz and the highest value of f_s = 5 kHz = 0.005 MHz, then $f_c + f_s$ = 1.005 MHz and $f_c - f_s$ = 0.995 MHz, as in Fig. 18.07(*b*). The *bandwidth* needed to transmit a.fs up to 5 kHz is therefore 10 kHz and since the medium waveband extends from about 500 kHz to 1.5 MHz, the 'space' is limited if interference between stations is to be avoided. In fact it is restricted to 9 kHz.

(b) Tuned radio frequency (TRF) or straight receiver

The various elements are shown in the block diagram of Fig. 18.08. The wanted carrier from the aerial is selected and amplified by the r.f. amplifier, which should have a bandwidth of 9 kHz to accept the sidebands. The a.f. is next separated from the r.f. carrier by the detector or demodulator and amplified first by the a.f. (voltage) amplifier and then by the a.f. power amplifier which drives the loudspeaker. The bandwidths of the a.f. amplifiers need not exceed 4.5 kHz, i.e. the highest a.f. frequency.

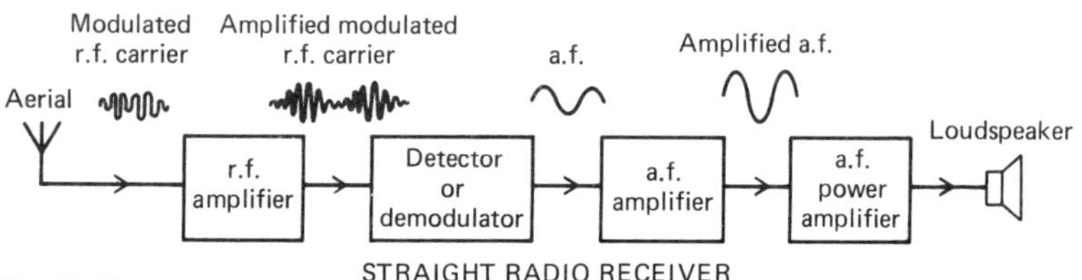

Fig. 18.08

The basic demodulation circuit is shown in Fig. 18.09(*a*). If the AM r.f. signal V of Fig. 18.09(*b*) is applied to it, the diode produces rectified pulses of r.f. current I, Fig. 18.09(*c*). These charge up C_1 whenever V exceeds the voltage across C_1, i.e. during part of each positive half-cycle, as well as flowing through R_1. During the rest of the cycle when the diode is non-conducting, C_1 partly discharges through R_1. The voltage V_{C1} across C_1 (and R_1) is a varying d.c. which, apart from the slight r.f. ripple (much exaggerated in Fig. 18.09(*d*), has the same waveform as the modulating a.f. C_2 blocks its d.c. component but allows the a.f. component to pass to the next stage.

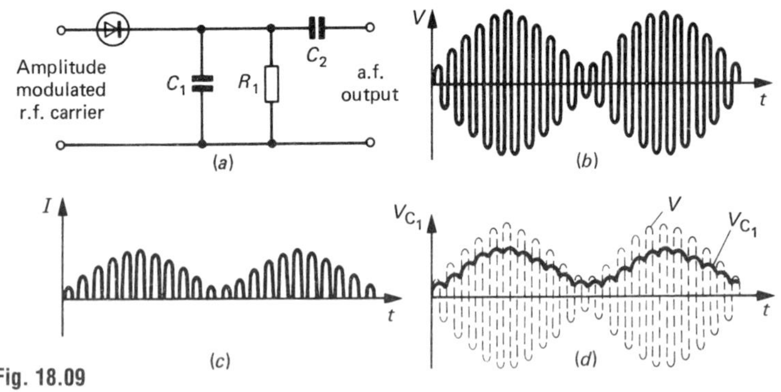

Fig. 18.09

The time constant $C_1 \times R_1$ must be between the time for one cycle of r.f., e.g. 10^{-6} s, and one cycle of a.f. (average), e.g. 10^{-3} s. If it is too large, C_1 charges and discharges too slowly and V_{C1} does not respond to the a.f. modulation; too small a value allows C_1 to discharge so rapidly that V_{C_1} follows the r.f. Typical values are $C_1 = 0.01$ μF and $R_1 = 10$ kΩ giving $C_1 \times R_1 = 10^{-4}$ s. Point-contact germanium diodes are used as explained in section 7.6.

(c) Superheterodyne receiver (superhet)

This type is much superior to the TRF receiver and is used in commercial AM radios. Fig. 18.10 shows a typical block diagram.

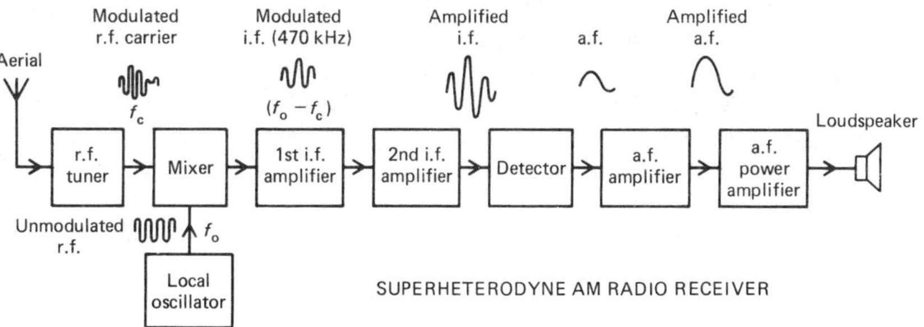

Fig. 18.10

The modulated r.f. carrier at the wanted frequency f_c is fed via an r.f. tuner (which may not be an amplifier but simply a tuned circuit that selects a band of r.fs) to the mixer along with an r.f. signal from the local oscillator which has a frequency f_o. The output from the mixer contains an r.f. oscillation at frequency $(f_o - f_c)$, called the *intermediate frequency* (i.f.), having the original a.f. modulation. Whatever the value of f_c, f_o is always greater than it by the same amount. The i.f., i.e. $(f_o - f_c)$ is therefore fixed, and in an AM radio is usually about 470 kHz. This is achieved by ganging the two variable capacitors in the tuned circuits of the r.f. tuner and local oscillator so that their resonant frequencies change in step, i.e. *track*. Long, medium and short wavebands are obtained by switching a different set of coils into each tuned circuit. Longer waves (smaller frequencies) require greater inductance, i.e. more turns on the coil. The i.f. amplifiers are double-tuned r.f. amplifiers (see section 12.4) with a resonant frequency of 470 kHz. They amplify the modulated i.f. before it goes to the detector which, like the a.f. stages, acts as in the TRF receiver.

The frequency changing of f_c to $(f_o - f_c)$ is achieved in the mixer by *heterodyning*. This is similar to the production of beats in sound when two notes of slightly different frequencies f_1 and f_2 $(f_1 > f_2)$ create another note, the beat note, of frequency $(f_1 - f_2)$. In radio, the 'beat' note must have a supersonic frequency (or it will interfere with the sound on the a.f. modulation), hence the name *super*(sonic) *heterodyne* receiver.

By changing all incoming carrier frequencies to one lower, fixed frequency (the i.f.) at which most of the gain and selectivity occurs, we obtain:

(i) greater stability because positive feedback is less at lower frequencies;
(ii) greater sensitivity because more i.f. amplifiers can be used (but two are usually enough) as a result of (i), so giving greater gain;
(iii) greater selectivity because more i.f. tuned circuits are possible without creating the tracking problems that arise when several variable capacitors are ganged.

18.4 Frequency modulated (FM) radio

(a) Frequency modulation

This is the other common type of modulation. It is used for v.h.f. radio and for u.h.f. TV sound signals. The frequency of the r.f. carrier, not the amplitude, is changed by the a.f. signal. The change or *deviation* is proportional to the amplitude of the a.f. at any instant and is expressed in kHz V^{-1}.

For example, if a 100 MHz carrier is modulated by a 1 V 1 kHz sine wave, the carrier frequency might swing 15 kHz either side of 100 MHz, i.e. from 100.015 MHz to 99.985 MHz and this would happen 1 000 times a second. A 2 V 1 kHz signal would cause a swing of ±30 kHz. at the same rate; for a 2 V 2 kHz signal the swing remains at ±30 kHz but it occurs 2 000 times a second. By international agreement, the maximum deviation allowed is ±75 kHz. Fig. 18.11 shows frequency modulation; note that when the a.f. signal is positive, the carrier frequency increases but it decreases when the a.f. signal is negative.

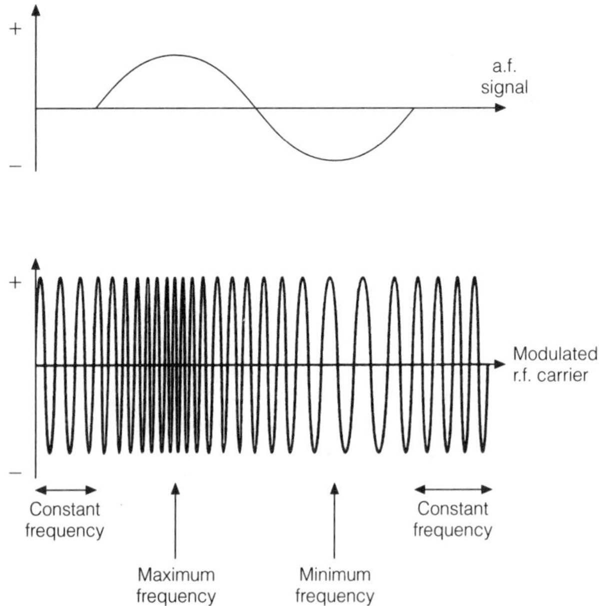

Fig. 18.11

In FM, each a.f. modulating frequency produces a large number of side frequencies (not two as in AM) but their amplitudes decrease the more they differ from the carrier. In theory, therefore, the bandwidth of an FM system should be extremely wide but in practice the 'outside' side frequencies can be omitted without noticeable distortion. The bandwidth may be taken as roughly $\pm(\Delta f_c + f_m)$ where Δf_c is the deviation and f_m the highest modulating frequency. The BBC uses a 250 kHz bandwidth which is readily accomodated in the v.h.f. band and also allows f_m to have the full range of audio frequencies. This accounts for the better sound quality of FM.

An FM transmitter is similar to the AM one shown in Fig. 18.05, except for the modulation process.

(b) FM receiver

The superheterodyne principle is also used in FM receivers. The block diagram (Fig. 18.12) is similar to that of the AM superhet. To ensure adequate amplification of the v.h.f. carrier, the first stage is an r.f. amplifier with a fairly wide bandwidth. The mixer and local oscillator change all carrier frequencies to an i.f. of 10.7 MHz. These three circuits often form a single module, called the FM tuner.

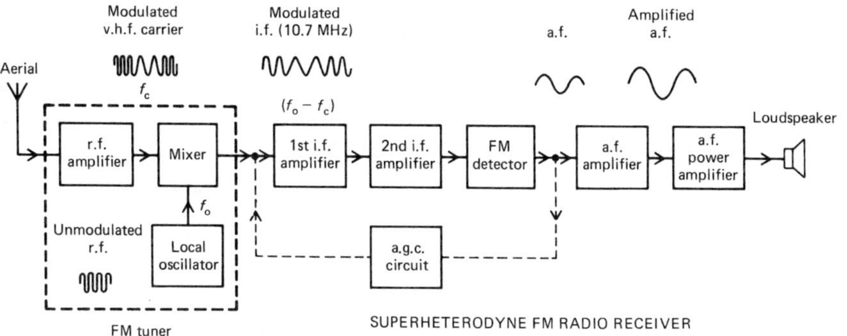

Fig. 18.12

The FM detector is more complex and quite different from an AM detector and consists of a *discriminator*. As well as extracting and passing on the a.f., it removes any amplitude changes in the carrier arising from unwanted 'noise'—see section 11.10. 'Quiet' reception is another good feature of FM.

(c) Automatic gain control (a.g.c.)

In any receiver, FM or AM, if the strength of the signal at the aerial varies because of fading or a change of station, the output will also vary. To overcome this and avoid the need for continual adjustment of the volume control, automatic gain control (a.g.c.) is used. It is achieved by feeding back the d.c. component in the detector output (which is proportional to the amplitude of the carrier), as the d.c. bias voltage to the base of the transistor forming the first i.f. amplifier, as shown by the dashed line in Fig. 18.12. The feedback circuit is such that a stronger aerial signal makes the bias more negative, so reducing the gain of the amplifier. Conversely a weaker aerial signal increases the gain.

(d) Automatic frequency control (a.f.c.)

Mistuning of an FM receiver operating in the v.h.f. or u.h.f. bands can give a distorted a.f. output. It occurs because the i.f. bandwidth is only a small fraction of the carrier frequency and even a slight error in the local oscillator frequency may lead to partial rejection of the wanted signal by the i.f. stages. As a result, many FM receivers include automatic frequency control (a.f.c.). It works by applying the i.f. output to a frequency comparator circuit which produces a direct voltage if the i.f. is not correct. This voltage, whose polarity depends on whether the i.f. is too high or too low, is fed back to a varicap (varactor) diode (see section 7.8) in the local oscillator tuned circuit. Therefore, the capacitance of the diode is changed so altering the frequency of the local oscillator sufficiently to correct the i.f.

(e) FM/AM receiver

The FM/AM superhet shown in the block diagram of Fig. 18.13 uses ICs and discrete components—the ceramic filter (see section 12.4) for the AM i.f. amplifier being an example. The a.f. outputs of both detectors are connected by a two-way switch to the common a.f. amplifier (IC3).

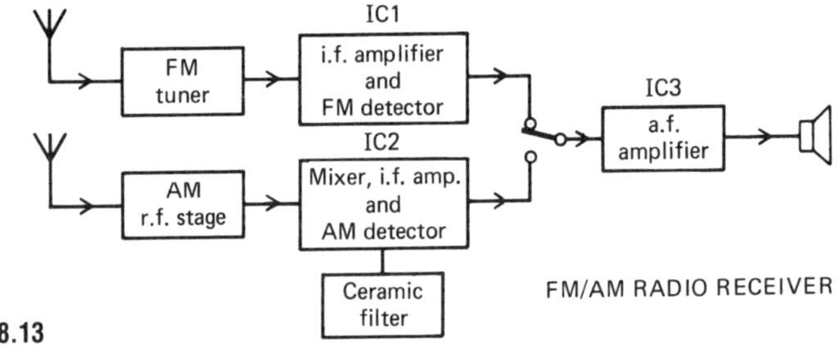

Fig. 18.13

18.5 Black-and-white television

Television is concerned with the transmission and reception of visual (video) information, usually as a picture.

(a) TV camera

This changes light into electrical signals. The *vidicon* tube, shown simplified in Fig. 18.14(*a*), consists of an electron gun, which emits a narrow beam of electrons, and a target of photoconductive material (see section 9.4) on which a lens system focuses an optical image.

The action of the target is complex but it behaves as if the resistance between any point on its back surface and a transparent aluminium film at the front, depends on the brightness of the image at the point. The brighter it is, the lower the resistance. The electron beam is made to scan across the target and the resulting beam current (i.e. the electron flow in the circuit consisting of the beam, the target, the load resistor *R*, the power supply and the gun) varies with the resistance at the spot where it hits the target. The beam current thus follows the brightness of the image and *R* turns its variations into identical variations of voltage, shown in Fig. 18.14(*b*), for subsequent transmission as the video signal.

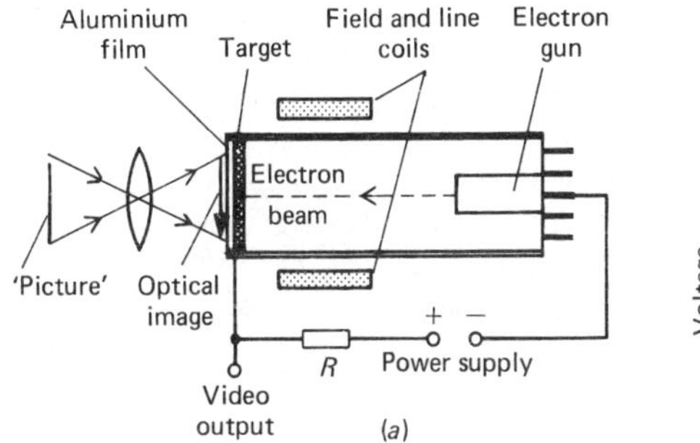

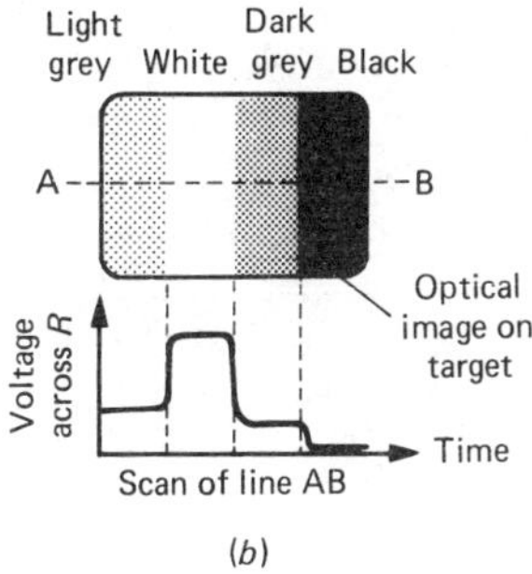

Fig. 18.14

(b) Scanning

The scanning of the target by the electron beam is similar to the way we read a page of print, i.e. from left to right and top to bottom. In effect the picture is changed into a set of parallel lines, called a *raster*; two systems are needed to deflect the beam horizontally and vertically. The one which moves it steadily from left to right and makes it 'flyback' rapidly, ready for the next line, is the *line* scan. The other, the *field* scan, operates simultaneously and draws the beam at a much slower rate down to the bottom of the target and then restores it suddenly to the top. Magnetic deflection is used in which relaxation oscillators (see section 12.8), called *time bases*, generate currents with sawtooth waveforms, Fig. 18.15(*a*), at the line and field frequencies. These are passed through two pairs of coils mounted round the camera tube.

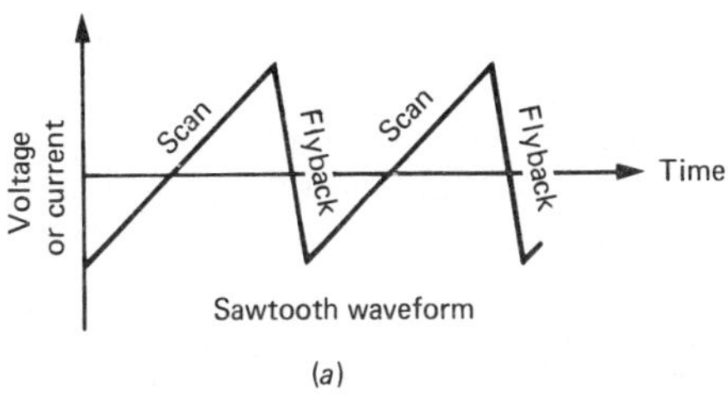

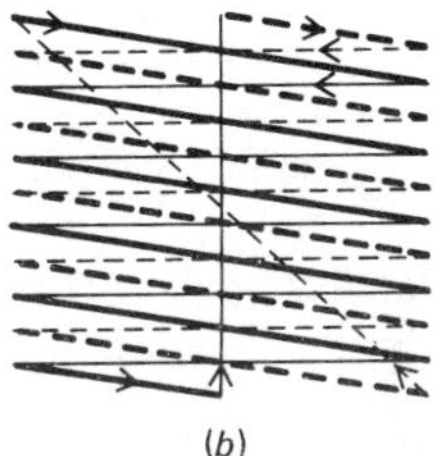

Fig. 18.15

Two conditions are necessary for the video signal to produce an acceptable picture at the receiver. First, the raster must have at least 500 scanning lines (or it will seem 'grainy') and second, the total scan should occur at least 40 times a second (or the impression of continuity between successive scans due to persistence of vision of the eye will cause 'flicker'). The European TV system has 625 lines and a scan rate of 50 Hz.

It can be shown that for such a system the video signal would need the very wide bandwidth of about 11 MHz, owing to the large amount of information that has to be gathered in a short time. This high value would make extreme demands on circuit design and for a broadcasting system would require too much radio wave 'space'. However, it can be halved to 5.5 MHz by using *interlaced* scanning in which the beam scans alternate lines, producing half a picture (312.5 lines) every $\frac{1}{50}$ s, and then returns to scan the intervening lines, Fig. 18.15(*b*). The complete 625 line picture or *frame* is formed in $\frac{1}{25}$ s, a time well inside that allowed by persistence of vision to prevent 'flicker'.

(c) Synchronization

To ensure that the scanning of a particular line and field starts and ends at the same time in the TV receiver as in the camera, synchronizing pulses are also sent. These are added to the video signal during the flyback times when the beam is blanked out. The field pulses are longer and less frequent (50 Hz compared with 312.5 × 50 = 15 625 Hz) than the line pulses. Fig. 18.16 shows a simplified video waveform with line and field sync pulses.

Fig. 18.16

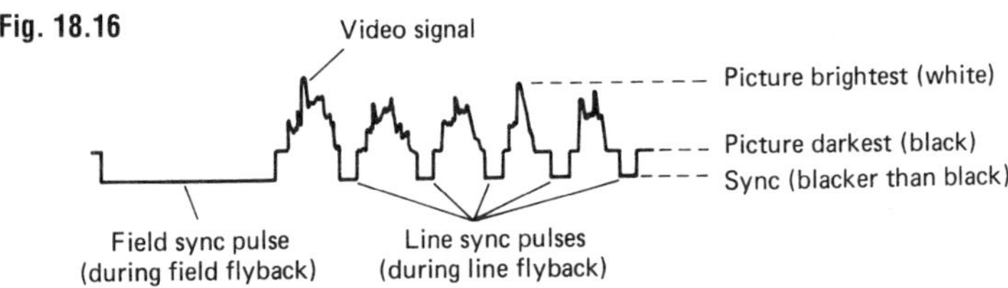

(d) Transmission

In broadcast television the video signal is transmitted by amplitude modulation of a carrier in the u.h.f. band (400 to 900 MHz). Since the video signal has a bandwidth of about 5.5 MHz, the bandwidth required for the transmission would be at least 11 MHz owing to the two sidebands on either side of the carrier. In practice a satisfactory picture is received if only one sideband and a part (a vestige) of the other is transmitted. This is called *vestigial sideband* transmission. The part used contains the lowest modulating frequencies closest to the carrier frequency, for it is the video information they carry in both sidebands that is most essential. The video signal is therefore given 5.5 MHz for one sideband and 1.25 MHz for the other (Fig. 18.17).

The accompanying audio signal is frequency modulated on another carrier, spaced 6 MHz away from the video carrier. The audio carrier bandwidth is about 250 kHz and is adequate for good quality FM sound. The complete video and sound signal lies within an 8 MHz wide channel.

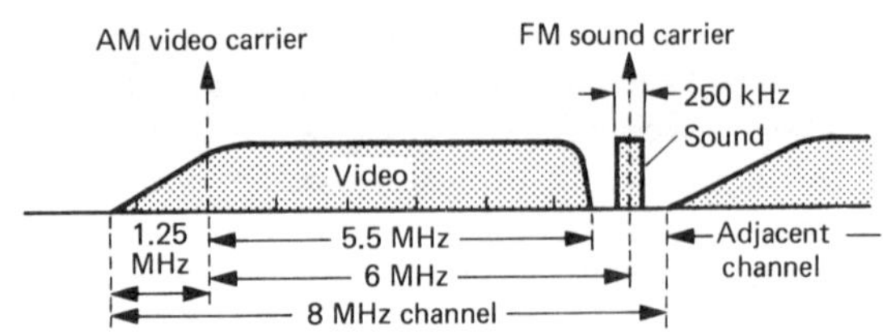

Fig. 18.17

(e) Receiver

In a TV receiver the incoming video signal controls the number of electrons travelling from the electron gun of a cathode ray tube (CRT—see section 9.8) to its screen. The greater the number, the brighter the picture produced by interlaced scanning as in the camera.

The block diagram in Fig. 18.18 for a broadcast receiver shows that the early stages are similar to those in a radio superhet, but bandwidths and frequencies are higher (e.g. the i.f. = 39.5 MHz). The later stages have to demodulate the video and sound signals as well as separate them from each other and from the line and sync pulses.

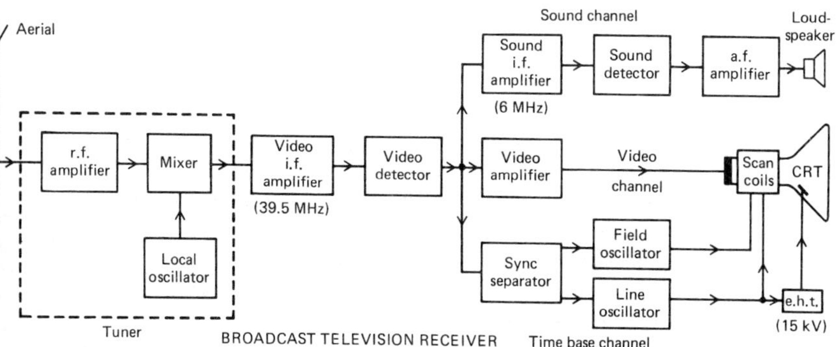

Fig. 18.18

Separation occurs at the output of the video detector. There, the now demodulated AM video signal is amplified by the video amplifier and applied to the modulator (grid) of the CRT to control the electron beam. The still-modulated FM sound signal, having been heterodyned in the video detector with the video signal to produce a sound i.f. of 6 MHz (i.e. the frequency difference between the audio and visual signals, shown in Fig. 18.17) is fed into the sound channel where, after amplification and FM detection, it drives the loudspeaker.

The mixed sync pulses are processed in the time base channel by the sync separator which produces two sets of different pulses at its two outputs. One set is derived from the line pulses by a differentiating circuit with a small time constant (see section 5.8) and triggers the line oscillator. The other set is obtained from the field pulses by an integrating circuit with a long time constant and synchronizes the field oscillator. The oscillators produce the deflecting sawtooth waveforms for the scan coils. The line oscillator also generates the extra high voltage or tension (e.h.t.) of about 15 kV required by the final anode of the CRT.

In a closed circuit television (CCTV) system the video input is connected directly to the video amplifier; the earlier stages of a broadcast receiver are unnecessary.

18.6 Colour television

Colour TV uses the fact that almost any colour of light can be obtained by mixing the three primary colours (for light) of red, green and blue in the correct proportions. For example, all three together give white light; red and green give yellow light.

The principles of both transmission and reception are similar to those for black-and-white (monochrome) TV but the circuits are more complex.

(a) Transmission

A practical requirement is that the colour signal must produce a black-and-white picture on a monochrome receiver—this is called *compatibility*—and, therefore, in a broadcast system, the bandwidth must not exceed 8 MHz despite the extra information to be carried.

A monochrome picture has only brightness or *luminance* variations ranging from black through grey to white. In a colour picture there are also variations of colour or *chrominance*. The eye is more sensitive to changes of luminance than to changes of chrominance and so less detailed information about the latter is necessary. Thus, the luminance of a colour picture is transmitted as it would be in a monochrome system using a 5.5 MHz bandwidth, so ensuring compatibility. In the PAL colour transmission system, which is used in Europe (except France and Russia) but not in the USA, the chrominance signal is amplitude modulated on a subcarrier, 4.43 MHz from the luminance carrier, its sidebands extending about 1 MHz on either side, as shown in Fig. 18.19(*a*).

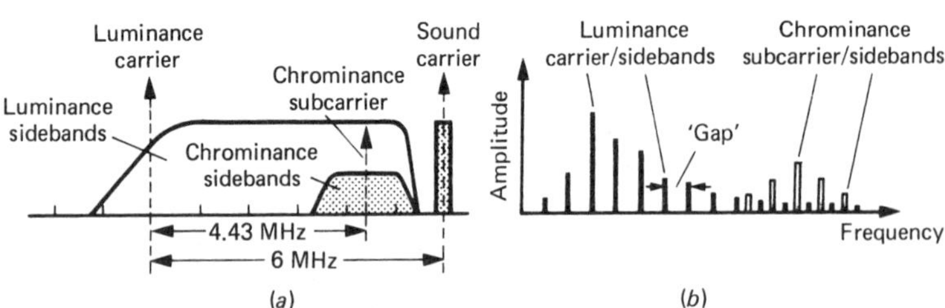

Fig. 18.19

A monochrome video signal consists of groups of frequencies with 'gaps' between, each group being centred on a multiple of the line scan frequency. The chrominance subcarrier is also made up of groups of frequencies and is chosen so that its groups fall into the luminance (monochrome) 'gaps', as in Fig. 18.19(*b*). In this way the luminance and chrominance signals are combined without affecting each other or requiring extra bandwidth.

In a colour TV camera, three vidicon tubes are required, each viewing the picture through a different primary colour filter. The 'red', 'green' and 'blue' electrical signals so obtained provide the chrominance information, which is modulated on the subcarrier by encoding circuits, and, if added together correctly, they give the luminance as well, Fig. 18.20. Alternatively the luminance may be recorded by a fourth tube (without a colour filter) as in black-and-white TV.

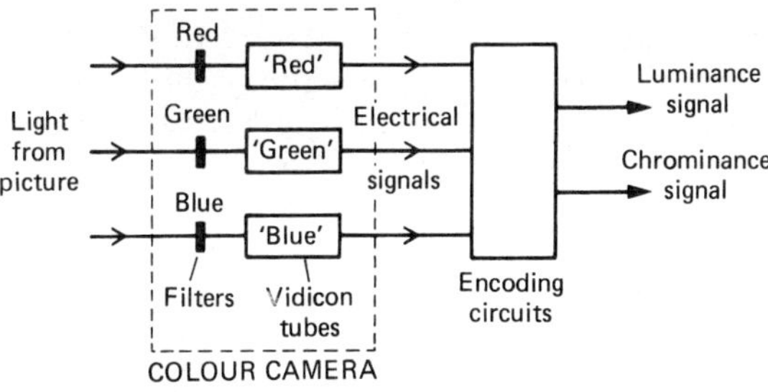

Fig. 18.20

(b) Receiver

In a colour TV receiver, decoding circuits are needed to convert the luminance and chrominance carriers back into 'red', 'green' and 'blue' signals and a special CRT is required for the display.

One common type of display is the shadow mask tube which has three electron guns, each producing an electron beam controlled by one of the primary colour signals. The principle of its operation is shown in Fig. 18.21. The inside of the screen is coated with many thousands of tiny dots of red, green and blue phosphors, arranged in triangles containing a dot of each colour. Between the guns and the screen is the shadow mask consisting of a metal sheet with about 400 000 holes (for a 25 inch diagonal tube).

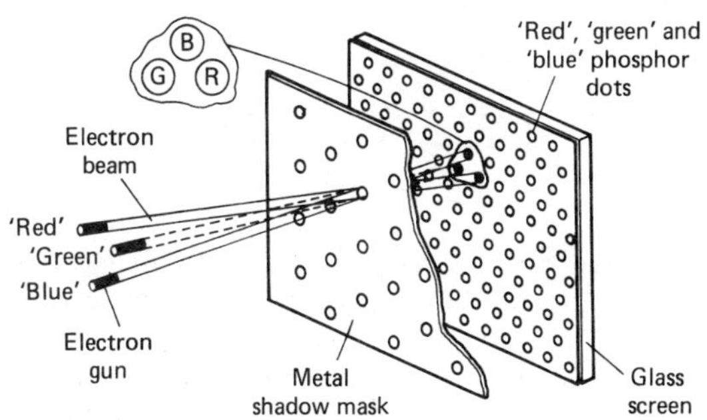

Fig. 18.21

As the three electron beams scan the screen under the action of the same deflection coils, the shadow mask ensures that each beam strikes only dots of one phosphor, e.g. electrons from the 'red' gun strike only 'red' dots. Therefore, when a particular triangle of dots is struck, it may be that the red and green electron beams are intense but not so the blue. In this case the triangle would emit red and green light strongly and so appear yellowish. The triangles of dots are excited in turn and since the dots are so small and the scanning so fast, the viewer sees a continuous colour picture.

The holes in the shadow mask occupy only 15% of the total mask area; 85% of the electrons emitted by the three guns are stopped by the mask. For this reason, the beam current in a colour tube is much greater than that in a monochrome tube for a similar picture brightness. Also, the final anode voltage in the tube is higher (about 25 kV).

More recent colour TV tubes than the shadow mask type are: (i) the Sony (Japan) Trinitron tube which produces three beams from one electron gun and has an 'aperture grille' with vertical slits that match up to vertical phosphor stripes on the screen; (ii) the RCA (USA) precision-in-line (PI) tube, having elongated holes in the shadow mask; and (iii) the Mullard-Philips (UK-Holland) tube which makes circuit design easier. If present research is successful, much flatter tubes, based on LCDs and other technologies, will become available (see section 20.3).

18.7 Sound recording

(a) Tape

Sound can be stored by magnetizing plastic tape coated with tiny particles of iron oxide or chromium oxide. Every particle contains thousands of little magnets each with a north and a south pole. In unmagnetized tape the little magnets point in all directions and cancel out one another's magnetic fields. Under the influence of an external magnetic field, they can be made to turn so that their north poles all point in the same direction. The tape is then magnetized, the strength of the magnetization depending on the number of little magnets turned into line.

The principle of tape recording is shown in Fig. 18.22. The *recording head* is an electromagnet with a very small gap between its poles.

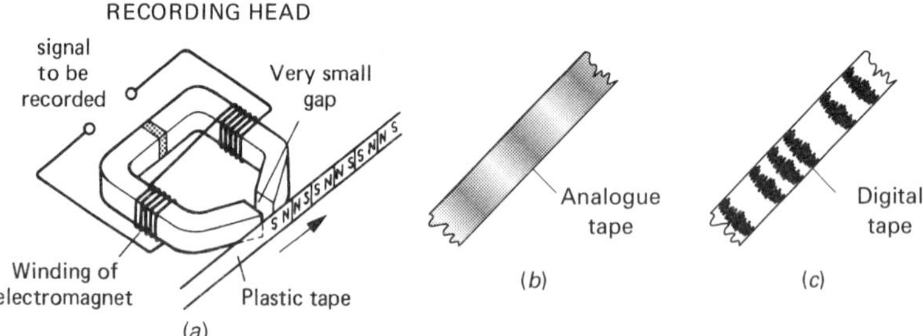

Fig. 18.22

If the input to the recording head is a.c., i.e. an analogue a.f. signal, an *analogue* tape is produced. The strength of the magnetization of any part of it varies continuously from a minimum to a maximum, depending on the value of the input current. This is represented in Fig. 18.22(*b*) by the shading of different density on the tape.

If the input to the head is a digital electrical signal, a *digital* tape is produced, Fig. 18.22(*c*). In this the sound is stored as a sequence of regions of 'high' and 'low' magnetization that represents the 1s and 0s of the digital input.

We will consider in more detail now analogue recording and playback.

(i) Recording When the a.c. due to the a.f. signal to be recorded (e.g. from a microphone) passes through the windings of the electromagnet, it causes an alternating magnetic field in and just outside the gap. As unmagnetized tape passes close to the gap at constant speed, a chain of magnets is produced in it, in a magnetic pattern which represents the original sound. The north pole of each magnet points in the opposite direction to that of its two neighbours owing to the a.c. in the windings reversing the direction of the field in the gap once every cycle.

The *strength* of the magnetization produced at any part of the tape depends on the value of the a.c. when that part passes the gap and this in turn depends on the loudness of the sound being recorded. (A magnet has a maximum strength when all its little magnets are lined up, i.e. when it is saturated; currents exceeding the saturation value cause distortion on playback.) The *length* of a magnet depends on the frequency of the a.c. (and the sound) and on the speed of the tape. High frequencies and low speeds create shorter magnets and very short ones tend to lose their magnetism immediately. Each part of the tape being magnetized must have moved on past the gap in the recording head before the magnetic field reverses. High frequencies and good quality recording require high tape speeds.

Fig. 18.23 shows tapes for two-track and stereo recordings. In the former, half the tape width is used each time the tape is turned over. In the latter, the recording head has two electromagnets, each using one-quarter tape width.

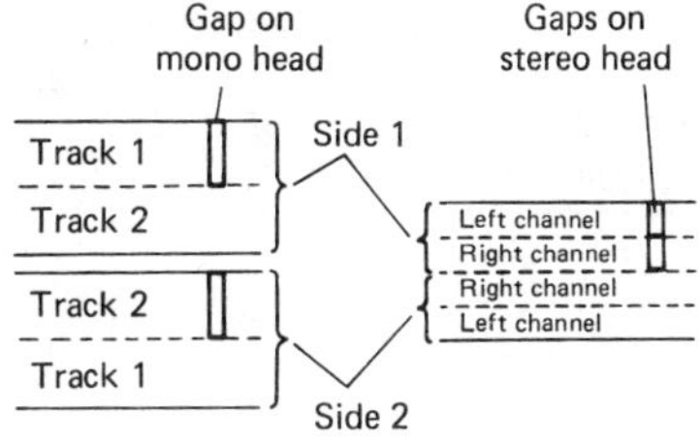

Fig. 18.23

(ii) Bias Magnetic materials are only magnetized if the magnetizing field exceeds a certain value. When it does, doubling the field does not double the magnetization. That is, the magnetization is not directly proportional to the magnetizing field, as the graph in Fig. 18.24(*a*) shows. Therefore, signals fed to the recording head would be distorted when played back. To prevent this, *a.c. bias* is used. A sine wave a.c. with a frequency of about 60 kHz is fed into the recording head along with the signal. The a.c. bias, being supersonic, is not recorded but it enables the signal to work on linear parts of the graph, Fig. 18.24(*b*).

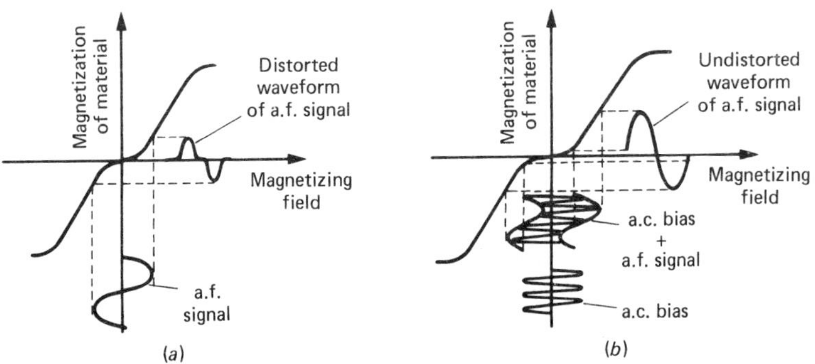

Fig. 18.24

(iii) Playback On playback, the tape runs past another head, the *playback head*, which is similar to the recording head—sometimes the same head records and plays back. As a result, the varying magnetization of the tape induces a small a.f. voltage in the windings. After amplification, this voltage is used to produce the original sound in a loudspeaker.

(iv) Erasing The a.c. bias, after boosting, is also fed during recording to the *erase head*, which is another electromagnet but with a larger gap so that its magnetic field covers more tape. It is placed before the recording head and removes any previous recording by subjecting the passing tape to a decreasing, alternating magnetic field. This allows the little magnets in the magnetic particles to point in all directions again and demagnetize the tape.

A simplified block diagram of a tape recorder is given in Fig. 18.25.

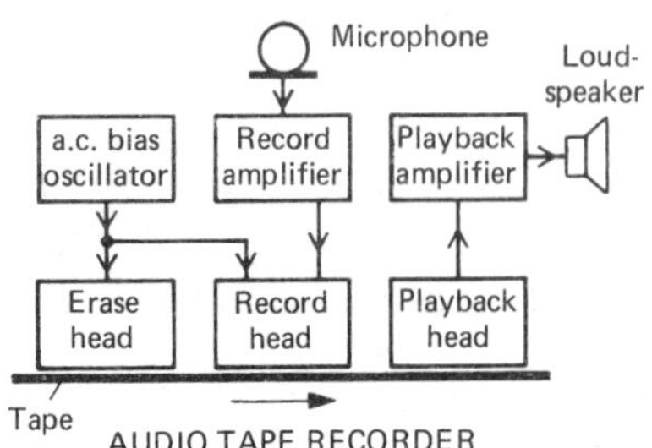

Fig. 18.25

(v) Cassette recorder This is now considered a hi-fi (high fidelity, or high quality of reproduction) system. At the quite low tape speed of 4.75 cm s^{-1}, it can handle frequencies up to 17 kHz. Its improvement is due to better head design and tape quality and to the use of noise reduction circuits, such as the *Dolby* one, which improve higher frequency recording by reducing the level of 'hiss' that occurs during quiet playback spells.

(b) Compact disc (CD)

An audio CD is a plastic (polycarbonate) disc, 120 mm in diameter, which can store about 70 minutes of sound in digital form as a pattern of tiny 'pits' of various lengths and spacing, which form a spriral track about 3 miles long.

(i) Recording During recording, the sound (e.g. music) is converted from an analogue to a digital electrical signal and used to control a very intense, narrow beam of light from a powerful laser which, in effect, 'burns' out the track of 'pits' on the disc's surface as it revolves. The disc is then coated with a film of aluminium to make it reflective and lacquered for protection.

(ii) Playback During playback on a CD player, the pick-up in the form of a semiconducting laser sends a fine beam of light to the disc via a partly reflecting mirror and a lens system. If the beam falls on the rough bottom of a 'pit', Fig. 18.26(*a*), it is scattered, no reflected beam or electrical signal is produced in the photodiode detector and this represents a '0'. If the beam falls on a smooth space between 'pits', Fig. 18.26(*b*), there is a reflected beam and signal which gives a '1'. The photodiode converts the interruptions of the beam into a *digital* electrical signal before it undergoes digital-to-analogue conversion and operates an amplifier and loudspeaker. The pick-up follows the sound-carrying spiral track by moving across the CD as it revolves.

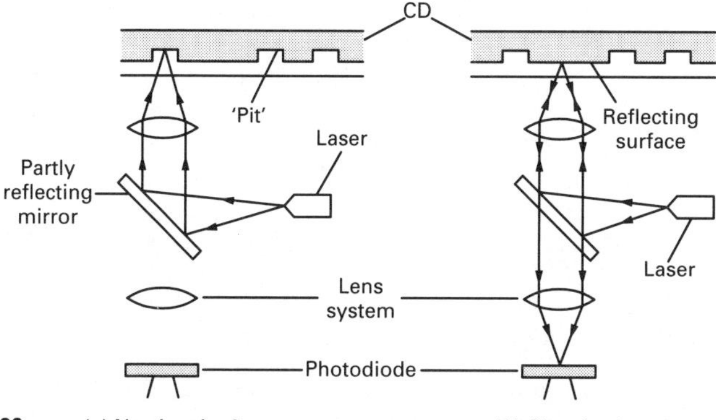

Fig. 18.26 (*a*) No signal = 0 (*b*) Signal = 1

A CD is not subject to the wear that its predecessor, the vinyl record, suffered since there is no physical contact between it and the pick-up. Recording and retrieving sound digitally is less subject to *distortion* and *noise* than analogue methods but is more difficult technically.

18.8 Video recording

(a) Tape

In theory there is no reason why video signals should not be recorded on magnetic tape in the same way as audio signals. In practice, because frequencies of the order of 5 MHz are involved, the very high tape speeds required would create difficult problems. However, the same effect can be obtained if a slow tape speed is used and the head (recording or playback) moves at high speed.

In a video cassette recorder (VCR), the tape passes round a drum with a slit in its side and containing two heads on an arm rotating at 25 revolutions per second, as shown in Fig. 18.27(*a*). The mechanical arrangement is such that each head scans one complete TV field (picture) in a series of parallel diagonal lines during the half-revolution it is in contact with the tape, as in Fig. 18.27(*b*). A track on one edge of the tape carries synchronizing pulses for control purposes and sound is recorded on the other edge; both use stationary heads.

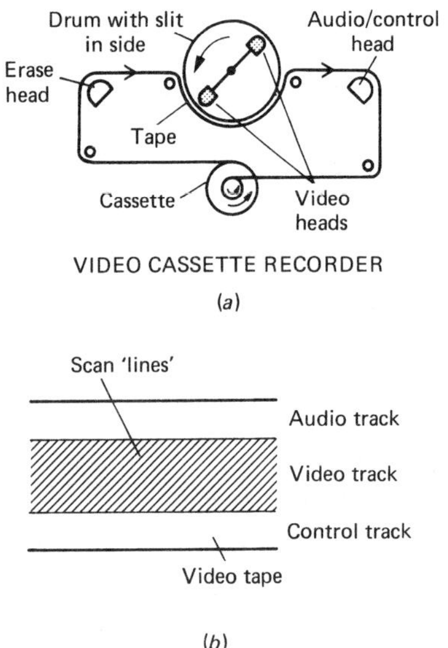

Fig. 18.27

(b) Compact disc (CD)

Video CDs are based on the same technology as audio CDs but in this case the spiral track of 'pits' in the disc are the digital representation of video signals arising from, for example, drawings (graphics), photographs or animation (moving figures). Single video CDs can store 100 colour photos and in the near future a full-length film lasting over 2 hours will be available by compressing information on a DVD (Digital Versatile Disc—see section 20.4).

'Rewritable' CDs are now available that can be 'written' on and erased many times—like video cassette tapes, which they may eventually replace.

18.9 Digital electronics in communications

(a) The communications revolution

The revolution now occurring in communications is largely due to the use of digital electronics. Digital technology has taken over in public telephone systems Digital radio and TV, which will give more channels and better quality sound and pictures, are planned. One of the most far-reaching developments in recent years has been the establishment and rapid growth of the Internet as a way in which people communicate with each other worldwide using the digital language of computers (see section 20.6). This interactive global network may become the much talked-about 'Information Superhighway'.

In the near future, communication satellites are likely to make a further significant contribution to the communications revolution. With synchronous satellites (see section 18.1) there is a time delay in two-way communication because of the large distances involved (36 000 km). As a result, some companies are planning to launch several hundreds of fairly inexpensive satellites with low-level polar orbits at about 750 km above the Earth. Unlike synchronous satellites they would move relative to the Earth, hence the need for a large number to ensure that at least one was orbiting overhead at all times in a given region.

It is not easy to predict what the full impact of the communications revolution will be on our daily lives in the future but already it is clear that the world has become a much smaller place with the availability of services such as videoconferencing, mobile phones and electronic mail. It has been suggested that the implications for human culture may turn out to be as profound as those that arose from the invention of printing in the 15th century.

(b) Pulse code modulation (PCM)

In pulse code modulation, which gives us another way of representing information electrically and is at the heart of the communications revolution, an analogue signal is converted into a digital one by an analogue-to-digital (A/D) converter—see section 17.3. The amplitude of the analogue signal is measured at regular intervals (Fig. 18.28(*a*)) and, each time, it is represented in binary code by the number and arrangement of the pulses in the digital output from the converter. The three-bit code in Fig. 18.28(*b*) can represent eight (0 to 7) voltages; four bits would allow sixteen (0 to 15) voltages to be encoded.

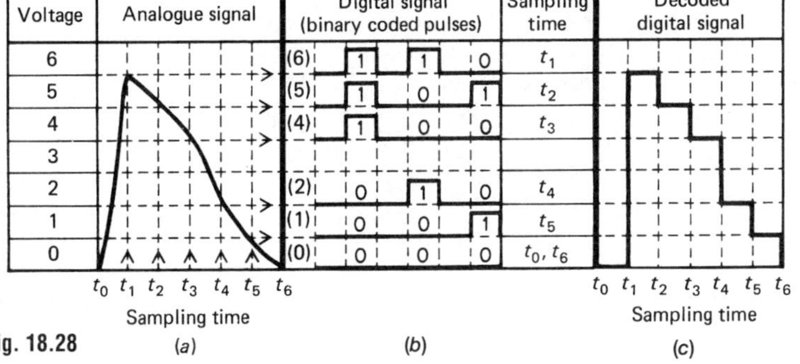

Fig. 18.28 (*a*) (*b*) (*c*)

The accuracy of the representation also increases with the sampling frequency, which has to be greater than twice the highest frequency of the analogue signal to be sampled. The highest frequency needed for intelligible speech in a telephone system is about 3500 Hz and a sampling frequency of 8000 Hz is chosen, i.e. samples are taken at 125 μs intervals, each sample lasting for 2 to 3 μs. An eight-bit code (giving $2^8 = 256$ levels) is used and so the number of bits that have to be transmitted is $8000 \times 8 = 64\,000$, i.e. the *bit-rate* is 64 kbit/s and is given by:

$$\text{bit-rate} = \text{sampling frequency} \times \text{no. of bits in code}$$

For good quality music where frequencies up to about 16 000 Hz must be transmitted, the sampling frequency is 32 000 Hz and a sixteen-bit code ($2^{16} = 65\,536$ levels) is used. The bit-rate required is $32\,000 \times 16 = 512$ kbit/s. If the music was to be stored on an audio compact disc playing for one hour, then:

$$\text{number of bits stored} = 512 \times 10^3 \text{ bit/s} \times 3600\text{s}$$

or, since 1 byte = 8 bits,

$$\begin{aligned}\text{number of bits stored} &= 512 \times 36 \times 10^5/8 \\ &= 230 \times 10^6 \text{ bytes} \\ &= 230 \text{ Mbytes}\end{aligned}$$

For television signals, which carry much more information, a bit-rate of 70 000 000 = 70 Mbit/s is needed.

The analogue voltage shown in Fig. 18.28(*a*) would be represented in digital form by the train of pulses in Fig. 18.29 using a three-bit code.

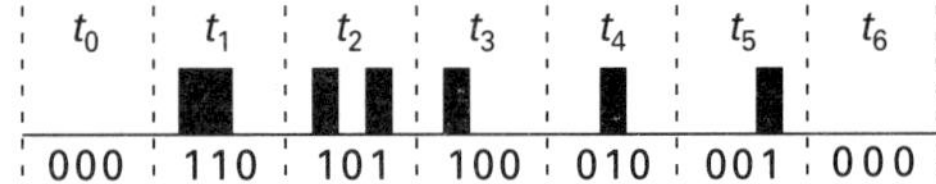

Fig. 18.29

Information in digital form (with just the two voltage levels of 'high' and 'low') has certain advantages over that in analogue form (with voltages that vary continuously) when it has to be transmitted, by cable or r.f. carrier, or processed (e.g. amplified, recorded). First, distortion may occur and in an analogue signal this is difficult to remove but in a digital one it does not matter because the pulse still exists. Second, noise can be picked up and while it is easily 'cleaned off' a digital signal by 'clipping', Fig. 18.30, it causes problems in an analogue system. Therefore, although most transducers produce analogue signals, digital signals are much more reliable information carriers.

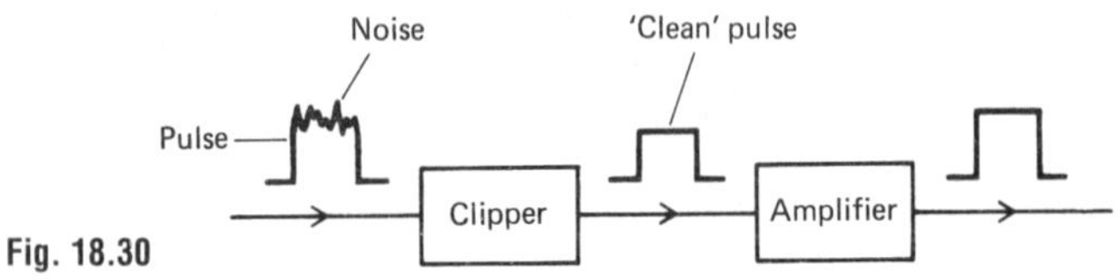

Fig. 18.30

When a digital signal has to be changed back to an analogue one, a digital-to-analogue (D/A) converter (see section 17.2) decodes it to give a stepped waveform like that in Fig. 18.28(*c*). If a high sampling frequency is used during encoding (e.g. 1 MHz), the steps are very small and the decoded signal is a good copy of the original analogue signal.

(c) Representing text

To be processed in a digital system such as computer, words have to be digitized. This means a binary code pattern of 1s and 0s has to be agreed for the letters of the alphabet. The *American Standard Code for Information Interchange* (ASCII) is an eight-bit code that allows $2^8 = 256$ characters to be coded in binary. This is adequate for all letters of the alphabet (capitals and small), the numbers 0 to 9, punctuation marks and other common symbols. Capital A is represented by the number 65 (0100 0001 in binary), capital B by 66 (0100 0010) and so on. Lower-case letters start with 97 and the space between words is represented by 32 (0010 0000). An average (six-letter) word therefore needs about 48 bits, i.e. 6 bytes or roughly 1 byte per character.

(d) Compressing digital data

The number of bits to be processed in even a short piece of text can be huge. To reduce the time needed to transmit large numbers, the technique of digital data compression is used. The idea is to omit redundant information. For example, the letter *u* is redundant when it comes after the letter *q* because we know that a *u* will follow a *q*. It is not unusual for compression to halve the amount of text to be sent, thereby allowing it to be sent twice as fast.

Compression can also be applied to sound and video data, even reducing it to one-tenth of its original size. For instance, suppose that ten picture elements (i.e. the separate dots or *pixels* making up the picture) are in one row on the screen of a cathode ray tube. It takes fewer bits to represent the colour once and state that it should be repeated ten times than it does to indicate the colour ten times. That is, compression occurs by describing how the colour *changes*.

(e) Digital bandwidth

The 'bandwidth' of a transmitting medium for digital signals is a measure of the bit-rate, i.e. the number of bits it can transmit per second (1 bit per second = 1 *baud*). It gives the rate of transfer of information (in bits per second), something that is difficult to do exactly for analogue signals.

Twisted-pair copper cable that connects homes to the local telephone system is a 'narrow-band' medium (bandwidth up to about 150 kilobits per second). It carries text and voice data and services such as e-mail (see section 18.10) and computer games based on the Internet.

Coaxial copper cable (Fig. 18.04(*b*)) has a much higher bandwidth (up to about 100 megabits per second) and can carry graphics, animation and TV pictures.

Optical fibres (see section 18.11), which now form the backbone of the telephone network and also carry cable TV to homes, have broad-band capacity (up to a few gigabits per second) and can carry good quality audio and video signals.

Digital transmission via microwaves is a rapidly developing medium with narrow-band capacity (less than that of coaxial cable).

18.10 Some digital communication systems

(a) Digital telephone system

By the early 1990s the whole of Britain's trunk telephone network (i.e. between cities) and about half of local telephone lines were connected by digital exchanges. British Telecom's digital communication system, *System X*, uses pulse code modulated digital signals, electronic switching circuits, and computers to control the routing of telephone calls and faxes, as well as *e-mail* and the transmission of computer data (see below).

(b) Electronic mail (e-mail)

Electronic mail or *e-mail* allows users instant communication, using the ordinary telephone network, to carry text from the sender's microcomputer and be stored in an electronic 'mailbox' in the service provider's mainframe computer. The receiver can open the mailbox using a personal password and either file the message, send an immediate reply using his or her microcomputer or make a hard copy using a printer.

(c) Modems

While most of the telephone network is digital, ordinary copper telephone lines from telephones to local exchanges are designed to carry audio frequency signals in the range 300 Hz to 3400 Hz. They are not meant to handle long strings of fast rising and falling d.c. pulses which would be affected by capacitive and inductive effects.

However, digital data can be sent over the telephone network so long as there is a device called a *modem* (*mo*dulator–*dem*odulator) at each end. The modem has to do three things:

(i) at the sending end it changes digital data into audio tones; in one system it encodes a 1 as a burst of high tone (2400 Hz) and a 0 as a burst of low tone (1200 Hz);
(ii) at the receiving end it decodes the tones into digital data;
(iii) at both ends it electrically isolates the low voltage levels of the digital equipment (e.g. a computer) from the high voltage levels of the telephone system and so prevents damage to either.

Fax and e-mail are among the services that require a modem when transmission is via the ordinary telephone line. A typical modem processes data at a rate of 33.6 kilobits per second, but this rate is ever-increasing.

ISDN (*Integrated Services Digital Network*) is a development which enables all digital communications—high quality speech, text, data, drawings, photographs, music—to be carried from computer to computer using a *terminal adaptor* instead of a modem. The computer and digital telephone are connected directly to a digital local exchange by an ISDN line which can carry two independent digital communications on a pair of high quality copper wires. Transmission is much faster and more reliable than via a modem.

(d) Mobile (cellular) phones

Cellphone networks like *Cellnet* and *Vodafon* allow hand-held phones to send and receive messages by microwave radio. If the user travels from one 'cell' in the country to another, connection to the new switching centre occurs automatically by computer.

The first cellphone systems, launched in the early 1980s, broadcast on a frequency of 900 MHz using analogue technology and were only meant to carry speech. The more recent, more secure, second generation systems are all-digital; they suffer less from interference and give longer battery life. They also operate at 900 MHz but with separate channels for speech and data, with a bandwidth of 9.6 kilobits per second.

A system at present under development is known as UMTS (*Universal Mobile Telephone System*). It is also an all-digital system, designed to handle mainly data, giving access to a company's computer network from a mobile office. When working at full power to carry calls over a wide area, the data rate is 150 kilobits per second (twice that obtainable on ISDN lines). UMTS uses a frequency in the 2 GHz range (near to the 2.45 GHz of microwave ovens).

(e) Teletext

This is a system which, aided by digital techniques, displays on the screen of a modified domestic TV receiver, up-to-the-minute facts and figures on news, weather, sport, travel, entertainment and many other topics. It is transmitted along with ordinary broadcast TV signals, being called *Ceefax* ('see facts') by the BBC and *Oracle* by ITV.

During scanning, at the end of each field (i.e. 312.5 lines), the electron beam has to return to the top of the screen. Some TV lines have to be left blank to allow time for this and it is on two or three of these previously blank lines in each field (i.e. four or six per frame of 625 lines) that teletext signals are transmitted in digital form.

One line can carry enough digital signals for a row of up to 40 characters in a teletext page. Each page can have up to 24 rows and takes about $\frac{1}{4}$ second (i.e. $12 \times 1/50\,\text{s}$) to transmit. The pages are sent one after the other until, after about 25 seconds, a complete magazine of 100 pages has been transmitted before the whole process starts again.

The teletext decoder in the TV receiver picks out the page asked for (by pressing numbered switches on the remote control keypad) and stores it in a memory. It then translates the digital signals into the sharp, brightly coloured words, figures and symbols that are displayed a page at a time on the screen.

18.11 Optical fibre systems

The suggestion that information could be carried by light sent over long distances in thin fibres of very pure (optical) glass was first made in the 1960s. In just eleven years, the world's first optical fibre telephone link was working in Britain. Now nearly all of the trunk 'traffic' is via optical fibre cable and local telephone lines are also being replaced. Undersea optical fibre links are in operation between Britain and the Continent and the USA.

(a) Outline of system

A simplified block diagram is shown in Fig. 18.31(*a*) and a system (and fibres) in Fig. 18.31(*b*). The electrical signals representing the *information* (speech, television pictures, computer data, for instance) are pulse code modulated in an *encoder* and then changed into the equivalent digital 'light' signals by the *optical transmitter*. This is either a miniature laser or an LED bonded on to the end of the fibre. The 'light' used is infrared radiation in the region just beyond the red end of the visible spectrum (with a wavelength of 0.85, 1.3 or 1.5 μm) because it is attenuated less by absorption in the glass than 'visible' light.

The *optical fibre* is 123 μm (0.125 mm) in diameter and has a glass core of higher refractive index than the glass cladding around it. As a result the infrared beam is trapped in the core by total internal reflection at the core–cladding boundary and bounces off it in zig-zag fashion along the length of the fibre (Fig. 18.32).

The *optical receiver* is a photodiode and converts the incoming infrared signals into the corresponding electrical signals before they are processed by the *decoder* for conversion back into *information*.

(b) Advantages

A digital optical fibre system has important advantages over other communication systems.

(i) It has a high information-carrying capacity, typically 560 Mbits s^{-1} at present (i.e. about 9000 telephone channels or over 1000 music channels or 8 television channels), which is over 5 times that of the best copper cable and 8 times that of microwaves. (Capacities as high as 2 Gbit s^{-1} have been achieved in laboratory tests.)
(ii) It is free from 'noise' due to electrical interference.
(iii) Greater distances (e.g. 50 km) can be worked without regenerators (see section 18.12(b) below); copper cables require repeaters to be much closer.
(iv) An optical fibre cable is lighter, smaller and easier to handle than a copper cable.
(v) Optical fibre cable is cheaper to produce than copper cable, does not suffer from corrosion problems and is generally easier to maintain.
(vi) Crosstalk between adjacent channels is negligible.
(vii) It offers greater security to the user.

Information → Encoder → Optical transmitter — Optical fibre — Optical receiver → Decoder → Information

(*a*)

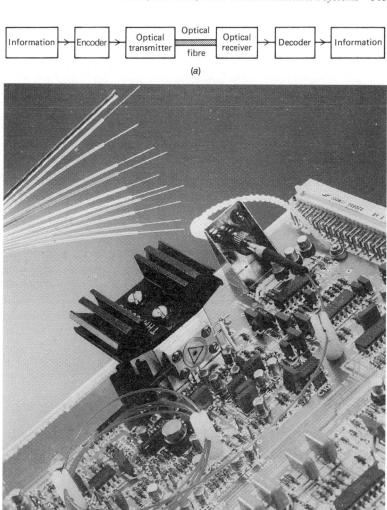

Fig. 18.31 (*b*)

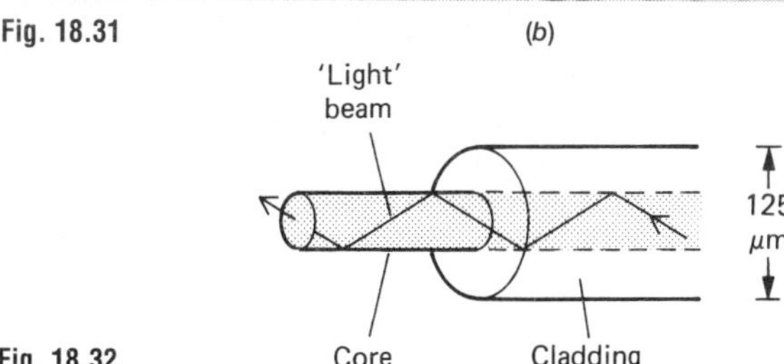

Fig. 18.32

18.12 Optical fibres

(a) Types

There are two main types—multimode and monomode.

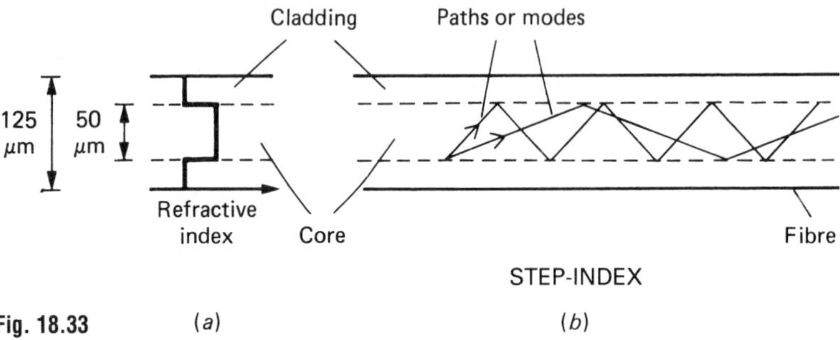

Fig. 18.33 (*a*) (*b*)

In the *step-index multimode* type, the core has the relatively large diameter of 50 μm and the refraction index changes abruptly at the cladding, Fig. 18.33(*a*). The wide core allows the infrared radiation to travel by several different paths or modes. Paths that cross the core more often are longer, Fig. 18.33(*b*), and signals in those modes take longer to travel along the fibre. Arrival times at the receiver are therefore different for radiation from the same pulse, 30 ns km^{-1} being a typical maximum difference. The pulse is said to suffer *dispersion*, i.e. it is spread out, Fig. 18.34. In a long fibre separate pulses may overlap and at the receiving end, errors and loss of information will occur.

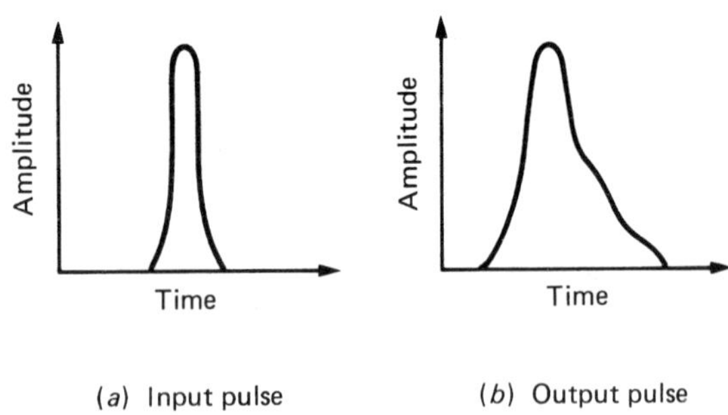

(*a*) Input pulse (*b*) Output pulse

Fig. 18.34

In the *graded-index multimode* type, the refractive index of the glass varies continuously from a high value at the centre of the fibre to a low value at the outside, so making the boundary between core and cladding indistinct, Fig. 18.35(*a*). Radiation following longer paths travels faster on average since the speed of light is inversely proportional to the refractive index. The arrival time for different modes are then about the same (to within 1 ns km^{-1}) and all arrive more or less together at the receiving end, Fig. 18.35(*b*). Dispersion is thereby much reduced.

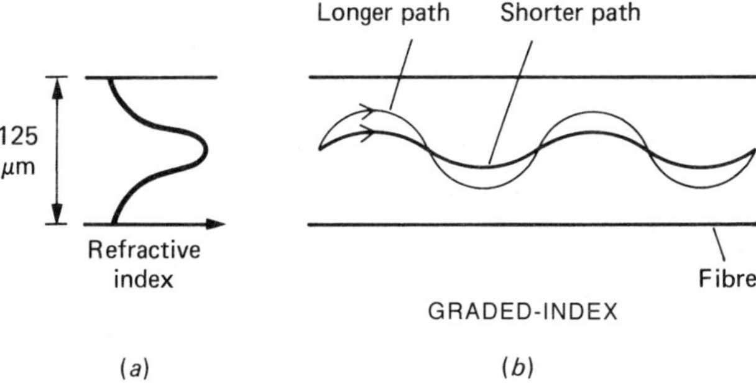

Fig. 18.35

In the *monomode* fibre the core is only 5 μm in diameter, Fig. 18.36, and only the straight-through transmission path is possible, i.e. one mode. This type, although more difficult and expensive to make, is being increasingly used. For short distances and low bit-rates, multimode fibres are quite satisfactory.

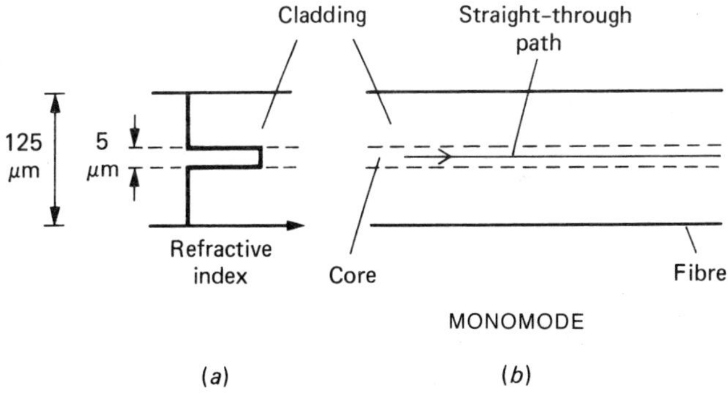

Fig. 18.36

(b) Attenuation

Absorption of infrared radiation occurs when it travels through glass, being less for longer wavelengths than for shorter ones. The *optical power* decays exponentially with fibre length x according to the equation

$$P = P_o e^{-\alpha x}$$

where P and P_o are the input and output powers respectively and α is a constant called the *attentuation coefficient* of the fibre. It is expressed in km^{-1} and is given by

$$\ln\left(\frac{P}{P_o}\right) = -\alpha x$$

As well as loss due to absorption by impurity atoms in the glass, scattering of the radiation at imperfect joints in the fibre also adds to the attenuation.

Attenuated pulses have their size and shape restored by the use of *regenerators* at intervals along the optical cable.

Increased absorption and scattering of the 'light' signal occurs as an optical fibre ages due to hydrogen diffusing into the glass. Hydrogen is liberated by the slow decomposition of the protective plastic cover on the fibre. It can also be formed by water seeping in and causing electrolytic action among traces of metals that are added to the glass to change its refractive index.

18.13 Optical transmitters

(a) Light-emitting diode (LED)

A LED is a junction diode made from the semiconducting compound gallium arsenide phosphide; its action was considered earlier—see section 9.6(*a*). Those used as optical fibre transmitters emit infrared radiation at a wavelength of about 850 nm (0.85 μm). Pulse code modulated signals from the coder supply the input current to the LED which produces an equivalent stream of infrared pulses for transmission along the fibre. The spectral spread of wavelengths in the output is 30 to 40 nm.

LEDs are a cheap, convenient 'light' source. They are generally used only with multimode fibres because of their low output intensity and in low bit-rate digital systems (up to 30 Mbit s^{-1} or so) where the 'spreading' of output pulses due to dispersion is less of a problem. A lens between the LED and the fibre helps to improve the transfer of 'light' energy between them.

(b) Lasers

A laser, named from the first letters of *l*ight *a*mplification by the *s*timulated *e*mission of *r*adiation, produces a very intense beam of light or infrared radiation which is:

(i) *monochromatic*, i.e. consists of one wavelength;
(ii) *coherent*, i.e. all parts are in phase;
(iii) *collimated*, i.e. all parts travel in the same direction.

Those used in optical fibre systems are made from gallium arsenide phosphide and one the size of a grain of sand can produce a continuous power output of around 10 mW. The speed at which a laser can be switched on and off by the digital pulses of input current, is much faster than for an LED. Spectral spreading of the radiation emitted is also smaller (1 to 2 nm or less) and as a result, dispersion is not such a problem (since refractive index and so speed depend on wavelength). Lasers are therefore more suitable for use with monomode, high bit-rate fibre systems. However they are more complex and currently more expensive.

18.14 Optical receivers

The receiver converts 'light' signals into the equivalent stream of electrical pulses. A reverse biased photodiode—section 9.5(a)—is used to do this both at the end of the system and in regenerators along the cable.

The *p-i-n photodiode* has a low doped intrinsic (i) depletion layer between the p and n regions, Fig. 18.37. When 'light' photons (bundles of light energy) are absorbed in this i-region, the resulting electrons and holes then move in opposite directions under the applied voltage to form the current through the external circuit of the diode. Reverse bias depletes the i-region completely and produces an electric field high enough to cause rapid motion of the charge carriers. This ensures they respond rapidly to changes of 'light' intensity on the photodiode.

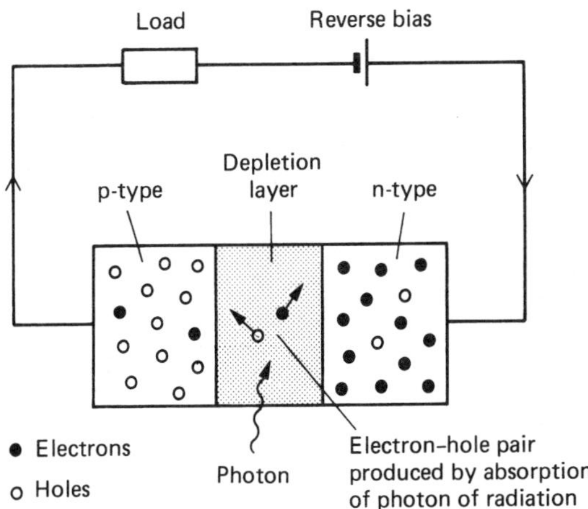

Fig. 18.37

18.15 Multiplexing

Multiplexing involves sending several different information signals along the same communication channel so that they do not interfere. Two methods are outlined.

(a) Frequency division (for analogue signals)

The signals (e.g. speech in analogue form) modulate carriers of different frequencies which are then transmitted together at the same time, a greater bandwidth being required.

The principle is shown in Fig. 18.38 for cable transmission. The information signal contains frequencies up to 4 kHz and the carrier frequency is 12 kHz. *Each* sideband contains *all* the modulating frequencies, i.e. all the information, and so only one is really necessary. Here, a 'bandpass filter' allows just the upper sideband to pass.

The multiplexing of three 4 kHz wide information signals 1, 2 and 3, using carriers of 12, 16 and 20 kHz is shown in Fig. 18.39.

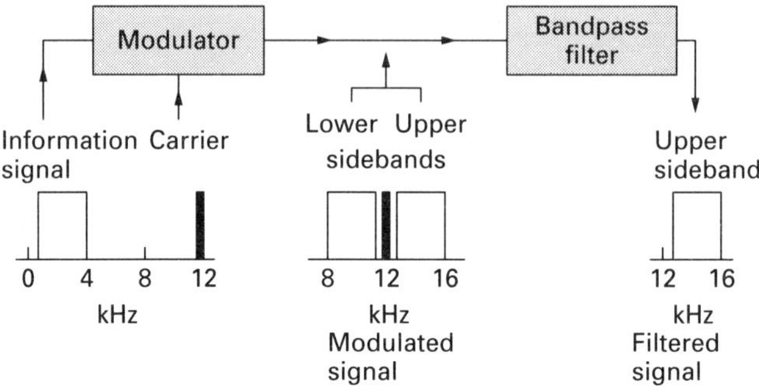

Fig. 18.38

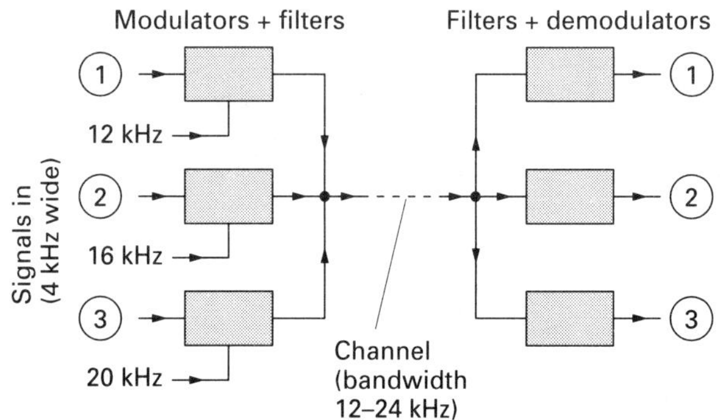

Fig. 18.39

(b) Time division (for digital signals)

This method is shown in Fig. 18.40 for three signals. An electronic switch, i.e. a multiplexer (section 15.10 and here 3 : 1 line), samples each signal in turn (for speech 8000 times per second). The encoder converts the samples into a stream of pulses, representing, in binary, the level of each signal (as in PCM). The transmissions are sent in sequence, each having its own time allocation. The decoder and a de-multiplexer (1 : 3 line), which are synchronized with the multiplexer and encoder, reverse the operation at the receiving end.

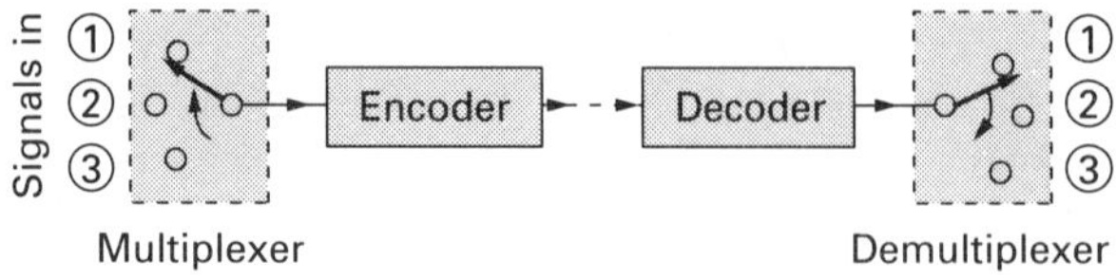

Fig. 18.40

(c) Example

Consider the transmission of data from a database to a computer. If a page (screen) of data consists of 20 lines, each with 50 characters, then:

$$\text{number of characters per page} = 20 \times 50 = 1000$$

Suppose each character of data comprises 1 start bit, 8 data bits and 2 stop bits, then:

$$\text{total number of bits to be transmitted} = 1000 \times 11$$

If the sending rate is 1100 baud, then:

$$\text{time to transmit one page} = \frac{1000 \times 11}{1100} = 10\text{s}$$

If several sets of separate signals had to be sent to different destinations on the same telephone line, time-division multiplexing would be used, each terminal receiving the data being allotted a time slot. Frequency division multiplexing would be restricted by the bandwidth of the line.

(d) DWDMs

New technologies are being developed that allow the number of channels carried over, for example, an existing single optical fibre to be increased. One such system is based on *Dense Wavelength Division Multiplexers*, whereby more efficient use can be made of the optical fibre bandwidth.

18.16 Revision questions and Problems

1. a) State three ways in which radio waves travel.
b) Explain the terms: ionosphere, critical frequency, fading.

2. In what frequency bands do the following operate:
(i) a 27 MHz citizen's band radio,
(ii) a 500 kHz ship's radio telephone link,
(iii) a 500 MHz TV broadcast station,
(iv) a 96 MHz FM radio transmission,
(v) a communication satellite?

3. a) Calculate the length of a dipole aerial suitable for operation at 600 MHz.
b) If a dipole has a length of 1 m, at what frequency is it a half-wavelength long?

4. Explain the action of
a) the reflectors and directors on a dipole aerial,
b) a ferrite rod aerial.

5. a) Why must sound modulate an r.f. carrier to be transmitted?
b) Explain with the help of diagrams the terms: modulation depth, side frequency, sidebands, bandwidth.
c) What is the bandwidth when frequencies in the range of 100 Hz to 4.5 kHz *amplitude modulate* a carrier?

6. a) In Fig. 18.41, what do the numbered blocks represent if it is for
(i) an AM,
(ii) an FM, superhet radio receiver?
b) What advantages does a superhet have over a straight radio receiver?

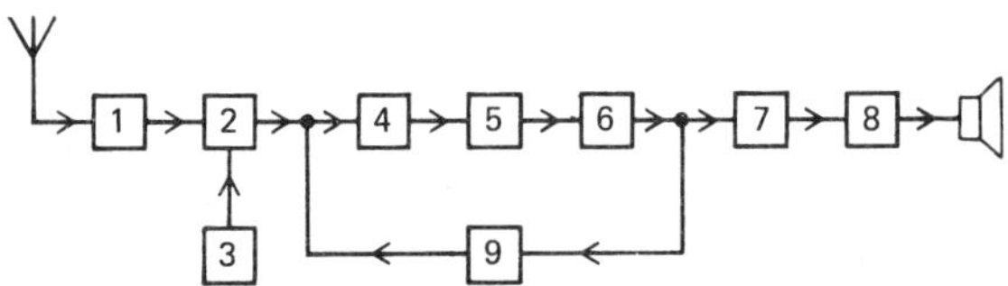

Fig. 18.41

7. a) What is the effect of increasing the amplitude of the modulating frequency in an FM transmitter?
b) If a carrier is frequency modulated by a pure sine wave, how many side frequencies are produced?
c) Why is AM used rather than FM for medium wave radio broadcasts?

8. Explain the terms: a.g.c., a.f.c.

9. a) In TV what type of modulation is used for
(i) sound signals,
(ii) video signals?
b) What is the bandwidth of a TV video amplifier?
c) Explain the following terms in connection with TV: raster, line scan, field scan, interlacing, synchronization, vestigial sideband transmission.
d) In the simplified block diagram of a black-and-white TV receiver in Fig. 18.42, state what the numbered blocks represent.

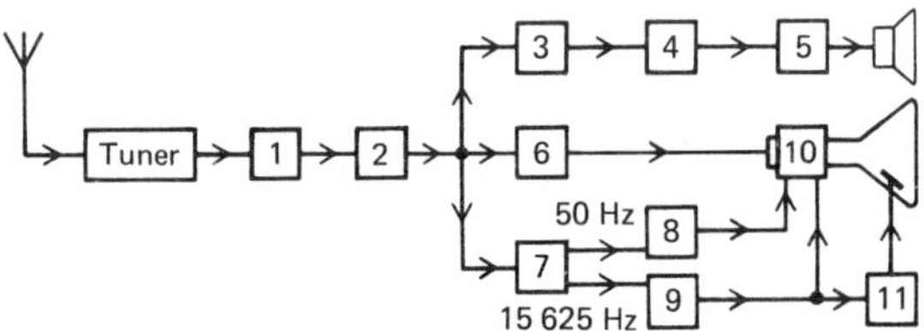

Fig. 18.42

10. a) Explain the following terms used in colour TV: compatibility, luminance, chrominance.
b) Name four types of colour TV tube.

11. Explain the principles of sound recording on
a) magnetic tape,
b) compact disc.

12. a) Give two reasons in favour of
(i) the digital system,
(ii) the analogue system,
of handling information.
b) Explain briefly the meaning of the following:
(i) ASCII,
(ii) digitization of data,
(iii) digital data compression,
(iv) digital bandwidth.

13. a) Give *three* reasons why a *modem* must be used to send computer data along a telephone link.
b) If a modem has a bit-rate of 33 600 baud, estimate how long it takes to send, in ASCII code, a page of text that contains 50 lines, each line having on average 100 characters.

14. a) Draw a block diagram for an optical fibre communication system.
b) State seven advantages optical fibre cables have over copper cables.
c) Explain the terms *pulse spreading* and *attenuation* and state their causes in optical fibres.
d) Distinguish between
(i) multimode and monomode fibres,
(ii) step-index and graded-index fibres.

15. a) Compare the spectral output and radiation characteristics of two types of optical transmitter.
b) Describe the structure and operation of a p-i-n photodiode.

16. a) Why does *multiplexing* make the use of telephone cables more cost effective?
b) Explain what is meant by
(i) time-division multiplexing,
(ii) frequency-division multiplexing.

UNIT 19

Digital systems

19.1 Designing digital systems

There are no rule-of-thumb methods for finding the best and most economical design for a digital system built from logic circuits. However, once it is known what the system has to do, it is often a good starting point to draw up a truth table. This technique will be used to design some simple systems. The first three systems control lamps which have to flash in certain sequences and are driven by 'clock' pulses of constant frequency (e.g. from an astable) applied to an appropriate binary counter.

(a) System 1

During each cycle of four pulses a lamp has to flash during the fourth one, i.e. it has to be off for the first three pulses and on for the fourth. A logic circuit is required for the system which has a 'high' (a 1) output only during the fourth pulse of each cycle.

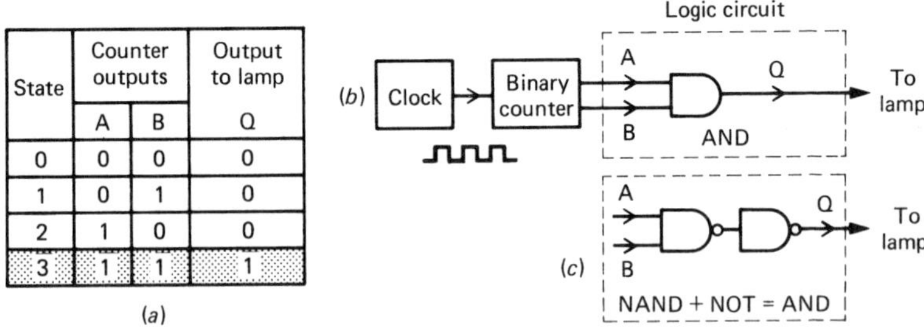

State	Counter outputs		Output to lamp
	A	B	Q
0	0	0	0
1	0	1	0
2	1	0	0
3	1	1	1

Fig. 19.01

The cycle is a four-state one requiring a modulo-4 counter—see section 16.7. Such a counter has two outputs, A and B, and, as the clock pulses are applied to it, it will go through the four binary combinations in turn in the truth table of Fig. 19.01(a). (Note that the first stage is state 0, the second is state 1, and so on.) The output Q from the counter is 'high' only when both inputs A and B are 'high'. Using *Boolean algebra*, where a dot (.) represents the AND logic operation,

$$Q = A \, . \, B$$

The logic circuit for the system therefore consists simply of a two-input AND gate whose output feeds the lamp, as in Fig. 19.01(*b*). In practice, for economic reasons, logic circuits are usually constructed from NAND gates because of their availability and cheapness: two are required here, the second one in Fig. 19.01(*c*) has one input and acts as an inverter—see section 15.4.

(b) System 2

A lamp has to light for the fourth and fifth pulses every eight pulses.

State	Counter outputs			Output to lamp
	A	B	C	Q
0	0	0	0	0
1	0	0	1	0
2	0	1	0	0
3	0	1	1	1
4	1	0	0	1
5	1	0	1	0
6	1	1	0	0
7	1	1	1	0

(*a*)

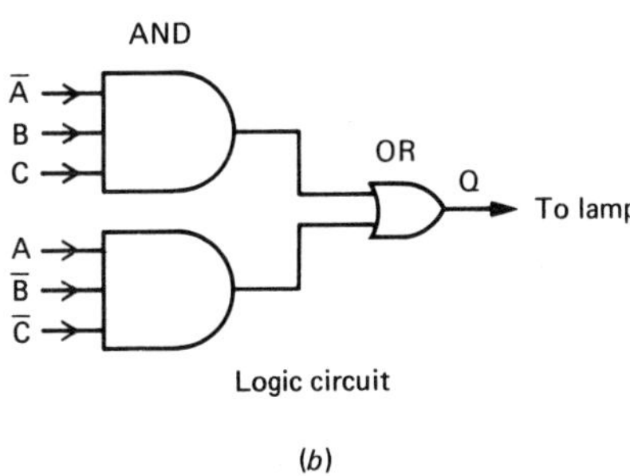

(*b*)

Fig. 19.02

This is an eight-state cycle and a modulo-8 counter with three inputs is needed. From states 3 and 4 of the truth table Fig. 19.02(*a*) we see that Q is 1 when:

(i) A is 0 *and* B is 1 *and* C is 1
or
A is 1 *and* B is 0 *and* C is 0

Now, if an input (or output) is 0, its complement is 1, for example if A is 0, $\overline{A}$ (not A) is 1. Rewriting (i) so that all inputs are 1s, the statement is still the same and we get Q is 1 when:

(ii) $\overline{A}$ is 1 *and* B is 1 *and* C is 1
or
A is 1 *and* $\overline{B}$ is 1 *and* $\overline{C}$ is 1

In Boolean notation, where a plus (+) indicates the OR logic operation, we get:

$$Q = \overline{A}.B.C + A.\overline{B}.\overline{C}$$

In terms of logic gates this means the control circuit has two three-input AND gates feeding a two-input OR gate as in Fig. 19.02(*b*). To implement it with only NAND gates we use the facts explained in section 15.4:

AND = NAND followed by NOT

OR = NOT of each input followed by NAND

The circuit becomes that of Fig. 19.03(*a*) but inverting an input twice gives the original input, i.e. two successive NOT gates cancel each other so only the NAND gates 1, 2 and 3 are needed. The equivalent NAND gate circuit is given in Fig. 19.03(*b*); note that the complements $\overline{A}$, $\overline{B}$ and $\overline{C}$ are obtained by inverting A, B and C with one-input NAND gates.

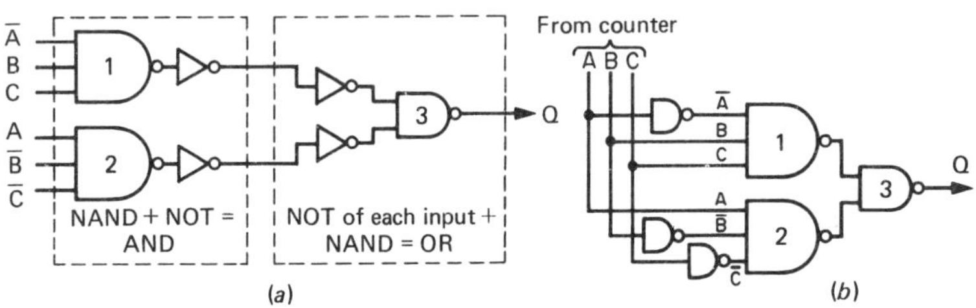

Fig. 19.03

(c) System 3 (traffic lights)

The problem is to make three lamps (red R, yellow Y and green G) light in the order of British traffic lights.

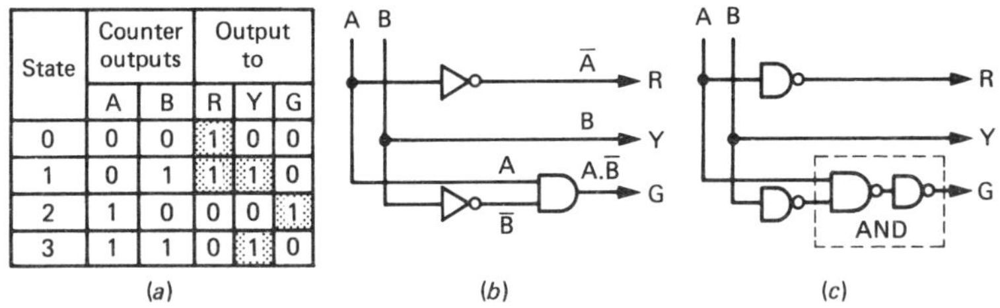

State	Counter outputs		Output to		
	A	B	R	Y	G
0	0	0	1	0	0
1	0	1	1	1	0
2	1	0	0	0	1
3	1	1	0	1	0

Fig. 19.04

As with system 1, this is a four-state cycle which needs a modulo-4 counter. From the truth table in Fig. 19.04(*a*), we can write:

(i) R is 1 when:

A is 0 *and* B is 0 (i.e. $\overline{A}$ is 1 *and* $\overline{B}$ is 1)
or
A is 0 *and* B is 1 (i.e. $\overline{A}$ is 1 *and* B is 1)

(ii) Y is 1 when:

A is 0 *and* B is 1 (i.e. $\overline{A}$ is 1 *and* B is 1)
or
A is 1 *and* B is 1

(iii) G is 1 when:

A is 1 *and* B is 0 (i.e. A is 1 *and* $\overline{B}$ is 1)

Rewriting (i), (ii) and (iii) in Boolean notation and factorizing as in ordinary algebra:

$$R = \overline{A}.\overline{B} + \overline{A}.B = \overline{A}(\overline{B} + B)$$

$$Y = \overline{A}.B + A.B = B(\overline{A} + A)$$

$$G = A.\overline{B}$$

Now if we OR an input and its complement, one input is 1, making the output 1, i.e. $\overline{A} + A$ is 1 and $\overline{B} + B$ is 1, hence we get:

$$R = \overline{A} \qquad Y = B \qquad G = A.\overline{B}$$

Thus, in the logic circuit, the red light is fed from input A via an inverter, the yellow light goes directly to input B and green is supplied by the output of an AND gate, with A and the complement of B as its inputs, Fig. 19.04(*b*). Using only NAND gates, the circuit is shown in Fig. 19.04(*c*).

(d) System 4 (two-to-one line multiplexer or data selector)

Basically this is a two-way switch which connects its output Q to digital input A when the command input C is 'low' and to digital input B when C is 'high', Fig. 19.05(*a*). In the truth table, Fig. 19.05(*b*), Q has the same logic state as A when C = 0 and is the same as B when C = 1. We can write the Boolean equation from lines 1, 3, 6 and 7 as:

$$Q = A.\overline{B}.\overline{C} + A.B.\overline{C} + \overline{A}.B.C + A.B.C$$

Factorizing we get:

$$Q = A.\overline{C}(\overline{B} + B) + B.C(\overline{A} + A)$$

$$= A.\overline{C} + B.C \quad \text{since } \overline{B} + B = \overline{A} + A = 1$$

In Fig. 19.05(*c*) the circuit is implemented with various gates and in Fig. 19.05(*d*) with NAND gates only.

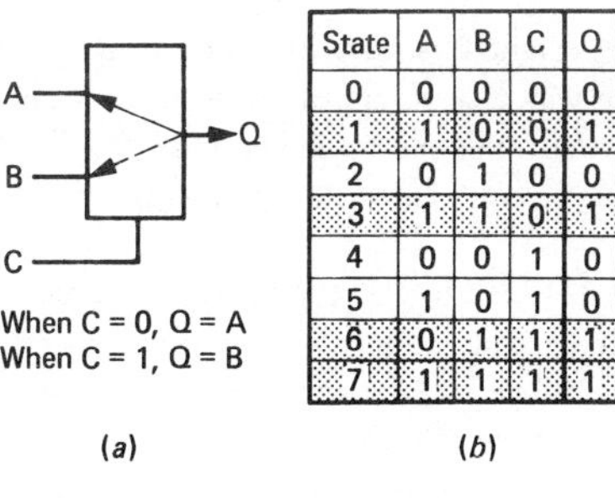

State	A	B	C	Q
0	0	0	0	0
1	1	0	0	1
2	0	1	0	0
3	1	1	0	1
4	0	0	1	0
5	1	0	1	0
6	0	1	1	1
7	1	1	1	1

(*a*) (*b*)

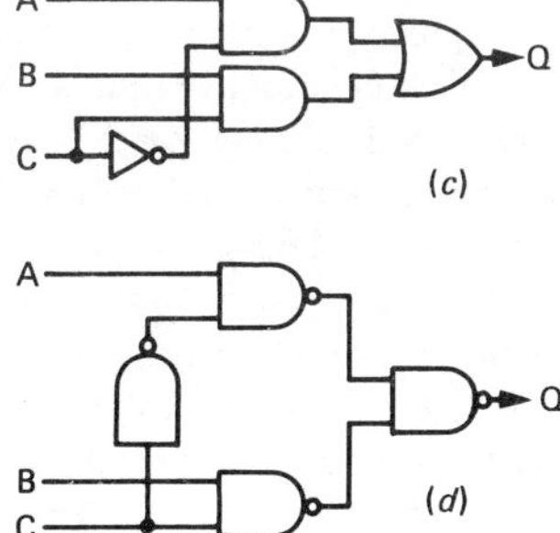

(*c*) (*d*)

Fig. 19.05

(e) System 5 (simple ALU)

Suppose the ALU (see section 15.11) has two inputs A and B and that:

(i) when the select input C = 0, it is to be in the *arithmetic mode* and give an output Q = 1 when A is greater than B (i.e. it is a magnitude comparator);
(ii) when C = 1, it is to be in the *logic mode* and perform the AND operation on A and B, i.e. Q = 1 when A = B = 1.

The truth table is given in Fig. 19.06(*a*) and from it the Boolean equation is:

$$Q = A.\overline{B}.\overline{C} + A.B.C$$

In Fig. 19.06(*b*) the system is implemented with various two-input gates. Note how the need to use three-input gates is avoided by 'ANDing' A and $\overline{B}$ first, then 'ANDing' their output $(A.\overline{B})$ with $\overline{C}$. The 'ANDing' of A, B and C is similarly treated in two stages. Fig. 19.06(*c*) gives the equivalent NAND gate system.

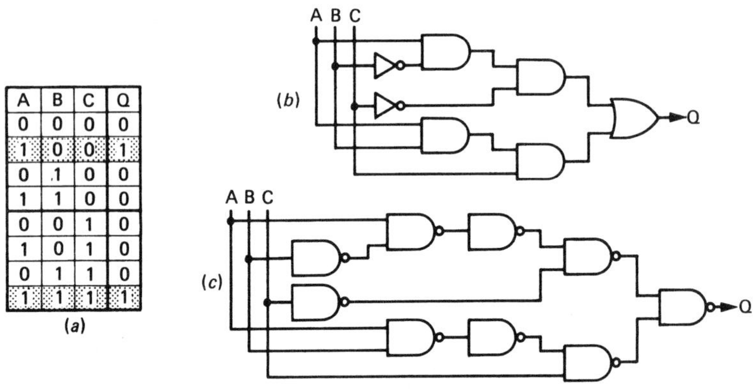

A	B	C	Q
0	0	0	0
1	0	0	1
0	1	0	0
1	1	0	0
0	0	1	0
1	0	1	0
0	1	1	0
1	1	1	1

Fig. 19.06

19.2 More Boolean algebra

In the previous section logic circuits were designed using truth tables to describe the logical operations required. Some use was also made of Boolean notation. Now we will develop Boolean algebra more formally as a quicker and neater way of summarizing the action of a digital logic circuit. The notation also often indicates how simplifications can be made.

Boole, who invented his form of algebra in 1847, showed that logical statements which are either 'true' or 'false' have a binary nature applicable to symbols as well as words. However it was not until nearly 100 years later that the approach was used in digital electronics to analyse and predict the behaviour of telephone switching circuits. They were concerned with on-off signals that can be considered as at logic level 1 if 'on' and at logic 0 if 'off'.

(a) Boolean expressions for logic gates

These are a form of shorthand which describes what a gate does in logical terms. Certain conventions mentioned earlier are used and are restated again:

(i) a plus (+) represents an OR operation and a dot (.) represents an AND operation;

(ii) a bar above the top of a symbol, e.g. $\overline{A}$, means the value of the symbol is inverted or complemented so that $\overline{1} = 0$ and $\overline{0} = 1$;

(iii) a double bar, e.g. $\overline{\overline{B}}$, is a double inversion and cancels out to give the original symbol, i.e. $\overline{\overline{B}} = B$.

The expressions for the various gates can then be written as in Table 19.1, where A and B are inputs that can have logic levels of either 0 or 1, as can the output, usually denoted by Q (or F) in Boolean algebra.

Remember that in electronic circuits logic levels 0 and 1 represent 'low' and 'high' voltages and are not the same as mathematical 0 (zero) or 1.

Table 19.1 Boolean expressions for two-input gates

Gate	Expression	Logic operation
NOT	$Q=\overline{A}$	Q is *not* A, i.e. if A=1, $Q=\overline{1}=0$
OR	$Q=A+B$	Q is the result of 'ORing' A and B, and is 1 if A *or* B is 1
AND	$Q=A.B$	Q is the result of 'ANDing' A and B, and is 1 if A *and* B are both 1
NOR	$Q=\overline{A+B}$	Q is the complement of the result of 'ORing' A and B, and is 1 if neither A *nor* B is 1
NAND	$Q=\overline{A.B}$	Q is the complement of the result of 'ANDing' A and B, and is 1 if both A *and* B are *not* 1

(b) Boolean identities

These are statements that arise from the properties of OR and AND operations and frequently allow Boolean equations (and so also logic circuits) to be simplified. Table 19.2 gives the identities that depend on the OR operation.

Table 19.2 OR operation identities

No.	Identity	Explanation
1	$A+0=A$	For an OR gate if A is 1, the output is 1 since one input is 1; if A is 0, the output is 0 since both inputs are 0, i.e. the output always has the same logic level as A
2	$A+1=1$	If one input is 1 (as here), the output is 1 whether A is 1 or 0
3	$A+\overline{A}=1$	If A is 1, the output is 1; if A is 0, $\overline{A}$ is 1 making one input 1 and giving a 1 output
4	$A+A=A$	If A is 0, the output is 0 but if A is 1, the output is 1, i.e. the output is always A

Table 19.3 sets out the identities that arise from the AND operation.

Table 19.3 AND operation identities

No.	Identity	Explanation
5	$A.0=0$	For an AND gate if one input is 0 (as here), the output is 0, whatever the other input
6	$A.1=A$	If A is 1, both inputs are 1 so the output is 1; if A is 0, the output is 0, i.e. the output is always A
7	$A.\overline{A}=0$	If A is 1, $\overline{A}$ is 0, so the output is 0; if A is 0, $\overline{A}$ is 1, so the output is again 0
8	$A.A=A$	If A is 1, the output is 1; if A is 0, the output is 0, i.e. the output is always A

Some other combinations can be treated as in ordinary algebra. For example:

$$A.B + A.C = A.(B + C)$$

and

$$A + A.B = A(1 + B) = A$$

since from identity 2 in Table 19.2, B + 1 = 1.

(c) De Morgan's theorems

These are two useful statements, due to De Morgan, a contemporary of Boole, that enable conversions to be made from one kind of gate to another. They allow circuits to be built from just one type—usually combinations of the commoner and more easily manufactured NAND or NOR gates (which we saw earlier could be done).

The *1st theorem* states in symbols that:

$$\overline{A+B} = \overline{A}.\overline{B} \tag{1}$$

and asserts that the *complement of the output* of 'ORing' A and B is equivalent to the output of 'ANDing' the *complements* of A and B (i.e. $\overline{A}$ and $\overline{B}$).

The *2nd theorem* in symbols is:

$$\overline{A.B} = \overline{A} + \overline{B} \tag{2}$$

and states that the *complement of the output* of 'ANDing' A and B equals the output of 'ORing' the *complements* of A and B (i.e. $\overline{A}$ and $\overline{B}$).

Both theorems can be proved by writing down the truth table for each side of the expression, as in Table 19.4.

The fifth and sixth columns are the same which proves the 1st theorem, and the 2nd theorem follows since the seventh and eighth columns are identical.

The theorems still apply when there are three or more symbols, viz:

$$\overline{A+B+C} = \overline{A}.\overline{B}.\overline{C}$$

and

$$\overline{A.B.C} = \overline{A} + \overline{B} + \overline{C}$$

Table 19.4 Proof of De Morgan's theorems

A	B	$\overline{A}$	$\overline{B}$	$\overline{A+B}$ NOR	$\overline{A}.\overline{B}$	$\overline{A.B}$ NAND	$\overline{A}+\overline{B}$
0	0	1	1	1	1	1	1
0	1	1	0	0	0	1	1
1	0	0	1	0	0	1	1
1	1	0	0	0	0	0	0

Inspection of expressions (1) and (2) for De Morgan's theorems shows that to get from the expression on the left-hand side of the equals sign to that on the right-hand side, you have to first '*break the bar*' over the left-hand expression and then '*change the logical operation*' connecting its variables.

For example, for equation (1), 'breaking the bar' over $\overline{A+B}$ gives $\overline{A}+\overline{B}$, then 'changing the connecting operation' from OR (+) to AND (.) gives $\overline{A}.\overline{B}$, hence $\overline{A+B} = \overline{A}.\overline{B}$. Equation (2) can be treated in the same way.

19.3 Universal gates

Since any combinational logic circuit can be built from either NAND gates only or NOR gates only, these two types are called *universal gates*. Three circuits, which were considered earlier, will now be analysed using De Morgan's theorems.

(a) OR gate from NAND gates

The circuit in Fig. 19.07 is for an OR gate made from only NAND gates (i.e. NOT of each input followed by NAND), as can be checked by drawing up a truth table. However, it can be proved more easily using Boolean algebra and De Morgan's 2nd theorem.

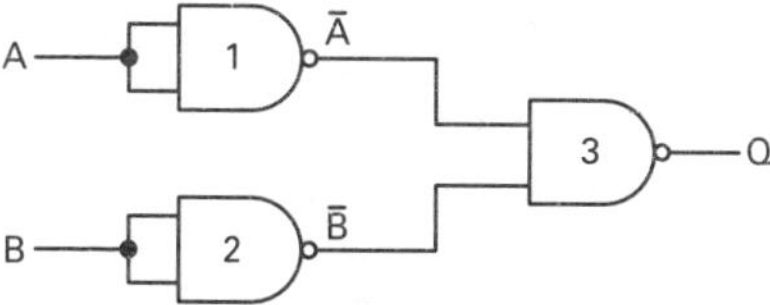

Fig. 19.07

The inputs to NAND gate 3 are $\overline{A}$ and $\overline{B}$ since A and B are inverted by the one-input NAND gates 1 and 2. Hence, for NAND gate 3, we have the Boolean expression:

$$Q = \overline{\overline{A}.\overline{B}}$$

Applying the 'break the bar and change the logic' rule we get:

$$Q = \overline{\overline{A}} + \overline{\overline{B}}$$

$$= A + B \quad \text{since } \overline{\overline{A}} = A \text{ and } \overline{\overline{B}} = B$$

This is the Boolean expression for an OR gate.

(b) AND gate from NOR gates

The circuit is shown in Fig. 19.08 with one-input NOR gates 1 and 2 acting as NOT gates which provide inverted inputs A and B to NOR gate 3. The output Q from NOR gate 3 is given by the Boolean expression:

$$Q = \overline{\overline{A}+\overline{B}}$$

Applying De Morgan's 1st theorem we get:

$$Q = \overline{\overline{A}}.\overline{\overline{B}}$$

$$= A.B \quad \text{since } \overline{\overline{A}} = A \text{ and } \overline{\overline{B}} = B$$

This is the expression for an AND gate.

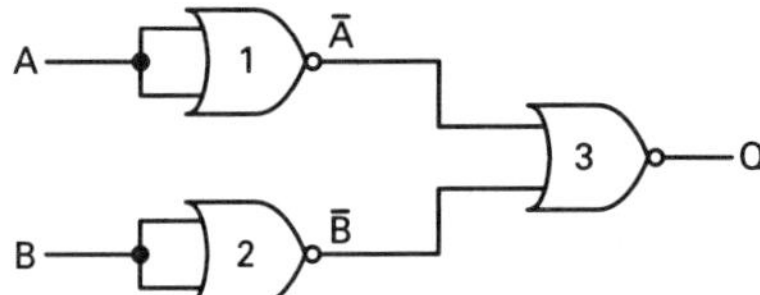

Fig. 19.08

(c) Exclusive-OR gate from NAND gates

The ex-OR gate circuit implemented with only NAND gates was derived earlier (section 15.4) from its truth table; both are shown again in Fig. 19.09. The Boolean expression was deduced from the truth table to be:

$$Q = \overline{A}.B + A.\overline{B}$$

Using the Boolean expression for a NAND gate and De Morgan, this can be proved as follows. The inputs to each NAND gate are shown in Fig. 19.09, from which we can say:

$$Q = \overline{\overline{\overline{A}.B}.\overline{A.\overline{B}}}$$

'Breaking the bar and changing the connecting logic' gives:

$$Q = \overline{\overline{\overline{A}.B}} + \overline{\overline{A.\overline{B}}}$$

But double inversions cancel out, i.e. $\overline{\overline{\overline{A}.B}} = \overline{A}.B$ and $\overline{\overline{A.\overline{B}}} = A.\overline{B}$, hence:

$$Q = \overline{A}.B + A.\overline{B}$$

A	B	Q
0	0	0
0	1	1
1	0	1
1	1	0

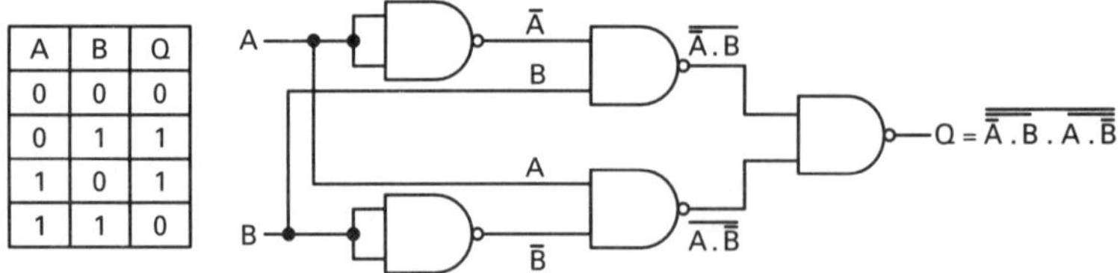

Fig. 19.09

19.4 Microelectronic options

When designing an electronic system using large-scale integrated circuits, two broad approaches are possible. In the first, the job done by the system is controlled by its circuit and cannot be changed without altering or modifying the circuit. This is the *circuit-controlled* option. In the second, the job done is decided by a program of instructions (software) which tells the system exactly how to perform the task. To change the job done, the program is changed. This is the *program-controlled* option.

In general, the first approach would seem best when the system has just one job to do and the second when it has several, but there are cases needing both approaches. Often the choice is a matter of technical and commercial judgement. Within the circuit-controlled option there are three further choices and these will be outlined in this section. The program-controlled option uses a microprocessor which will be discussed in section 20.8.

(i) Wired logic The system is wired together on a suitable board using standard, off-the-shelf SSI and MSI chips and, perhaps, discrete components. The chips might range from simple logic gates to an arithmetic unit. It is often the simplest and cheapest system to design and develop but it is inflexible. The traffic lights system of section 19.1(c) could readily be constructed by this method using three SSI chips (an astable, a binary counter and a quad two-input NAND gate) and a few discrete components.

(ii) Custom chips In this case all functions demanded of the system are done by one 'dedicated' LSI chip containing the exact number of components to do just the job required. It has the advantage of maximum efficiency, minimum size and minimum power consumption. A digital watch chip is an example, which, though designed initially as a custom special, found so large a market that it became a 'standard' chip. Otherwise, custom chips can be costly.

(iii) Uncommitted logic array (ULA) This is a halfway house between (i) and (ii). It is a chip containing an array of standard logic 'cells', each complete in itself but not yet interconnected into any circuit pattern. Together the 'cells' provide all the constituents (NOT, AND, OR, half-adder, full-adder, etc.) normally required in an LSI chip of average complexity. Connections are made subsequently by the chip manufacturer after a logic diagram has been constructed for the customer's application. Computer-assisted design techniques are used to decide which 'cells' and connections are required.

ULAs are cheaper than custom chips and can be designed in about one-third of the time. They are also more versatile, within limits. Two ULA chips are used in one type of fully automatic camera; one chip deals with analogue functions, the other with digital.

19.5 Digital watch

The block diagram of Fig. 19.10 shows the three main parts of a digital watch. They are the *quartz crystal* (the time keeper), the *integrated circuit* (the 'brain') and the *display* (the time indicator).

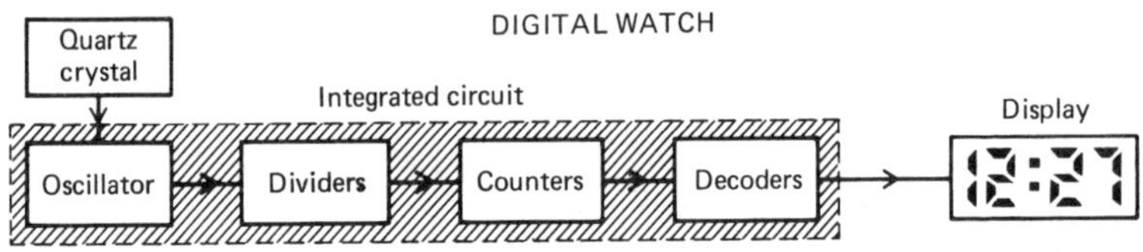

Fig. 19.10

When activated by a battery, the quartz crystal oscillator produces extremely stable electrical pulses with a frequency of exactly 32 768 Hz (2^{15} Hz). These are fed to a chain of frequency dividers which reduces them eventually to one pulse per second. After counting and decoding by the logic circuits in the decoders, they are sent to the appropriate electrodes of the seven-segment LCD or LED decimal displays to show the time or the date.

The custom chip contains over 2 000 transistors and incorporates all of the system's major circuits.

19.6 Calculator

The basic difference between a simple calculator and a digital computer is that the calculator requires the numbers and instructions to be entered during the calculation. This is less true of more sophisticated models. The three main parts of a simple calculator are:

(i) the *keyboard* or input device which enables us to put numbers into the calculator and to tell it what do do;
(ii) the '*brain*' which consists of an arithmetic and logic unit (ALU—see section 15.11) to store data, together with control circuits for encoding, decoding and providing 'clock' pulses to synchronize operations;
(iii) the *display* or output device for giving the answer.

In many calculators a microprocessor (see section 20.8) is now used as the 'brain' but the principles involved can be shown for the addition of two decimal digits using the very simple system of Fig. 19.11. It has a four-bit full-adder (see section 15.8) as its ALU. Two points have to be remembered. First, whilst the keyboard and display are decimal number devices, the adder works in binary. Therefore, encoding of the keyboard output and decoding of the input to the display are necessary. Second, only one decimal digit can be entered at a time from the keyboard and so one digit has to be stored until the other is entered. Storage is the role of the shift registers—see section 16.6.

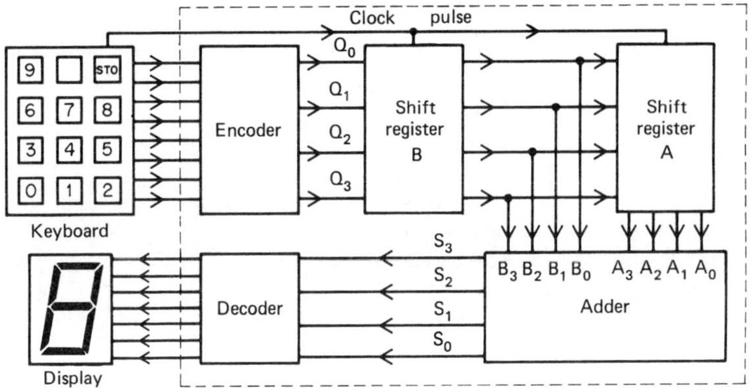

Fig. 19.11

Suppose the two numbers to be added are 3 and 5. When key '3' is pressed on the keyboard, the four-bit binary form of decimal 3 is produced by the encoder at its four outputs Q_0, Q_1, Q_2 and Q_3. In this case the output values would be Q_3(m.s.b.) = 0, Q_2 = 0, Q_1 = 1 and Q_0(l.s.b.) = 1. The binary number 0011 is thus applied to the four inputs of shift register B. If the 'store' key (STO) on the keyboard is pressed next, a 'clock' pulse causes shift register B to shift 0011 from its inputs to its outputs, where it is stored and becomes the input to shift register A.

The second number can now be entered and if key '5' is pressed, the binary number applied to the inputs of shift register B becomes 0101, i.e. $Q_3 = 0$, $Q_2 = 1$, $Q_1 = 0$ and $Q_0 = 1$. On applying a second 'clock' pulse to the shift registers, the first number (0011 = 3) is shifted from the inputs of shift register A to its outputs where it becomes the $A_3A_2A_1A_0$ input to the adder. At the same time, the second number (0101 = 5) is shifted from the inputs of shift register B to its outputs and becomes the $B_3B_2B_1B_0$ input to the adder.

The adder adds the two binary numbers $A_3A_2A_1A_0$ and $B_3B_2B_1B_0$ immediately and produces their binary sum $S_3S_2S_1S_0$ at its four outputs. Here the sum is 1000 (= 8) so that $S_3 = 1$, $S_2 = S_1 = S_0 = 0$. These outputs are applied to the four inputs of the decoder which, if it was driving a seven-segment decimal display, would create seven outputs each capable of driving one segment. In this example all seven segments would light up and give 8 as the sum of 3 + 5.

19.7 Revision questions and Problems

1. Write down the truth table for, and design, a system, driven by a clock and a suitable binary counter, which makes a lamp light during
a) state 0 of a four-state cycle,
b) states 0 and 5 of an eight-state cycle.

2. a) Design a system of traffic lights in which the sequence is red, yellow, green, red. What would be the modulo of the counter required to control it?
b) (i) Design an exclusive-OR gate (see section 15.4) using various gates and then using NAND gates only.
(ii) How would you modify the circuit, using NAND gates only, to make it a half-adder (see section 15.8)?
c) Design a logic circuit for a *two-bit decoder* (see section 15.7) that implements this truth table. (A is the l.s.b.)

Truth table for a two-bit binary decoder

	Inputs		**Outputs**			
State	B	A	Q_0	Q_1	Q_2	Q_3
0	0	0	1	0	0	0
1	0	1	0	1	0	0
2	1	0	0	0	1	0
3	1	1	0	0	0	1

d) Write the truth table for a magnitude comparator (see section 15.9) that compares two bits and gives a 'high' output when they are equal. Design the system, using NAND gates only. (It is called an *exclusive-NOR* or *equivalence gate*.)

3. Explain the following terms used in connection with microelectronic systems: circuit-controlled, program-controlled, wired logic, custom chip, ULA.

4. Write the Boolean expression for NOT, OR, AND, NOR and NAND gates.

5. State the result of performing the following logical operations:

a) $A + 0$ e) $A + \overline{A}$ i) $0 + 1$
b) $A.1$ f) $\overline{\overline{A}}$ j) 1.0
c) $A + 1$ g) $0 + 0$ k) $1 + 1$
d) $A.A$ h) 0.0 l) 1.1

6. a) What are the Boolean expressions for De Morgan's two theorems?
b) State the rule used to apply them.

7. Prove the following Boolean identities.

a) $A.(A + B) = A$

b) $(A + B).(B + C) = A.C + B$

c) $\overline{A+B+C} = \overline{A}.\overline{B}.\overline{C}$

d) $\overline{A}.\overline{B}.C + \overline{A}.B.C + A.C = C$

e) $\overline{(A+ B).(B+ C)} = \overline{B}.(\overline{A} + \overline{C})$

f) $\overline{\overline{A}.B+B.\overline{C}} = \overline{B} + A.C$

8. a) What are the Boolean expressions for the logic circuits in Fig. 19.12 (i), (ii), (iii), (iv), (v), (vi)?

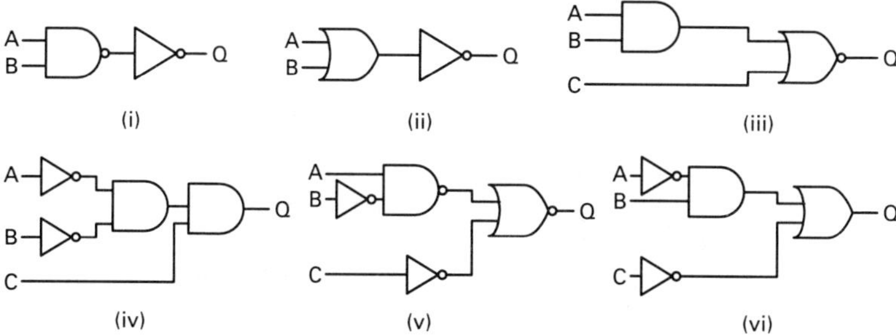

Fig. 19.12

For each circuit state the input combination(s) that give an output of Q = 1.

b) Draw up truth tables for circuits (iii) and (iv) and check that they agree with your predictions about when Q = 1 in each case.

9. a) If $Q = \overline{A}.B + A.\overline{B}$ is the Boolean expression for an exclusive-OR gate, show that for an exclusive-NOR gate $Q = \overline{A}.\overline{B} + A.B$.

b) Work out the Boolean expression for the circuit in Fig. 19.13. What kind of gate is it?

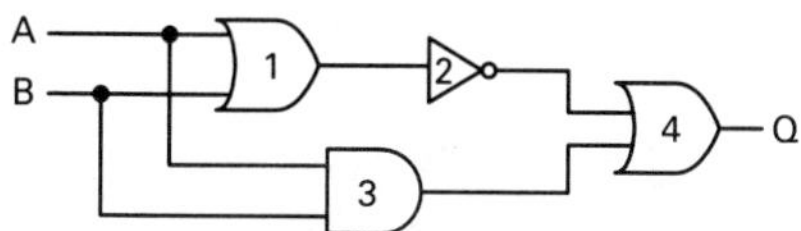

Fig. 19.13

UNIT 20

Computers and microprocessors

20.1 Digital computer

(a) Introduction

Computers are playing an increasing role in our everyday lives at home, at work, in education, and in our leisure. This is due to the speed with which they can undertake almost any task involving the processing of information (data).

The power of a computer depends on its speed of working and the amount of data it can handle at one time. There are three very broad classes.

(i) *Mainframe computers* are usually the most powerful and contain several large units, housed in an air-conditioned room and operated by a team of people.
(ii) *Minicomputers* are made from smaller units and only require one or two operators.
(iii) *Microcomputers* are the desktop personal computers (PCs) found today in homes, schools and offices. They have been made possible by the development of microprocessors. The smallest portable types go under such names as laptops, palmtops, powerbooks, handhelds and notebooks.

The brief description which follows is based on the simplified block diagram of Fig. 20.01.

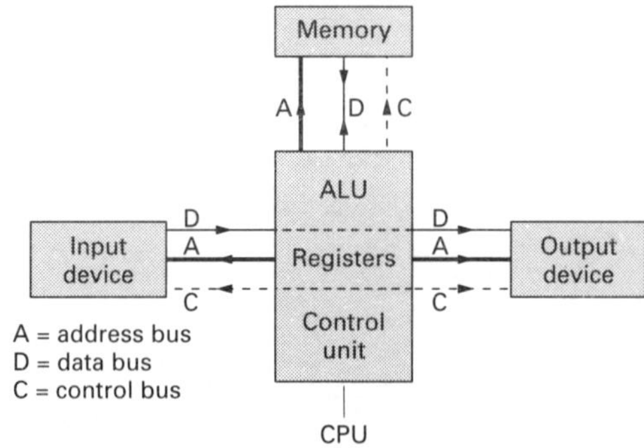

Fig. 20.01

(b) Central processing unit (CPU)

The CPU performs, organizes and controls all the operations the computer can carry out. It is sometimes called the 'brain' of the computer and what it can do depends on the set of instructions (typically 50 or more), called the *instruction set*, which it is designed to interpret and obey.

It consists of an *arithmetic and logic unit* (ALU see section 15.11), a number of *shift registers* (see section 16.6) and a *control unit*. The ALU performs arithmetic calculations and logical operations. The registers are temporary stores; one of them, the *accumulator*, contains the data actually being processed. Two important components of the control unit are the *clock* and the *program counter*. The clock is a crystal-controlled oscillator (see section 12.6) which generates timing pulses at a fixed frequency (called the *clock speed*, at present typically 120 MHz to 200 MHz) to synchronize the computer's operations and ensure they all occur at the right time. The program counter counts the timing pulses in binary as they initiate the next step in the program and points to the address of the next instruction.

The operation of the CPU and some of its other components will be described more fully when the microprocessor is considered in section 20.8.

(c) Memory

A computer needs a *working* memory to run the program of instructions (the software) it has to execute. If the program is fixed (as in a microprocessor-controlled washing machine or electronic game), the memory has only to be 'read' and a ROM (see section 16.8), programmed by the manufacturer, is used since its contents are retained when the power is switched off. A PROM, EPROM or EEPROM (see section 16.8(*f*)) would also be suitable if the user wanted to construct the program and keep it permanently in the memory.

If the program has to be changed, the 'read' and 'write' facilities of a RAM are required to allow instructions to be written in, read out and altered at will. A RAM is also needed to store the *data* for processing because it too may change. Sometimes a RAM is used for both the program of instructions and the data and so all is normally lost at switch off. As we saw earlier (section 16.8), each location in a memory has its own *address* which allows us to get directly to any instruction or item of data.

These *internal semiconductor* memories are, in general, temporary read/write devices (RAMs) that are just extensions of the CPU's registers. In a modern PC 16 megabytes of RAM is common. As well as these short-term memories, there is usually provision for permanent storage of programs and data not in current use, in *external magnetic* or *optical* memories (see section 20.4). The latter are often referred to as 'hard disk' to distinguish from the computer's working memory or 'RAM'.

(d) Input and output devices

The CPU accepts digital signals from the input device, e.g. a keyboard, a mouse or a modem, via its *input port*. After processing these are fed out via its *output port* to the output device, e.g. a *visual display unit* (VDU; usually called a monitor) or a *printer*, in a form we can understand.

(e) Buses

The CPU is connected to other parts by three sets of wires, called *buses* because they 'transport' information in the form of digital electrical signals. The *data bus* carries data for processing and is a two-way system with 4, 8 or 16 lines, each carrying one bit at a time. The one-way *address bus* conveys signals from the CPU which enable it to find data stored in a particular location of the memory. It has anything from 4 to 32 lines depending on the number of memory addresses there are (8 lines give $2^8 = 256$ addresses). The *control bus* transmits timing and control signals and could have 3 to 10 lines. In diagrams each bus is represented by one line.

To stop interference between parts that are not sending signals to a bus and parts that are, all parts have in their output a circuit called a *tristate gate*. Each gate is enabled or disabled by the control unit. When enabled, the output is 'high' or 'low' and pulses can be sent to the bus or received from it. When disabled, the output has such a large impedance that it is in effect disconnected. It can neither send or receive signals and does not upset those passing along the bus between other parts.

(f) Cache

The very rapid increase in the clock speed of processors during the last 25 years (from 0.1 MHz to 200 MHz, due to miniaturization) has not been matched by a decrease in the time it takes to access instructions from memory. As a result, processors are often kept waiting by a slow memory.

A *cache* is a small, high-speed memory (typically 256 kilobytes) incorporated in the process chip which reduces the problem by storing and allowing faster access to information that is used frequently. The slower main memory is then accessed only for information not in the cache.

(g) Operating system

This is the software which makes the computer work and controls such operations as how, for example, it reads programs. It is the language the computer 'speaks'.

Two popular systems are *Windows* (by Microsoft) and *Macintosh* (by Apple); PCs designed to use one system will not run on the software of the other. Recently a new system called *Java* has been developed which, it is claimed, can be run on almost any computer.

(h) Viruses

A computer virus is a program which reproduces itself and can become attached to other programs. There, it can cause the unwelcome changes planned by its originator. It is spread for instance via the Internet but may be countered using protective software.

20.2 Computer peripherals—Input devices

The input, output and external memory devices of a computer are called *peripherals*. They enable the computer to communicate with the outside world.

(a) Keyboard

This is similar to the one on a typewriter and contains the usual range of alphanumeric characters (A to Z and 0 to 9) as well as 'command' keys for giving instructions to the computer; some keys have specific functions within certain programs.

(b) Mouse

Controlling and giving commands to a computer can also be done by a small box called a 'mouse', which makes complex programs easier to use. It is especially useful, for example, when the computer has to draw.

The mouse is moved by hand over a mat on the desk top. As it does so, it rolls on a ball that causes electrical pulses to be sent to the computer. These indicate the mouse's exact location and the computer reacts by shifting a pointer (↖) on the screen of the monitor in the direction indicated by the mouse's movement. If the pointer points to one of the symbols, i.e. *icons*, on the screen, the computer can be given commands by 'clicking' or 'double clicking' a switch on the mouse. The command represented by the icon is then executed, e.g. to load a program or call up a file, or to draw.

When the ball rolls, it turns two slotted discs at right angles to one another. Each has two photodiodes and two LEDs and electrical pulses are produced in the diodes by light from the LEDs passing through the slots. These pulses give the horizontal and vertical coordinates of the printer on the screen.

A *trackball* works like a mouse but stays at rest when the ball is rotated by the fingers. Some portable computers have a *trackpad* consisting of an area of the keyboard on which the user presses to move the pointer to the desired location.

(c) Joystick

This allows the position of a cursor or graphics character to be controlled when the joystick handle is moved. Some are just four-position switches (left, right, up, down). Others are basically variable resistors that give instructions by applying different voltages. They are used mostly for fast response in computer games.

(d) Touch screen

Different systems are used but all consist of a clear panel (of glass or plastic) placed in front of, or incorporated into, the screen of a monitor. When the panel is touched, the x and y coordinates of the location of the finger are worked out, decoded and used to control the computer as would a mouse or keyboard. The system is useful for menu choice, i.e. obtaining a list of optional facilities available to the user.

The touch-sensitive panel contains a matrix of cells comprising devices which respond when the finger

(i) interrupts infrared beams that cross the panel, or
(ii) causes a pressure change where the panel is touched, or
(iii) alters the capacitance formed by tiny metal dots on the panel and the finger which connects it to earth, thereby causing charge flow.

(e) Optical readers

These are input devices used in data processing which read the data directly from the source of information.

The *bar code reader* (which may be a laser) is used, for example, in supermarkets for stock control. It works on the principle that a narrow beam of light from the reader is reflected back from the pattern of black lines of various widths on the bar code, and changed into a series of electrical pulses. These enable the computer to identify the item.

An *optical mark reader* (OMR) works on the same principle and detects marks on specially prepared documents, e.g. answer sheets for multiple-choice questions in an examination!

Optical character readers (OCR) read letters and figures; these are used in the preparation of gas and electricity bills.

Magnetic ink character readers (MICR) are used by banks for handling cheques. Some of the characters are printed in a special magnetic ink.

(f) Sensors

In industrial process control, in science, in medicine and in other areas, inputs come from transducers or sensors that produce analogue voltages. These may represent continuously varying quantities such as temperature, light intensity, sound level, pressure, fluid flow rate or movement, and must be changed to digital signals by an analogue-to-digital converter before being input to the computer.

(g) Scanners

These are able to transfer full-colour images from documents into a digital form which can be used by the computer in many different ways.

(h) Other input devices

Other methods of control that are claimed to be easier to use are emerging, one of the most natural being spoken commands of the *voice*. Within the next ten years we will be able to enter information into a computer by simply talking to it. Software already exists which can be 'trained' to recognize the voice of the user.

Another method is based on *pens* that write or draw on a sheet of paper next to the computer and simultaneously transfer the image to the screen, Fig. 20.02.

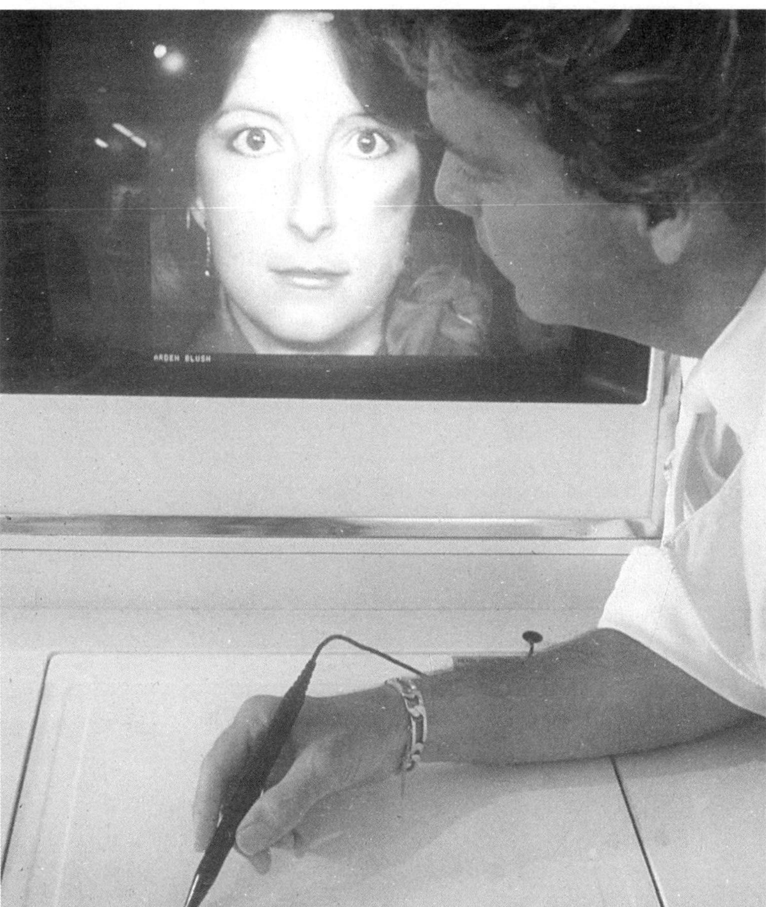

Fig. 20.02

20.3 Computer peripherals—Output devices

(a) Visual display unit (VDU) or monitor

This is usually a colour or black-and-white cathode ray tube (CRT) similar to that used in a domestic TV receiver but which makes direct use of the computer's video signals.

The screen is divided into picture elements, called *pixels*, each being the smallest size of dot (typically 0.3 mm) that can be controlled independently by the computer; these are the building blocks from which graphics are constructed. The original drawing, picture or photograph has to be 'digitized' (i.e. represented in binary code) so that it can be treated as a two-dimensional array of data and reproduced as a series of small dots—just like a newspaper photograph.

The *resolution* of a system measures the fineness of detail that it shows, i.e. the degree of sharpness of the image or printed character on the screen. It is usually stated in terms of the number of pixels. For example, a resolution of 320 × 200 (= 64 000) means that there are 320 pixels across the screen and 200 down, i.e. 64 000 altogether. The higher the resolution, the smaller the pixels and the more of them that will be displayed at one time. High resolution is considered to be 1 024 × 768 pixels, although a resolution as high as 1 600 × 1 200 is possible in some systems. There are many ways of controlling pixels by a progam.

Monitors using *non-interlaced scanning* produce displays with less flicker than those that use interlacing (see section 18.5) because of the higher rate at which they refresh the display.

Flat-screen black-and-white and colour displays are in use in portable computers and small TV receivers; various types are being developed. The most popular type at present is the *active matrix* liquid crystal display (LCD—see section 9.7), also known as the *thin film transistor* (TFT) display. It contains liquid crystals arranged in a matrix of cells that represent pixels, each colour of each pixel being turned on and off by a transistor, Fig. 20.03(*a*).

In the *gas plasma* display, Fig. 20.03(*b*), each pixel is a tiny fluorescent tube. When current passes through one 'column' electrode and one 'row' electrode, the tube at their intersection glows.

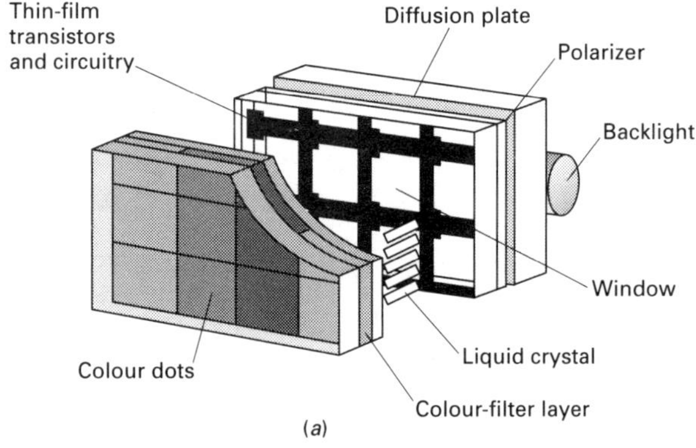

(*a*)

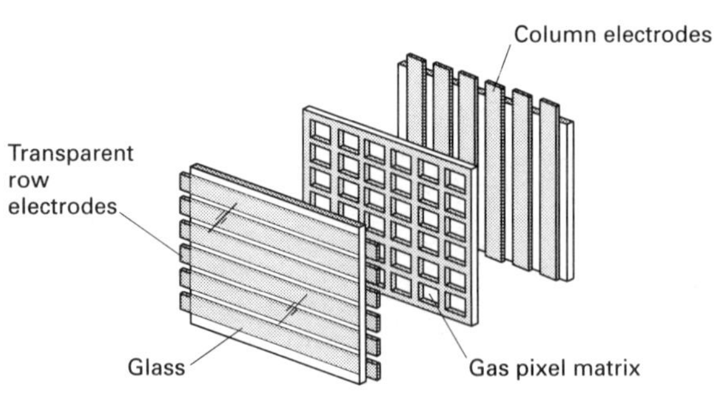

(*b*)

Fig. 20.03

Other promising new screen technologies include:

(i) the *field emission* display (FED) which uses a phosphor-coated screen like today's CRT monitors but with thousands of tiny cathode emitters instead of one huge tube to fire electrons at the screen;
(ii) the *polymer* display, made using a plastic called a 'conjugated' polymer with unusual properties; and
(iii) *laser semiconductor* screens based on gallium nitride which emits ultraviolet radiation that can cause certain materials to produce different colours of light.

The bulky CRT, which for many years has produced good quality images at low cost, clearly has rivals that will in the future hang like pictures on the wall. One firm recently exhibited a gas plasma monitor with a width of 106 cm (measured diagonally) and a depth of just 8.0 cm.

(b) Printers

A printer is used when a permanent record ('hard copy') of the output is required. There are several types.

(i) Dot-matrix printers These form letters and numbers from tiny dots. The computer sends the code for each character to a chip in the printer head which converts the code into a binary signal that turns electromagnets in the head on and off. The head contains a row of pins, some or all of which (depending on the character) are struck by a hammer when their corresponding electromagnets are switched on. As a result the pins strike an inked ribbon which makes dots on the paper. Good quality printers have a row of 24 pins in the head.

Stepper motors (see section 9.10) move head and paper to the correct positions.

(ii) Ink-jet and bubble-jet printers These give better quality printing than dot-matrix types but cost more. They work by firing small droplets of ink onto the paper from over 1000 jets, each half the thickness of a human hair, at a rate of 6000 drops a second from each.

(iii) Laser printers These combine extremely high quality black and colour printing with flexibility and speed but are expensive. They make up characters with dots, like a dot-matrix printer, but the dots are much smaller, giving very sharp printing at a typical speed of 12 pages (of A4) per minute (i.e. 12 ppm) which is several times faster than other types of printer. They work by melting fine plastic powder onto the page to give a permanent result.

(c) Activators

Outputs from the computer may be used to control activators, often electric motors, which in turn operate the switches, gears, valves, etc. that in turn control the machine or an industrial process. If an activator requires an analogue voltage to operate, the digital output from the computer has to be converted by a digital-to-analogue converter.

20.4 Computer peripherals—External memories

In addition to short-term internal memories such as semiconducting RAMs, external memories are required when programs and large amounts of data have to be stored permanently. Three types are common.

(a) Floppy disk

This is a flexible plastic disk, usually 3.5 inches in diameter, coated with a thin magnetic film, which stores data *magnetically*. When it is inserted into a floppy disk drive (FDD) in 'write' mode, digital electrical pulses (1s and 0s) from the computer are changed by the *write/read head* into small magnetized bands on the disk in either of two directions, one way for a 1, the other way for a 0. In 'read' mode, digital electrical signals are induced in the head (by the magnetized bands on the disk) for transfer to the computer.

The disk is divided into concentric circular *tracks* (e.g. 80 for 'double density' and 135 'for high density') and *sectors* (e.g. 10), Fig. 20.04(*a*). Each sector stores a certain number (e.g. 256) of bytes where one byte represents one stored character. The first two tracks near the edge of the disk store its 'catalogue'. This is the disk's filing system which contains the location (i.e. track and sector) of the required data. Setting up the catalogue, a process called *formatting* the disk, has to be done before the disk can be used. The necessary procedure is usually supplied with the computer.

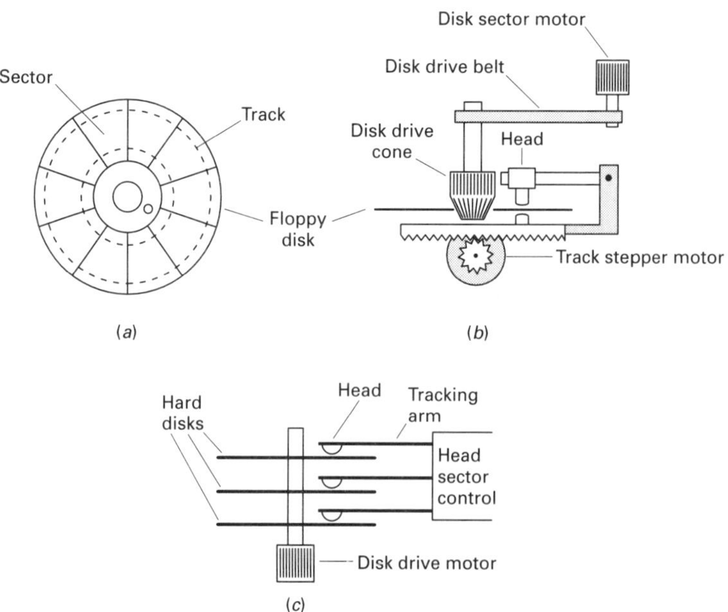

Fig. 20.04

A precision stepper motor (see section 9.10) moves the write/read head across the disk and locates the track while another motor simultaneously rotates the disk at high speed to locate the sector by timing from a reference point in the rotation, Fig. 20.04(*b*). In this way rapid, random access to data at any location on the disk is obtained.

Floppy disks are used in PCs and typically have a capacity of 1.44 megabytes. They are widely used as portable external memories for data storage.

(b) Hard disk

This is a rigid magnetic disk, similar to a floppy disk but with a much greater storage capacity, e.g. 2000 megabytes (2.0 gigabytes). It is driven by a hard disk drive (HDD) which often has several disks on the same shaft, Fig. 20.04(*c*).

Compared with semiconductor memories the *access time* (i.e. the time it takes to find and deliver a piece of data) for hard and floppy disks is slow, especially for the latter. Storage is permanent, i.e. non-volatile, unless the disk is demagnetized by a magnetic field or by excessive temperatures or by physical deformation.

Some systems now have a hard disk that can be removed simply by pulling out a cartridge, thus giving greater security and flexibility. Removable storage systems such as *Zip*, *Jaz* and *Syjet* can store between 100 megabytes and 1.5 gigabytes on a disk about the size of a floppy but thicker.

(c) CD-ROM

Based on the technology of the audio compact disc (see section 18.7), this is a 5 inch plastic disc that can store graphics, pictures, information and computer programs as well as sound. It is also called an *optical*, *video* or *laser* disc but more commonly a CD-ROM because initially such discs could be 'written' on only once, but 'read' many times.

Huge amounts of data can be stored. For example, one CD-ROM has a capacity of 650 megabytes and can store the contents of all 40 000 pages of the *Encyclopaedia Britannica*. Also, the rate of transfer of data from disc to computer by a CD-ROM is high. A portable computer having a CD-ROM drive as well as a floppy disk drive and a hard disk drive is shown in Fig. 20.05.

Fig. 20.05

As in audio, photo and video CDs, in CD-ROMs digital data is stored in a series of tiny 'pits' or 'craters' of different lengths and spacing on the disc and its reflecting surface. A narrow laser beam 'reads' the data as explained in section 18.7.

The latest generation of compact discs is the *Digital Versatile Disc* (DVD), known initially as the Digital Video Disc. It has a capacity of up to 25 times greater than that of a normal CD-ROM, depending on whether the format is single-side, single-layer (4.7 gigabytes) or double-side, double-layer (17 gigabytes). It marks a major advance for data storage and entertainment systems, being suitable for storing either computer data for a multimedia system (see section 20.6) with much improved graphics, or a full-length film (135 minutes) for playback on a digital video disc player.

Rewritable CD-ROMs can not only be 'read' many times but also repeatedly 'written' on and erased.

20.5 Interfacing chips

Data in its original form cannot usually be fed directly from an input device into the CPU. It must be presented in digital form. Similarly digital signals from the CPU may not be acceptable to an output device. Very often interfacing chips are required between the CPU and its peripherals. They may also have to compensate for differences of, for example, voltage levels, operating speeds and codes.

(a) Analogue-to-digital (A/D) and digital-to-analogue (D/A) converters

The need for chips to perform these operations for certain input and output devices was mentioned earlier and their action considered in sections 17.2 and 17.3.

(b) Encoders and decoders

To interface a keyboard to a computer, an encoder is required to produce a different pattern of binary bits, according to the ASCII code (American Standard Code for Information Interchange), for every key operated. The action of one kind of decoder was considered in section 15.7.

(c) Modems

A computer connected to an ordinary telephone line requires a modem at each end to convert digital signals to and from analogue signals—see section 18.10(c).

(d) PIO and SIO chips

A peripheral may be a serial or a parallel device. Monitors are serial types which produce and must receive a string of bits following one after the other. A seven-segment LED display and a printer require parallel interfacing so that the bits of the output are supplied together, i.e. in parallel.

Input and output devices are connected to the *input and output ports* of the computer which usually consist of eight lines, for handling eight-bit words. Interfacing a serial device therefore requires a serial input/output (SIO) chip; a parallel input/output (PIO) chip would enable a parallel device to communicate with a computer. Whatever method is used, synchronization is required since the rate at which a device such as a printer can receive data is generally slower than the rate at which it can be sent by a computer. The synchronizing signals are called *handshaking* signals.

20.6 Uses of computers

Some of the ever-increasing applications of computers are outlined below.

(a) Data processing

This accounts for about three-quarters of all computer use today. It includes payroll preparation for firms with large numbers of employees, stock control in large shops and factories, airline ticket booking, and preparation of invoices.

(b) Process control (automation)

Large-scale control applications using mainframe or minicomputers are common in process industries. Some examples are steel making, oil refining, brick making, chemical manufacture, paper making, sugar refining and float-glass production. Apart from achieving automation, the aim is to maintain quality of the end-product with maximum economy and output.

(c) Robotics

An industrial robot is like a human arm with several separately controllable joints. In small arms the joints are worked by stepper motors. These rotate through exact angles when electrical pulses are received from a computer that has the required movement program stored, often in a ROM chip. The 'hand' at the end of the arm may be a claw-like device which grips things or a special tool such as a paint sprayer or a spot welder.

(d) Computer-integrated manufacture (CIM)

At present this is being researched and hailed as the next great leap forward for industry. It is attempting to bring under computer control all the various stages of manufacturing, from designing the product to ordering materials, from quality control to issuing invoices, as well as actually making the product.

Traditional technologies such as production, mechanical and electrical engineering will be married with computer and information technology in the hope of creating a single completely integrated operation.

(e) Other computer users

These include the police for storing criminal and other records; banks for handling cheques, customers' accounts and processing 'cashless shopping' or Electronic Funds Transfer at Point of Sale (EFT-POS) transactions; the *Driver and Vehicle Licensing Agency* (DVLA) for keeping information about every vehicle in the country; newspaper publishers to speed up production; scientists, engineers, geologists and weather forecasters who can make models or simulations of real-life situations using a computer and predict likely outcomes; doctors and dentists for maintaining patients' records; farmers for recording production and planning the use of agriculatural chemicals, schools and colleges for educational and administrative purposes.

(f) The Internet

The Internet is a global *inter*connected collection of computer *net*works, to which all computer users have access via a modem and telephone line. *Internet Service Providers* (ISPs) use fast, powerful computers, known as servers, to give their subscribers (for a monthly fee) a temporary link to the Internet (after a password is exchanged) via their fast permanent Internet link.

Among the services available is the *World Wide Web* (*WWW*), known simply as the *Web*, which is rapidly becoming the most popular way to distribute information on the Internet. The information consists of millions of pages of text, graphics, photographs, sound and video residing on thousands of Web server computers distributed across the Internet. The pages are combined as an on-screen page and presented to the user (called a *browser*, who has sent his request using Web browser software) as formatted pages, known as *hypertext* documents. Each page on the Web has its own address, often starting with 'http://' (standing for *h*yper*t*ext *t*ransfer *p*rotocol). In addition, Web pages also contain link words, usually shown in a different style to the other text on the page, that when 'clicked' call up pages from other parts of the Web.

Another service available to Internet subscribers is *e-mail* (see section 18.10).

(g) Multimedia

This is the integration of text, colour graphics, animation, photographs, sound and video using a computer, for education, business and home entertainment. The software containing the large amounts of data required is stored in digital form mostly on CD-ROMs or on hard and floppy disks.

Encyclopaedias, atlases, instruction manuals, educational programs and games can be viewed and the user allowed to interact.

A typical multimedia system is shown in Fig. 20.06.

(h) Digital interactive TV (iTV)

It may be that in the future, the home computer, TV set and telephone will be combined. Computing, broadcasting and telecommunications will be brought together in what is called *digital interactive TV* using one piece of equipment and digital signals. Separate TV sets, video cassette recorders (VCRs), CD players, telephones and computers will be unnecessary.

The process has already started with several electronics companies having launched systems that combine the personal computer (PC) and the television receiver (TV) into a PCTV, using the TV as a full-screen monitor. Audio CDs, video CDs and CD-ROMs can all be played by the equipment. The Internet is also accessible as are fax and answerphone facilities. Letters can be written and sent as e-mail over telephone lines using the system's modem from up to 3 metres away by a remote wireless keyboard connected to the system by an infrared link.

Fig. 20.06

It is envisaged that a fully integrated multimedia system, controlled by a *Set Top Box* (a switching and memory device) will provide, possibly via the so-called Information Superhighway, hundreds of TV channels, films and computer games on demand as well as voice and visual telephone communication, computer data links and a host of interactive activities like home banking, tele-shopping, etc.

Some people in the computer industry predict that the desktop PC could in the future, in certain situations (e.g. in offices), be replaced by a slimmed-down device called a *Network Computer* (NC). This would take the form of a Set Top Box giving access to the Internet. In such a system large amounts of long-term memory (e.g. hard disks) and some processing power would be located centrally with the Internet Service Provider (ISP) but be available to the subscriber when required; similarly, all the applications software needed for most general purposes, e.g. word processing, could be downloaded. It is claimed the approach is much more economical, in that everyone does not require an expensive PC, and encourages teamwork computing.

(i) Virtual reality

This is an artificial world seen when computer-generated graphics are viewed on a monitor. Using the effect, an architect can 'walk' through a building that has yet to be built to detect any problems before construction starts. Doctors are experimenting with it to teach surgical techniques. In Fig. 20.07 an engineer is working on a virtual reality car engine. His headset provides 3D images and sounds of the equipment he is working on. By monitoring the engineer's movements, the system can provide computer-generated responses to his actions.

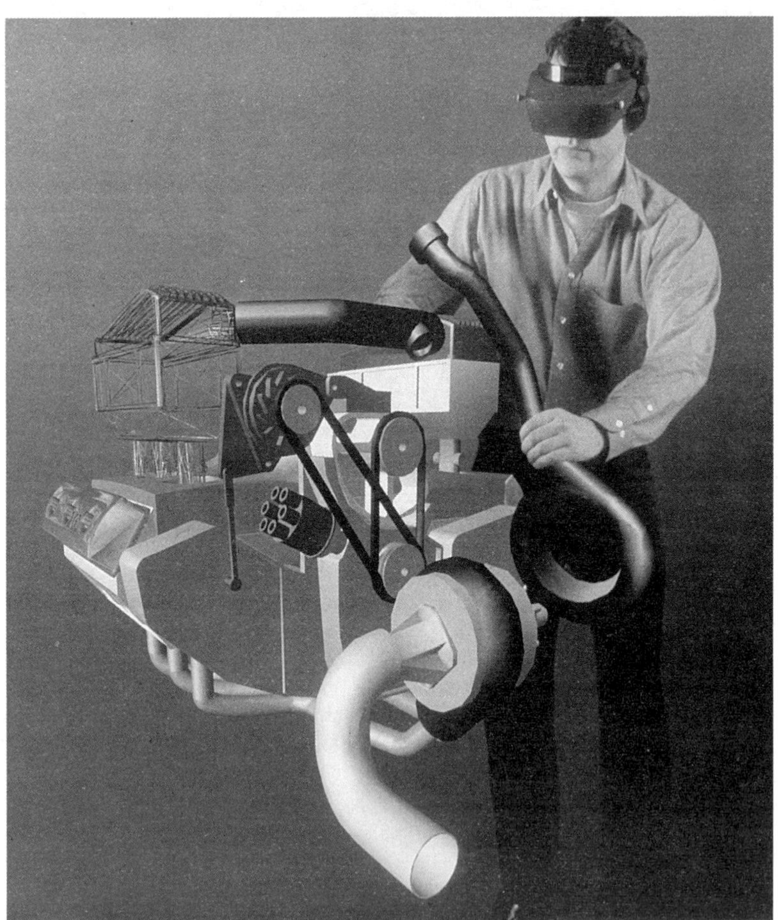

Fig. 20.07

20.7 Programs and flowcharts

(a) Programming

To communicate with a computer we must give it instructions, i.e. a *program*, so that it knows what it must do and how to do it. The program tells the CPU the order in which it has to perform the operations allowed by its instruction set. Each instruction is followed by an address for data and involves the computer in a 'fetch and carry' process. Different tasks require the operations to be done in a different order, so each task has a different program. Programs are called *software* (those stored in ROMs being *firmware*) whereas the electronic and mechanical devices making up the computer are *hardware*.

It is tedious and time-consuming to write a program of any length in the 0s and 1s of the binary system, called *machine code*, in which a computer has to work. Errors are likely and the program may be unintelligible to others. Program 'languages' have been developed to make the task easier. There are two main types.

(i) *Low-level* or *assembly* languages These are close to the binary system. Programs are written in mnemonics (i.e. memory aids) and by referring to the instruction set for the CPU we might find, for example, that the mnemonic for 'load data from memory into accumulator' is LDA. It would also give the binary code for this operation, say 1001 1110. This is more conveniently written in the program (beside LDA) and entered on the computer keyboard, in hexadecimal code—from section 15.5 you can deduce that the required code is 9E. Conversion of inputs into machine code is done by a special program stored in a ROM in the computer and called an *assembler*.

(ii) *High-level* languages These are more like everyday English and are easier to understand and work with since they use terms such as LOAD, ADD, FETCH, PRINT, STOP. However, they generally need more memory space and computer time because each statement converts into several machine code instructions, not just one as is usual in a low-level program. A *compiler* or an *interpreter* (the equivalents of an assembler and, like it, consisting of a program) does the translation into machine code. BASIC (Beginner's All-purpose Symbolic Instruction Code) is a popular high-level language; COBOL is designed for business use; FORTRAN, PASCAL and ALGOL are suitable for scientific and mathematical work.

(b) Flowcharts: binary multiplication

All programs, whether high- or low-level, must give absolutely clear instructions to the computer. Writing a program is easier if a diagram, called a *flowchart*, is drawn up initially to show the different steps required.

Suppose the problem to be solved is the binary multiplication of 6 and 5, i.e. 110 and 101. First consider how it is done by long-multiplication. The method is the same as for decimal multiplication, i.e. the multiplicand (6) is multiplied separately by each digit of the multiplier (5) in turn and the partial products properly shifted with respect to one another before they are added. The operation is simpler in binary because we have to deal with just two digits (1 and 0) and so at each stage there are only two possible answers, i.e. 110 × 1 = 110 or 110 × 0 = 0. It is performed as follows.

```
multiplicand        110 = 6
multiplier          101 = 5
                    -------
partial product     110←(110 × l.s.b. of 101, i.e. 110 × 1)
partial product    000 ←(110 × 2nd bit of 101, i.e. 110 × 0)
partial product   110  ←(110 × m.s.b. of 101, i.e. 110 × 1)
-----------------------
final product     11110 = 30 (sum of partial products)
-----------------------
```

The process is one of 'shift-and-add' in which the partial products are either zero or a shifted version of the multiplicand. When a digital computer is used for multiplication, it is more convenient to add each partial product as it 'accumulates', rather than adding them all at the end. The result is the same but the sequence is:

```
                      110
                      101
                     ----
partial product       110
partial product      000
                     ----
partial sum          0110
partial product     110
-------------------------
final product       11110
```

In practice, one shift register (X) is needed for the multiplicand, another (Y) for the multiplier and a third, the accumulator, stores the partial products and sums, and contains the final product when the multiplication is complete. This event is detected by a binary counter which records the number of 'shift-and-add' cycles to enable the whole operation to be stopped when complete.

A flowchart for binary multiplication is given in Fig. 20.08. The 'shift-and-add' cycle is repeated (as shown by the 'loop') until every bit in the multiplier has multiplied the multiplicand. Note that 'terminal' symbols (start, stop) are oval, 'operation' ones are rectangular, while those involving a 'decision' are diamond-shaped.

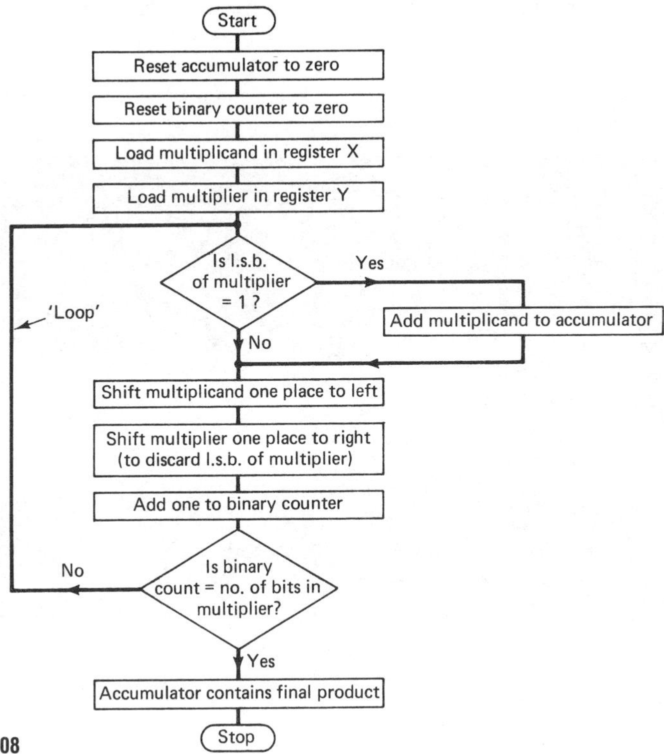
Start
Reset accumulator to zero
Reset binary counter to zero
Load multiplicand in register X
Load multiplier in register Y
Is l.s.b. of multiplier = 1 ?
Yes
Add multiplicand to accumulator
'Loop'
No
Shift multiplicand one place to left
Shift multiplier one place to right (to discard l.s.b. of multiplier)
Add one to binary counter
No
Is binary count = no. of bits in multiplier?
Yes
Accumulator contains final product
Stop

Fig. 20.08

20.8 Microprocessors

(a) Introduction

A microprocessor (MPU or μP) is a miniature version of the CPU of a digital computer, i.e. consisting of ALU, registers and control unit. It is an LSI chip containing thousands of transistors, Fig. 20.09, developed in the early 1970s when, as ICs became more complex and specialized, the need was felt for a general purpose device suitable for a wide range of jobs.

The versatility of a microprocessor is due to the fact that it is *program-controlled*. Simply by changing the program it can be used as the 'brain' not only of a microcomputer but of a calculator, a cash register, a washing machine, an electronic game or a petrol pump; it can control traffic lights or an industrial robot. *Smart cards*, which are taking us closer to the 'cashless' society, contain a microprocessor (and memory). They are being tried out by banks as a more secure alternative to magnetic strip cards used at present. The market for microprocessors with such diverse applications is much greater than for 'dedicated' chips that just do one job.

Microprocessors can now be designed only by using computers which probably contain microprocessors. Note that an MPU is not a computer itself—to be a microcomputer it needs memories and input/output units. However, the trend is to incorporate as many as possible of the peripheral support chips into the MPU package.

There are many MPUs on the market, with instruction sets for various tasks and which operate on words of different lengths (usually 4-, 8-, 16- or 32-bit). Some cost just a few pounds and are often the cheapest part of a system. They are often housed in 40 pin 0.6 inch wide d.i.l. packages and operate from 5 V and/or 12 V power supplies.

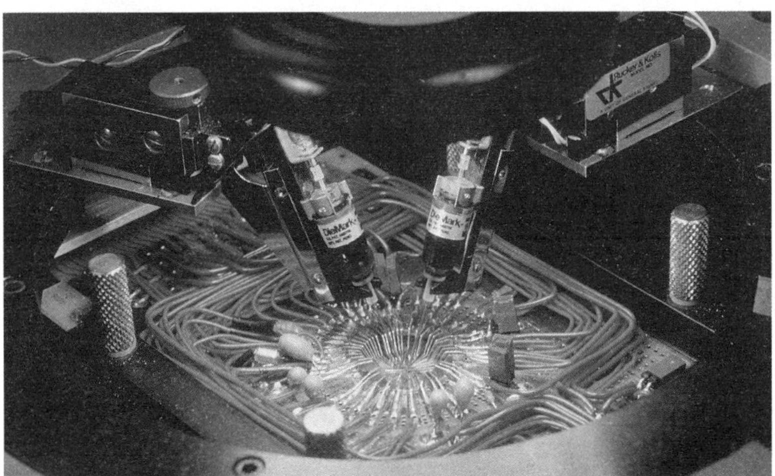

Fig. 20.09

(b) Architecture

The simplified block diagram in Fig. 20.10 of a typical microprocessor chip can be used to outline how it works. Assume it is programmed with the necessary instructions and data to add two numbers and that it has been 'reset', either manually by a switch or automatically when the power is applied, by a signal to its reset input. The *program counter* then reads zero, all *registers* contain 0s and the clock has stopped.

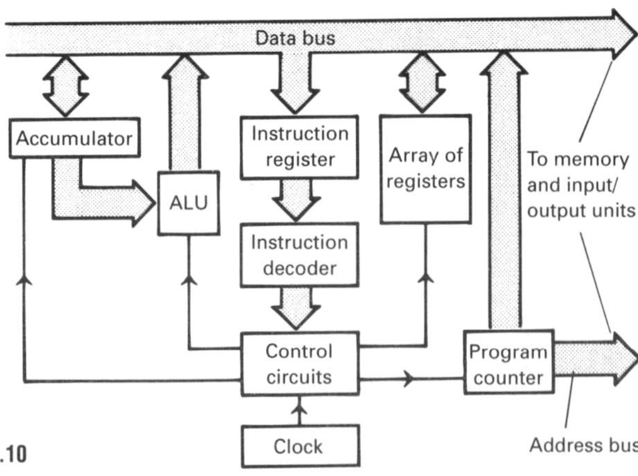

Fig. 20.10

(c) Action

The *program counter* is connected by the *address bus* to the *memory*, e.g. a ROM or PROM, in which the instructions are stored. The binary output (count) from the *counter* is the address input to the *memory* and is initially all 0s (e.g. 0000 0000 for an MPU with an eight-bit *address bus*). The instruction in this first address is thus 'read' out of the *memory* into the MPU via the *data bus*.

The instruction is held by the *instruction register* (until another is received) whose outputs are changed by the *instruction decoder* into a signal that goes to the *control circuits*. By opening and closing logic gates these set up the routes that enable the MPU to perform the operation required by the first instruction. Suppose it is LOAD.

If the *clock* is now started and advances the *program counter* by one count, the program advances by one step (line) and the data stored at the second address (e.g. 0000 0001) is 'read' out of the *memory* and loaded ('copied' is more exact) via the *data bus* into the *accumulator*. (The data will have been entered previously by the programmer and transferred into a RAM).

To obtain addition of the data (number) in the *accumulator* and the data (number) stored at another address in the RAM, the program must give the necessary instructions on succeeding clock pulses to enable the first data to be shifted from the *accumulator* to one of the internal *registers* in the MPU and for the second data to be copied into the *accumulator*.

If the ADD instruction is then given, the *instruction decoder* arranges for the *ALU* to perform the addition (i.e. act as a full-adder) and to store the result in the accumulator for subsequent transfer to the *output unit*.

(d) Register stack and subroutines

Sometimes instead of changing the program counter by one, it is useful to 'jump' from the step-by-step sequence of the main program to what is known as *a subroutine*. For example, multiplication involves a fairly long process of 'shift-and-add'. To save writing this out every time it is used in a program, it can be written just once as a subroutine, stored and recalled when needed, Fig. 20.11(*a*).

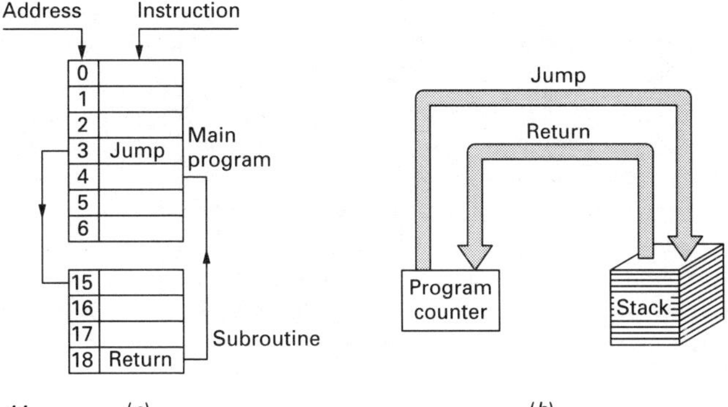

Fig. 20.11 (*a*) (*b*)

The store used is called a *register stack* because the data is 'stacked' on top of each other in order and then recovered from the top in reverse order, i.e. last-in, first-out, Fig. 20.11(*b*). Only one address is needed, that of the top and if the stack is an external RAM, a *stack pointer* is used to give the first location in the RAM chosen by the programmer as stack.

(e) Flags

A flag is a flip-flop which is set to the 'high' (1) state to show that a particular operation has been completed. For instance, in some MPUs the *carry flag* is set to 1 if there is a carry bit in the accumulator and the *zero flag* to 1 if an instruction puts a 0 in it.

(f) MMX technology

This important new technology is being built into the latest family of *Intel*'s 166 MHz microprocessors, making them 10 to 20% faster than non-MMX processors with an equivalent clock speed. Very significant improvements in multimedia and communications applications are obtained in the form of smoother and more realistic video, better graphics, CD quality sound and speech recognition.

The basis of the technology is an advanced technique called SIMD (Single Instruction, Multiple Data) which, as its name suggests, enables a single instruction on the PC to process multiple pieces of information. It has been achieved by incorporating into the architecture, among other things, 57 new instructions and eight 64-bit wide registers.

20.9 Computer abbreviations

These are common in technical and sales literature.

EDO RAM: *E*xtra *D*ata *O*utput RAM. A fast RAM.

MMX: *M*ulti*M*edia e*X*tensions. An advanced technology used in microprocessor architecture.

MPEG: *M*otion *P*ictures *E*xperts *G*roup. A compression technology which allows a PC to give full quality TV from a CD-ROM.

PCMCIA: *P*ersonal *C*omputer *M*emory *C*ard *I*ndustry *A*ssociation. This refers to a peripheral card of electronic circuits which provides additional features when it is plugged into an expansion slot on a PC. A *sound card* allows CD sound to be played. A *graphics card* or *accelerator* and a *video card* are needed for advanced computer games and for playing CD video. A *TV card* turns a PC into a TV. A *modem* (see section 18.10) connects a PC to telephone line facilities.

SVGA: *S*uper *V*ideo *G*raphics *A*daptor. A standard with which good quality monitor screens should be compatible for good colour pictures.

20.10 Analogue computer

(a) Electronic analogues

Analogue computers are used as electronic models or analogues of mechanical and other systems in cases where conducting experiments on the system itself would be costly or time-consuming or dangerous, or where the system has not yet been built. For example, when designing a bridge, an aircraft wing or any structure where motion can occur, the engineer must know beforehand how it will react to variable physical quantities such as wind speed and temperature.

A moving system can be described by a set of mathematical equations, called *differential equations*, which are derived from the laws of physics and take account of the quantities involved. Help with predicting the behaviour of the real system can be obtained by solving these equations, i.e. by finding how the quantities concerned vary with time, as most do. However, sometimes this is difficult or impossible by normal mathematical methods. Fortunately, an analogue computer will provide the solutions if it is set up to perform the necessary mathematical operations on voltages representing the quantities under investigation (and which may be varied continuously). One of the most important uses of an analogue computer is solving such differential equations.

In recent years analogue computers have become less popular because it is now possible to programme digital computers to simulate moving physical systems. None the less, for some problems it is easier and cheaper to use an analogue computer since all the operations occur at once and, no matter how complex the problem, the computing time required is the same. To perform the various mathematical operations, an analogue computer uses op amps in the form of inverting, summing and integrating amplifiers—see sections 13.4 and 13.5.

(b) Op amp as an integrating amplifier

The circuit is that of the inverting amplifier shown in Fig. 13.10 but feedback occurs not via a resistor but via a capacitor, as shown in Fig. 20.12. As we will see, it inverts, amplifies and integrates (adds up) the input voltage over any period of time. Basically its action is the same as the CR integrator circuit considered in section 5.8 but the op amp provides the greater output voltages required for analogue computing.

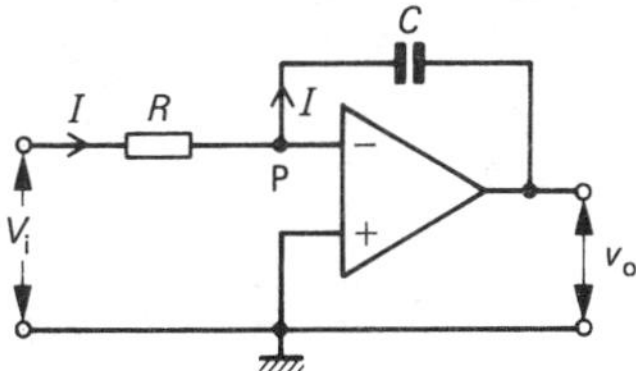

Fig. 20.12

Because of the very large open-loop gain of the op amp, point P is a virtual earth, i.e. at 0 V. Therefore the voltage across R equals the input voltage, V_i, and that across C is the output voltage, v_o. Assuming as before that a negligible fraction of the input current I enters the inverting (−) input of the op amp (partly because of its very high input impedance), then I also flows 'through' C and charges it up. If V_i is a constant positive d.c. voltage, v_o will be negative (owing to the action of the inverting input) and I will also be constant, with a value:

$$I = \frac{V_i}{R} \tag{1}$$

C charges at a constant rate and v_o falls, becoming more negative, at a constant rate. That is, the voltage of the output side of C (which is v_o, since its input side is at 0 V) changes so that the feedback path absorbs I. If q is the charge on C at time t and the voltage across it (the output voltage) changes from 0 to v_o in that time, then, from section 5.4, we have:

$$q = v_o \times C \tag{2}$$

But I is constant during t, therefore, from section 2.9, we have:

$$q = I \times t \tag{3}$$

From (2) and (3):

$$v_o \times C = I \times t$$

Substituting for I from (1), we get:

$$v_o \times C = \frac{V_i \times t}{R}$$

i.e.

$$v_o = -\frac{V_i \times t}{C \times R} = -\left(\frac{1}{C \times R}\right) V_i \times t$$

The negative sign is inserted because, when the inverting input is used, v_o is negative if V_i is positive and vice versa.

From (4) we see that the effective gain of the amplifier (v_o/V_i) equals $1/(C \times R)$. Taking $C = 1\ \mu\text{F}$ and $R = 100\ \text{k}\Omega$ makes the time constant $C \times R = 10^{-6}\ \text{F} \times 10^5\ \Omega = 10^{-1}\ \text{s} = 0.1$ s and gives a gain of 10. Therefore if $V_i = +0.1$ V, then $v_o/t = -V_i/(C \times R) = 0.1/0.1 = -1$ V per second, i.e. v_o falls steadily by 1 Vs^{-1} until it reaches the power supply voltage and the op amp saturates at, say, −9 V. This is shown by the graphs in Fig. 20.13(*a*). If $V_i = -2.0$ V, v_o rises steadily by 2 V s^{-1}, as in Fig. 20.13(*b*).

A more general mathematical treatment shows that if V_i varies then, using calculus notation:

$$v_o = -\frac{1}{C \times R}\int V_i\,\text{d}t$$

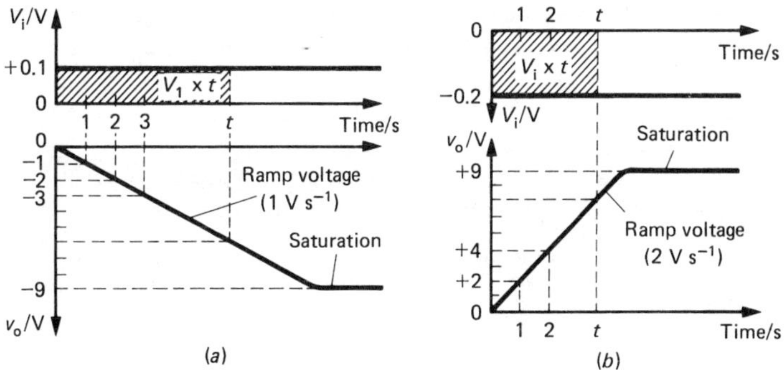

Fig. 20.13

$\int V_i \, dt$ stands for the 'integral of V_i with respect to time, t'. The integrating action on V_i is similar to that of a petrol pump which 'operates' on the rate of flow (in litres per second) and the delivery time (in seconds) to produce as its output, the volume of petrol supplied (in litres). (Those familiar with calculus will recognize $\int V_i \, dt$ as representing the area under the $V_i - t$ waveform, which; when V_i is constant, as in Figs. 20.13(*a*) and (*b*), equals $V_i \times t$.)

The output of an integrating amplifier with a constant d.c. input is a ramp voltage. It can be used as the time base for an oscilloscope (see section 21.8) and also for A/D conversion (see section 17.3). A square wave input produces a triangular wave output (Fig. 20.14).

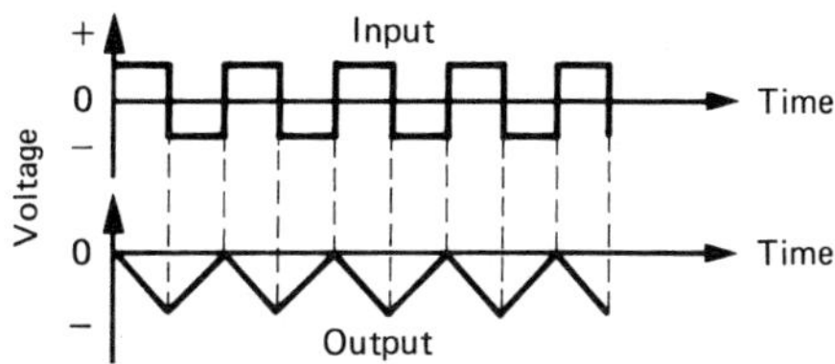

Fig. 20.14

(c) Solving a differential equation using an analogue computer

The variation of a quantity, y, with time t, i.e. its rate of change, is called its *first differential coefficient*, $\frac{dy}{dt}$. The rate of change of $\frac{dy}{dt}$ is the *second differential coefficient*, $\frac{d^2y}{dt^2}$, and so on. Differential coefficients arise in the differential equations used as mathematical models of real systems. For instance, in dealing with moving objects, velocity v and acceleration a are useful quantities. Velocity is rate of change of distance s (measured in, say, metres per second) while acceleration is rate of change of velocity (i.e. metres per second per second). They can be expressed in terms of the differential coefficients of s as $v = \frac{ds}{dt}$ and $a = \frac{d^2s}{dt^2}$.

To show how differential equations may be solved by analogue computing, we will consider the very simple case, easily soluble by mathematical methods:

$$\frac{d^2y}{dt^2} = 10$$

We wish to obtain an expression for y in terms of t. Integration can be regarded as the reverse process of differentiation (for both of which there are mathematical rules). For example, the integral of $\frac{dy}{dt}$ is y (i.e. $\int \frac{dy}{dt} dt = y$) and the integral of $\frac{d^2y}{dt^2}$ is $\frac{dy}{dt}$.

Therefore, if a constant d.c. voltage representing $\frac{d^2y}{dt^2}$ is supplied as the input to two integrating amplifiers in series, having $C \times R = 1$s for each stage, as in Fig. 20.15, integration occurs twice and the output voltage will represent y in terms of t. If the output is displayed on an oscilloscope, the graph produced shows how y varies with t. (In fact, one solution is $y = 5t^2$.) Knowing the scales of the voltage and time axes, numerical answers can be obtained.

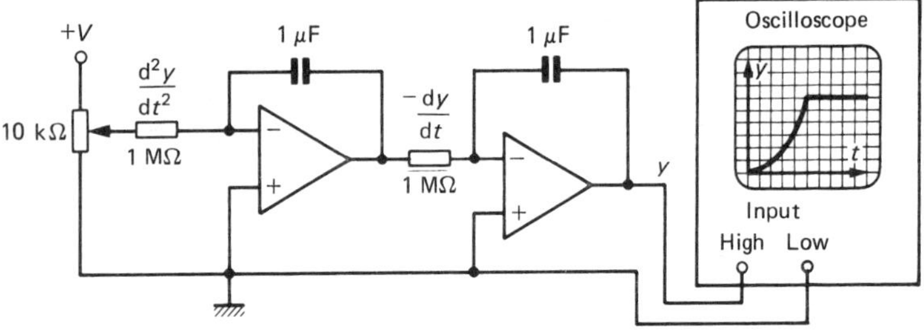

Fig. 20.15

The equation $\frac{d^2y}{dt^2} = 10$ represents the motion of an object falling from rest, neglecting air resistance, y being the distance fallen (metres) in time t (seconds). Equations for systems such as car suspensions, spacecraft travel and automatic pilots for aircraft and ships are much more complex but they are solved in basically the same way.

20.11 Brief history of computers

Charles Babbage, a mathematics professor at Cambridge University in the 1830s, is generally regarded as the father of the computer. He designed, but never completed, a mechanical device that could perform calculations. He called it an 'Analytical Engine' and believed it could process information that was represented by numbers. Furthermore, he saw his 'engine' as being able to follow different sets of instructions and so perform various tasks. That is, it would respond to what we now call 'software' and be a general-purpose device rather than one dedicated to a particular job.

One hundred years later, in the 1930s and 40s, *Alan Turing* and *Claude Shannon* did important theoretical work on calculating machines, logic circuits and information representation. Also, *John von Neumann* suggested that the architecture of general-purpose machines could be simplified if they had a memory to store different instructions—which is the principle of modern computers.

During World War II electronic computers were built in the UK, USA and Australia using thousands of thermionic valves (vacuum tubes) and relays as switches. These monsters occupied rooms 10 m × 5 m, weighed several tonnes, consumed vast amounts of power (e.g. 100 kW), had only a few kilobytes of memory and performed a mere 1000 operations per second. In 1948 *William Shockley* developed the transistor, which soon replaced the bulky thermionic valve. Computer electronics was then revolutionized with the arrival of integrated circuits in the early 1960s and of the microprocessor in the early 1970s. They heralded the birth of the personal computer.

The first MPU to be useful in a computer was *Intel's* '8080', introduced in 1974 and the direct ancestor of today's family of *Pentium* processors used in about 80% of modern PCs (i.e. IBM models and their clones). At the same time *Motorola* was promoting its '68 000' family and at present its *Power PC* processor is used in Apple computers. The latest MPUs have over 5 million transistors on one tiny silicon chip, a number which, according to *Moore's Law*, doubles every two years or so—as does the MPU clock speed. The microprocessor is unquestionably one of our greatest intellectual and technological accomplishments.

Current research suggests that MPU developments in the near future will be in three main directions. *First*, they will have faster clock speeds (over 500 MHz), *second*, they will offer extra on-chip multimedia functions more cheaply than today's arrangement of plug-in expansion cards and peripherals (giving better graphics, audio, video and telephone line access) and *third*, there will be chips that are used in network devices and are Internet-based.

20.12 Revision questions and Problems

1. a) Draw the block diagram for a simple digital computer and say what each part does.

b) Distinguish between the following pairs of terms: hardware and software; high-level and low-level languages; assembler and compiler; microprocessor and microcomputer.

c) Name and describe three kinds of external memory.

d) What is a flowchart? Construct one to get a microcomputer to

(i) switch off an electric kettle when it boils,

(ii) control the gas central heating system of Problem 5 in section 14.13.

2. a) When and why are analogue computers used as electronic models of mechanical and other systems?

b) What are the main differences between digital and analogue computers?

3. In the integrating amplifier circuit of Fig. 20.12, if $V_i = +1$ V and $C = 1\ \mu$F,

a) does v_o rise or fall and at what rate if $R = 10$ kΩ?

b) what value of R makes v_o change at a rate of 10 V s^{-1}?

UNIT 21

Power supplies and test instruments

21.1 Introduction

Electronic systems need a power supply in order to work. Test instruments are useful when electronic systems fail to work.

(a) Power supplies

The voltages for circuits containing semiconductor devices seldom exceed 30 V and may be as low as 1.5 V. Current demands vary from a few microamperes to many amperes in large systems. Usually the supply must be d.c.

Batteries are suitable for low power portable equipment such as radios, watches and calculators. In general however, power supply units operated from the a.c. mains are employed. In these, a.c. has to be converted to d.c., the process being called *rectification*. Apart from not requiring frequent replacement, as batteries may, power supply units are more economical and reliable and can provide larger powers. They can also supply very steady voltages when this is essential.

(b) Test instruments

These are required for circuit design, maintenance and fault-finding. The main features of multimeters, oscilloscopes and signal generators will be considered briefly.

21.2 Rectifier circuits

In most power supply units a transformer steps down the a.c. mains from 230 V to the required low voltage. This is then converted to d.c. using one or more junction diodes (usually silicon) in a rectifier circuit.

(a) Half-wave rectifier

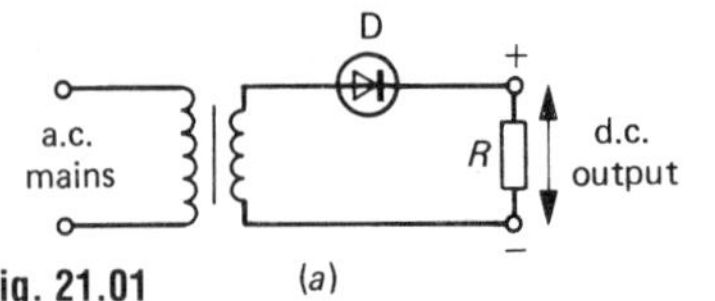

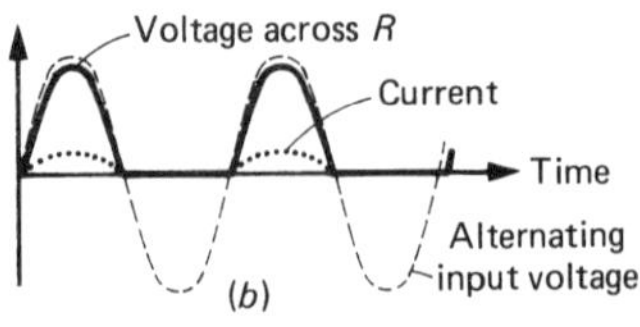

Fig. 21.01

In the simple circuit of Fig. 21.01(*a*) the load is represented by a resistor *R*, but in practice it will be a piece of electronic equipment. Positive half-cycles of the alternating input voltage forward bias diode D which conducts, creating a pulse of current. This produces a voltage across *R* of almost the same value as the input voltage if the forward resistance of D is small compared with the value of *R*. The negative half-cycles reverse bias D—there is little or no current in the circuit and the voltage across *R* is zero.

The various waveforms are shown in Fig. 21.01(*b*). The current pulses are unidirectional and so the voltage across *R* is direct, for, although it varies, it never changes direction.

(b) Centre-tap full-wave rectifier

In full-wave rectification, both halves of every cycle of input voltage produce current pulses.

In the circuit of Fig. 21.02(*a*), two diodes D_1 and D_2 and a transformer with a centre-tapped secondary are used. Suppose that at a certain time during the first half-cycle of the input, the voltage induced in AOB is 12 V. If we take the centre-tap O as the reference point at 0 V, then when A is at +6 V, B will be at −6 V. D_1 is then forward biased and conducts, giving a current pulse in the circuit AD_1CROA. During this half-cycle D_2 is reverse biased by the voltage across OB (since B is negative with respect to O). On the other half of the first cycle, B becomes positive relative to O and A negative. D_2 conducts to give current in the circuit BD_2CROB: D_1 is now reversed biased.

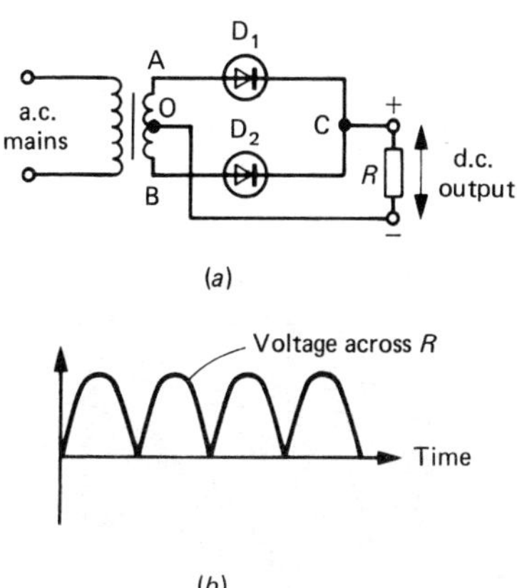

Fig. 21.02

In effect, the circuit consists of two half-wave rectifiers working into the same load on alternate half-cycles of the input. The current through *R* is in the same direction during both half-cycles and a fluctuating direct voltage is created across *R* like that in Fig. 21.02(*b*).

(c) Bridge full-wave rectifier

In the circuit of Fig. 21.02(*a*), only half of the secondary winding is used at any time and the transformer has to produce twice the voltage required. The problem does not arise if four diodes are arranged in a bridge network as in Fig. 21.03. If A is positive with respect to B during the first half-cycle, D_2 and D_4 conduct and current takes the path AD_2RD_4B. On the next half-cycle when B is positive, D_1 and D_3 are forward biased and current follows the path BD_3RD_1A. Once again, current through *R* is unidirectional during both half-cycles of input and a varying d.c. output is obtained. This is a popular circuit because diodes are cheap.

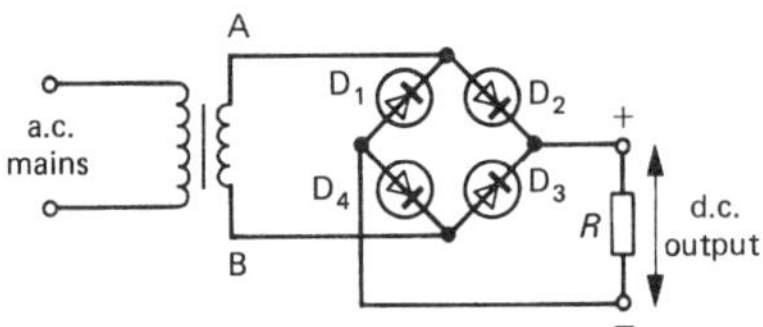

Fig. 21.03

All four diodes are available in one package, with two a.c. input connections and two output connections, classified, as with single diodes, according to the maximum reverse voltage V_{RRM} and average forward current I_F(av)—see section 7.5.

High current diode rectifiers are mounted on finned heat sinks (made of aluminium sheet and painted black), like that shown in Fig. 21.04, to keep their junction temperatures low. Manufacturers state what the *thermal resistance* of the heat sink should be for a particular rectifier. For example, if it is 2 °C W^{-1}, the temperature of the heat sink rises 2 °C above its surroundings for every watt of power it has to dissipate. To get rid of 10 W, the rise will be 20 °C.

Fig. 21.04

21.3 Smoothing circuits

The varying d.c. output from a rectifier circuit has to be smoothed to obtain the steady d.c. demanded by electronic equipment.

(a) Reservoir capacitor

The simplest smoothing circuit consists of a capacitor, called a reservoir capacitor C_1, connected as in Fig. 21.05(*a*) for half-wave rectification. The value of C_1 on a 50 Hz supply may range from 100 μF to about 10 000 μF (typically 1 000 μF), depending on the current and smoothing required.

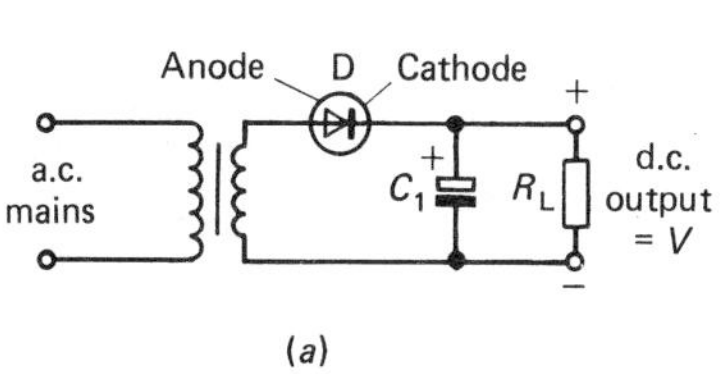

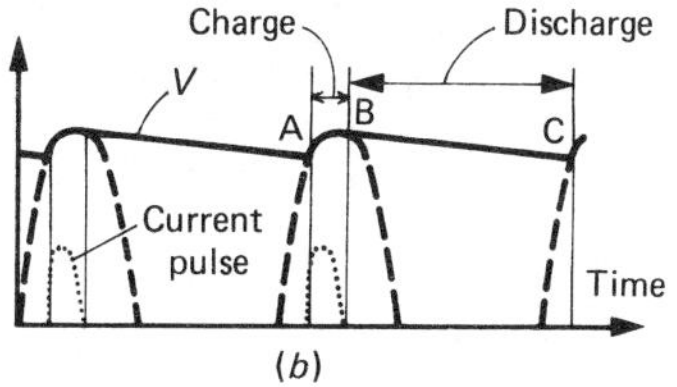

Fig. 21.05

The voltage and current waveforms for the circuit are given in Fig. 21.05(*b*). The solid line is the smoothed output voltage V developed across the load R_L. During the half-cycles of a.c. input (shown dashed) when D is forward biased, there is a pulse of rectified current (shown dotted) for part of the time (e.g. AB). This current charges up C_1 to near the peak value of the a.c. input. During the rest of the cycle (e.g. BC), C_1 keeps the load R_L supplied with current by partly discharging through it. While this occurs, V falls until the next pulse of rectified current tops up the charge on C_1. Hence for most of the time the load current is supplied by C_1 acting as a reservoir.

V is a d.c. voltage varying at mains frequency (50 Hz) but the amplitude of the variations, called the 'ripple' which reveals itself as 'mains hum', is much less than when C_1 is absent.

(b) Diode and capacitor ratings

The smoothing action of C_1 is due to its large value making the time constant $C_1 \times R_L$ large compared with the time for one cycle of the a.c. mains ($\frac{1}{50}$ s). The larger C_1 is, the better the smoothing but, as you can see from the waveform of V in Fig. 21.05(*b*), the smaller the fall in V, the briefer will be the rectified current pulse. Consequently, the greater will its peak value have to be to deliver a given amount of energy and damage may occur if this exceeds the peak current rating of the diode. Manufacturers sometimes state the maximum value of reservoir capacitor to be used with a diode rectifier.

Consideration has also to be given to the maximum reverse voltage V_{RRM} of the diode. When it is not conducting, Fig. 21.05(*a*) shows that the total voltage across it is the voltage across C_1 plus that across the transformer secondary. For example, if the lower plate of C_1 is at 0 V and the top is near the positive peak of the a.c. input, say +12 V, then the cathode of the diode is at +12 V. Its anode will be at −12 V, i.e. the negative value of the voltage across the transformer, so the total voltage across D is 24 V in this case. In practice it is wise to use a diode whose V_{RRM} is at least four times the r.m.s. output voltage of the transformer to allow for voltage 'spikes' picked up by the a.c. mains supply.

The reservoir capacitor should have a voltage rating at least equal to the peak value of the transformer output voltage, i.e. $\sqrt{2} \times$ r.m.s. value (see section 3.4).

(c) Capacitor-input filter

The smoothing produced by a reservoir capacitor can be increased and ripple reduced further by adding a filter circuit. This consists of a choke L or a resistor R and another large capacitor C_2, arranged as in Figs. 21.06(*a*) and (*b*). We can regard the varying d.c. voltage produced across C_1 as a steady d.c. voltage (the d.c. component) plus a small ripple voltage (the a.c. component). L or R offers a much greater impedance than C_2 to the a.c. component and so most of the unwanted ripple voltage is developed across L or R. For the d.c. component the situation is reversed and most of it appears across C_2. The filter thus acts as a voltage divider, separating d.c. from a.c., and producing a d.c. output voltage across C_2 with less ripple.

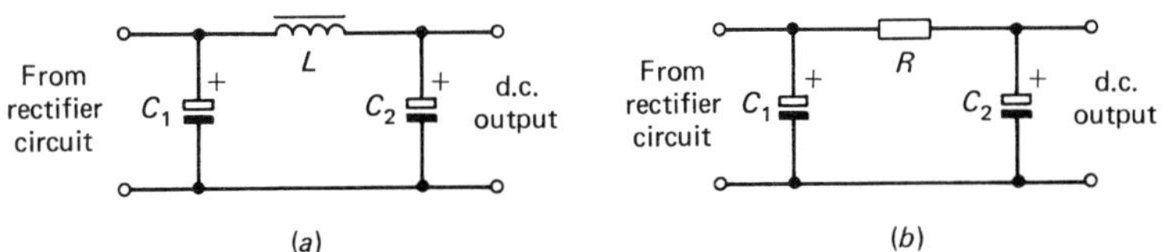

Fig. 21.06

Filter circuits have certain disadvantages which make them less suitable for semiconductor circuits which operate at low voltages and large currents. These are due to (i) the risk of large currents in chokes causing magnetic saturation of the iron cores, (ii) chokes being large (by present standards) and heavy, and (iii) resistors reducing the output voltage. Today when very steady d.c. voltages are required, a stabilized (regulated) power supply is used.

21.4 Stabilizing circuits

Because of its internal resistance, the d.c. output voltage from an ordinary power supply unit (and a battery) decreases from a maximum as the load current increases from zero. The greater the decrease the worse is said to be the *regulation* of the supply; Fig. 21.07(*a*) shows two regulation curves. There are many occasions, e.g. in circuits using TTL ICs, when a d.c. voltage is required which is not affected by load current changes. In these cases stabilizing or regulating circuits are added, as shown diagrammatically in Fig. 21.07(*b*).

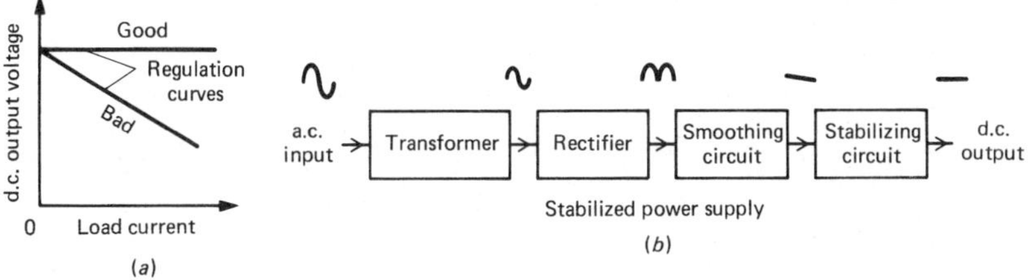

Fig. 21.07

(a) Zener diode stabilizer

The Zener diode was discussed in section 7.7 where we saw that, if it is reversed biased to the breakdown voltage, the voltage across it stays almost constant for a wide range of reverse currents. The characteristic is shown again in Fig. 21.08(*a*) and a simple voltage stabilizer circuit is shown in Fig. 21.08(*b*) where R_L is the load.

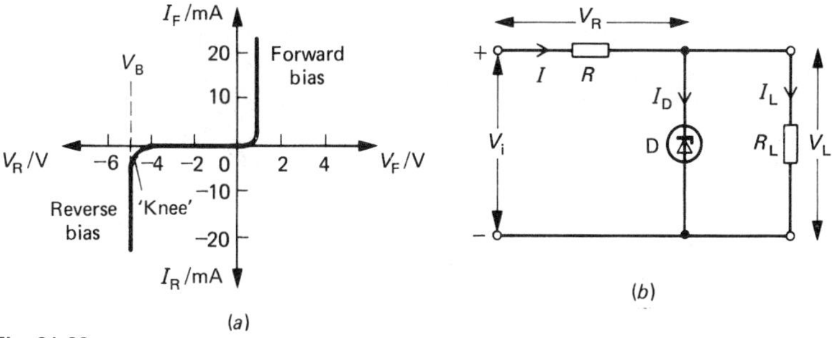

Fig. 21.08

Suppose that the unstabilized input voltage V_i is fixed. Then, since the voltage V_L across the diode, i.e. the d.c. output voltage, is also constant, it follows that the voltage V_R across resistor R will be constant as well, because $V_i = V_R + V_L$. The current I through R equals V_R/R and is therefore constant. But $I = I_D + I_L$, where I_D and I_L are the currents through the diode and the load respectively. Hence if I_L increases, I_D decreases by the same amount to keep I (and V_R) constant. V_L thus remains unchanged. Similarly if I_L decreases, there is an equal increase in I_D.

Two points should be noted. First, if $I_L = 0$, then $I_D = I$ and the diode has to be able to carry this current safely. Second, when I_L is a maximum, I_D has its minimum value which must be large enough (at least 5 mA for a small diode) to ensure the diode works on the breakdown part of its characteristic (not at or above the 'knee' where the voltage across the Zener falls to zero). This is achieved by making V_i 2 or 3 volts greater than the required output voltage.

Another reason for V_L not being steady is fluctuation of V_i due to ripple on it after smoothing or to a.c. mains variations. Stabilization helps to reduce both effects. If V_i rises, the increase in I is such that the rise in voltage occurs across R, leaving V_L about the same. The opposite action occurs when V_i falls.

An example showing how the value of the stabilizing resistor R can be calculated for a certain set of conditions was given in section 7.7.

(b) Emitter-follower stabilizer

Better stabilization is possible and higher load currents are supplied if a Zener diode is combined with an emitter-follower (see section 11.6) as shown in Fig. 21.09. The diode keeps the base voltage of the transistor fixed and the output voltage V_L is held constant at $V_D - V_{BE}$ for a wide range of load currents (V_D = Zener voltage, $V_{BE} = 0.6$ V). The transistor should be a power type (e.g. 2N3055) capable of carrying large load currents.

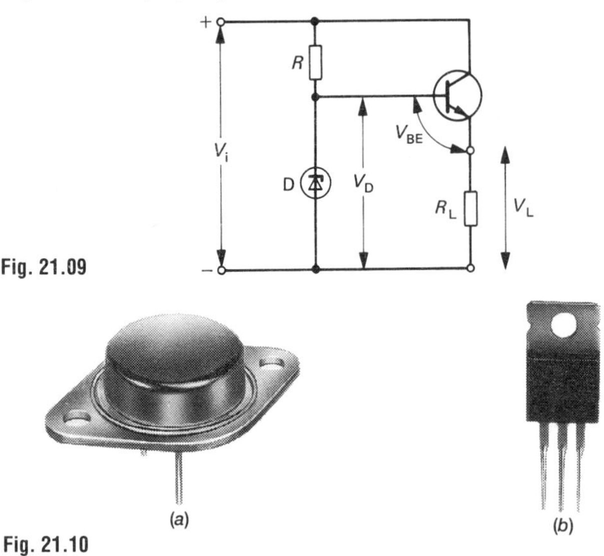

Fig. 21.09

Fig. 21.10

(c) IC regulators

A range of integrated circuit stabilizers, known as *voltage regulators*, are now available in a variety of packages, Fig. 21.10. They work on the principles just considered but their more complex circuitry provides protection against overloading and overheating. Many are designed for one voltage, e.g. 5 V for TTL circuits, or 15 V for operational amplifiers. Fig. 21.11 is the circuit for a stable 5 V 600 mA supply using the L005 regulator mounted on a 4 °C W^{-1} heat sink. C_1 is a reservoir capacitor, C_2 and C_3 are not essential but improve the performance of the regulator.

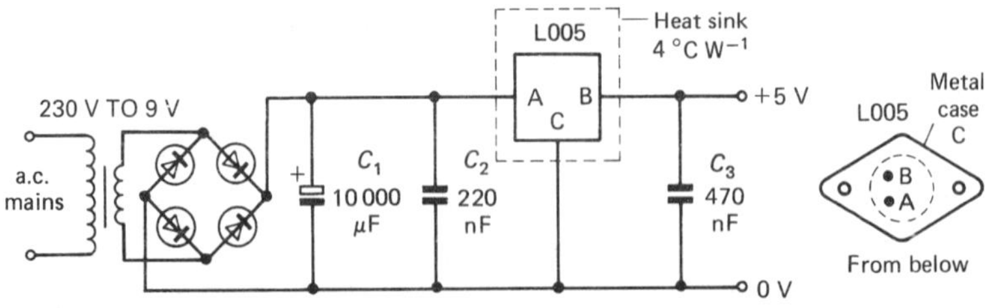

Fig. 21.11

21.5 Power control

(a) Thyristor

A thyristor is a four-layer, three-terminal semiconducting device, as shown in Figs. 21.12(*a*), (*b*) and (*c*). It used to be called a *silicon controlled rectifier* (SCR) because it is a rectifier which can control the power supplied to a load and in a way that wastes very little energy.

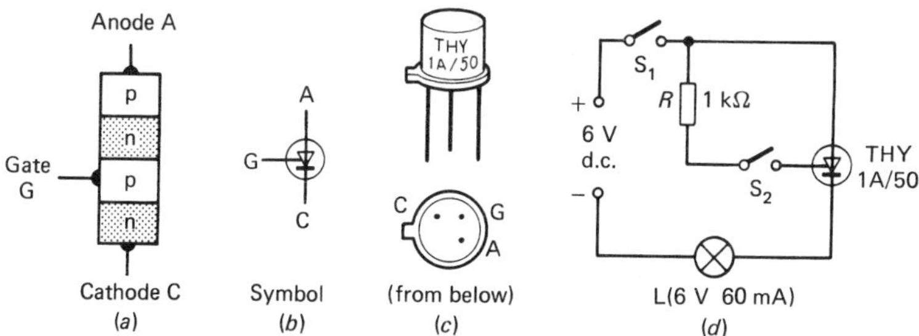

Fig. 21.12

When forward biased, a thyristor does not conduct until a positive voltage is applied to the gate. Conduction continues when the gate voltage is removed and stops only if the supply voltage is switched off or reversed or the anode current falls below a certain value. The circuit of Fig. 21.12(*d*) shows the action. When S_1 is closed, the lamp L stays off. When S_2 is closed as well, gate current flows and the thyristor switches on, i.e. 'fires'. The anode current is large enough to light L, which stays alight if S_2 is opened. This is a simple example of d.c. power control by the thyristor.

The control of a.c. power can be achieved by allowing current to be supplied to the load during only part of each cycle. A gate pulse is applied automatically at a certain chosen stage during each positive half-cycle of input. This lets the thyristor conduct and the load receives power. For the example shown in Fig. 21.13 the pulse occurs at the peak of the a.c. input. During negative half-cycles the thyristor is non-conducting and it does not conduct again till half-way through the next positive half-cycle. Current flows for only a quarter of a cycle but by changing the timing of the gate pulses, this can be decreased or increased. The power supplied to the load is thus varied from zero to that due to half-wave rectified d.c.

(b) Triac and diac

The thyristor, being a half-wave device like a diode, only allows half the power to be available. A triac consists of two thyristors connected in parallel but in opposition and controlled by the same gate. It is a two-directional thyristor which is triggered on both halves of each cycle of a.c. input by either positive or negative gate pulses. The power obtainable by a load can therefore be varied between zero and full-wave d.c. The connections are called *main terminals 1* and *2* (MT1 and MT2)

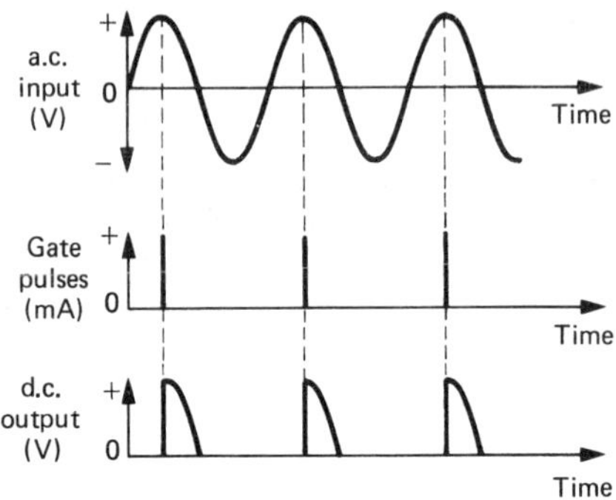

Fig. 21.13

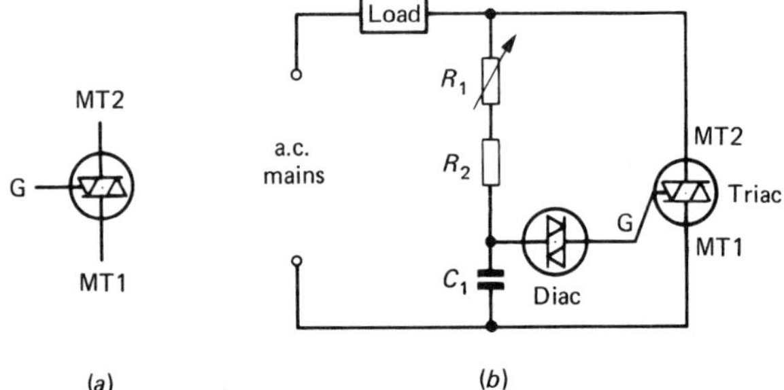

Fig. 21.14

and *gate* (G), as shown in Fig. 21.14(*a*). The triggering pulses are applied between G and MT1. The gate current for a triac handling up to 100 A may be no more than 50 mA or so.

Triacs are commonly used in lamp dimmer circuits and for motor speed control (e.g. in an electric drill). The basic dimmer circuit is given in Fig. 21.14(*b*). The diac is in effect two Zener diodes connected back to back. It conducts in either direction when the voltage across C_1 reaches its breakdown value of about 30 V. A current pulse from C_1 then triggers the triac and power is supplied to the load (lamp). The conduction time of the triac depends on the setting of the dimmer control R_1. The greater its value the more slowly C_1 charges up and the later in each half-cycle the triac is turned on, i.e. the dimmer the lamp.

21.6 Batteries

A battery consists of two or more cells connected together to produce a higher voltage or more electrical energy. Sometimes the term is used for a single cell. Batteries fall into two main groups depending on whether or not they can be recharged.

(a) Primary cells

These are non-rechargeable and have to be discarded after use. They are popularly known as 'dry' batteries. The more common types used today are listed in Table 21.1, and the graphs in Fig. 21.15 show approximately how their voltages for size AA (U12) vary when supplying moderate currents (e.g. 30 mA).

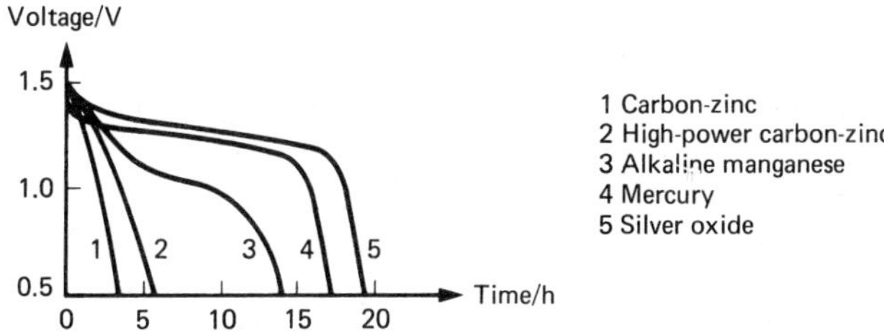

Fig. 21.15

Table 21.1 Types of primary cell

Cell	Voltage	Properties	Uses
Carbon-zinc	1.5	Most popular type; voltage falls as current increases; best for low currents or occasional use; cheap	Torch; radio
Alkaline-manganese	1.5	Voltage does not fall so much in use; long 'shelf' life; better for higher currents; last up to four times longer than same size carbon-zinc cell; medium price	Radio; calculator; photographic flash unit
Mercury	1.3	Voltage almost constant until discharged; best for low current use; large capacity for their size; made as 'buttons'; long 'shelf' life; expensive	Watch; calculator; camera; hearing aid
Silver oxide	1.5	Similar to mercury	As for mercury

(b) Secondary cells

Batteries in this group are often called accumulators. After discharging they can be recharged repeatedly (but not indefinitely) by sending a current through them in the reverse direction. They supply 'high' continuous currents depending on their *capacity*, which is measured in ampere-hours (A h) for a particular discharge rate. For example, a battery with a capacity of 30 A h at the 10 hour rate will sustain a current of 3 A for 10 hours, but while 1 A would be supplied for more than 30 hours, 6 A would flow for less than 5 hours.

Three other rechargeable types used in portable computers and mobile phones are the *Nickel Cadmium* (NiCad) cell (1.2 V) and the newer *Nickel Metal Hydride* (NiMH) and *Lithium Ion* cells. They are available in various sizes. The NiCad is cheap but should be fully discharged before recharging from a constant current source (sold for the purpose) to prolong its life. The NiMH has a greater capacity and no 'memory effect' like some NiCads so can be fully charged each time. The Lithium Ion cell is the most efficient but most expensive.

21.7 Multimeters

(a) Analogue

In this type of multimeter the continuous deflection of a pointer over a scale represents the value of the quantity being measured. Basically it consists of a *current-measuring* moving-coil meter with appropriate internal resistors brought into circuit by a range switch. Its operation was considered in section 2.7, where we saw that it can be used to measure current and voltage, both d.c. and a.c., as well as resistance. Tests can also be made on capacitors (see section 5.9) and transistors (see section 8.6).

A meter affects the circuit to which it is connected and alters the quantity it has to measure. A voltmeter causes least disturbance when its resistance is very large and only a small current is needed to produce a full-scale deflection (f.s.d.). For a good quality meter on a d.c. range, 50 μA is typical and in this case the *sensitivity* of the voltmeter is said to be 20 000 Ω V^{-1}, i.e. 1 V gives a current of 1 V/20 000 Ω = 1/20 000 A = 50 μA. An a.c. voltmeter with a sensitivity of 500 Ω V^{-1} would require a current of 1/500 A = 2 mA for a full-scale deflection.

The actual resistance of the meter depends on the range chosen as well as on the sensitivity. For example, if the sensitivity is 20 kΩ V^{-1}, then its resistance on the 1 V range is $20 \times 1 = 20$ kΩ and on the 10 V range it is $20 \times 10 = 200$ kΩ (the increase is due to extra resistors in the meter being connected by the range switch). For a particular sensitivity, the higher the range, the greater is the resistance of the meter and the less it 'loads' the circuit involved.

(b) Digital

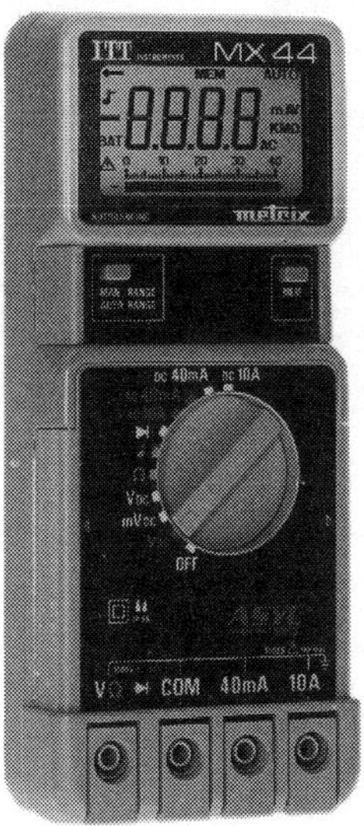

Fig. 21.16

An electronic multimeter with a digital decimal display (e.g. LCD) is shown in Fig. 21.16. The reading is produced by a *voltage-measuring* analogue-to-digital converter. Many instruments use the dual-slope ramp principle but a general idea of the action may be obtained by referring to the simpler single-slope method described in section 17.3. The binary output from the circuit in Fig. 17.07(*a*) would be applied to a decoder driving the multimeter display. When the input voltage is of the order of millivolts, it is amplified before being measured. For varying voltages, a latch circuit is included to hold the display steady at the last value while a further sample of the input is taken. With some extra internal components, currents and resistances can be measured and the various other multimeter jobs undertaken.

The input resistance of a digital multimeter on voltage ranges is high, of the order of 10 MΩ and unlike the analogue type, is the same on all ranges. It also has the advantage that errors are less likely to arise from reading the wrong scale or having to make an estimate when the pointer is not exactly over a mark on the scale. The highest operating frequency of an analogue multimeter using rectifiers on a.c. ranges is about 2 kHz; the upper limit for an electronic multimeter is much higher.

21.8 Oscilloscopes

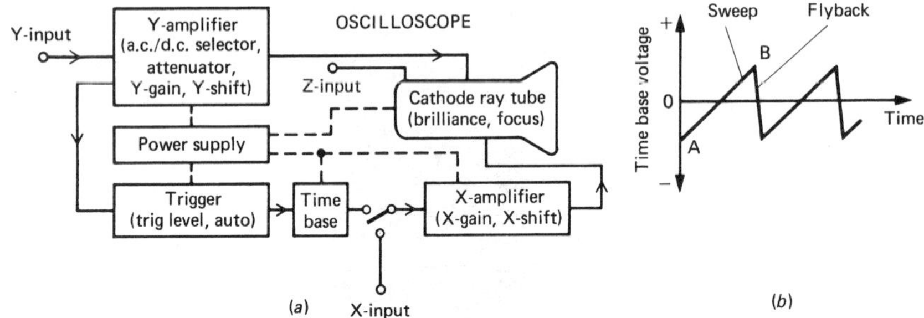

Fig. 21.17

(a) Introduction

The cathode ray oscilloscope (CRO) is one of the most useful test instruments in electronics. It produces a visual output, in the form of a graph or 'trace' on the screen of a cathode ray tube, of the voltage applied to its input. Non-electrical effects can be studied by using appropriate transducers.

The block diagram of a complete oscilloscope is shown in Fig. 21.17(*a*). A cathode ray tube (CRT) employing electrostatic deflection and whose operation was described in section 9.8, is at the heart of the instrument. The power supply provides the various voltages (high and low) and currents required by the CRT and other circuits. The input voltage is applied to the Y-plates, but usually it is first amplified by the Y-amplifier and its calibrated attenuator (a chain of resistors). Otherwise the vertical deflection of the electron beam is too small.

The voltage applied to the X-plates, also via an amplifier, the X-amplifier, is generally obtained from the time base, which is described in (b) below. Alternatively, an external voltage can be applied to the X-input terminal. X- and Y-shift controls allow the spot on the screen to be moved 'manually' in the X- or Y-directions respectively. They apply a positive or negative voltage to one of the deflecting plates according to the shift required.

The a.c./d.c. selector switch in the a.c. position connects a capacitor in the input so blocking d.c. but allowing a.c. to pass. Only a.c. or the a.c. part of the input is then displayed on the screen. With the switch in the d.c. position, d.c., a.c. and d.c. + a.c. inputs are shown.

The Z-input is connected to the cathode of the CRT. An external voltage applied to it brightens or blacks-out the trace depending on whether it makes Z negative or positive with respect to ground. Intensity modulation of the beam is thus possible.

(b) Time base

This is a relaxation oscillator (see section 12.8) whose job is to deflect the beam horizontally in the X-direction and make the spot 'sweep' across the screen from left to right at a speed which is steady but which can be varied by the time base controls. It must then make the spot fly back rapidly to its starting position, ready for the next sweep. The output voltage from the time base should therefore have a waveform like that in Fig. 21.17(*b*). Since AB is linear the distance moved by the spot is directly proportional to time and the horizontal deflection becomes a measure of time, i.e. a time axis or base. If the input voltage applied to the Y-plates represents a quantity which varies with time (most do that are studied with an oscilloscope), then its waveform is displayed.

To maintain a stable trace on the screen, each horizontal sweep must start at the same point on the waveform. The time base frequency should therefore be that of the input or a submultiple of it. Successive traces then coincide and, owing to the persistence of vision and the phosphor on the screen, the trace appears at rest. One way of obtaining synchronization of the time base and signal frequencies is to feed part of the Y-amplifier output to a trigger circuit. This gives a pulse at a chosen point on the Y-amplifier output, as selected by the trig level control, which starts the sweep of the time base sawtooth, i.e. it 'triggers' the time base. In many oscilloscopes automatic triggering occurs if the trig level control is set to 'auto'. The mean d.c. level of the input is detected and the time base starts when this is passed.

A negative blanking pulse applied to the modulator of the CRT during flyback cuts off the beam.

(c) Dual beam and dual trace oscilloscopes

These are useful for comparing two traces simultaneously, Fig. 21.18. The dual beam type has two electron guns, each with its own Y-plates but having the same time base. In the commoner dual trace type, electronic switching produces two traces from a single beam. The switching is done by a square wave having a frequency much higher than that of either input. The top of the wave connects one input (Y_1) to the Y-plates and the bottom connects the other (Y_2). This occurs so rapidly that there appears to be two separate traces, one above the other, on the screen.

Fig. 21.18

(d) Uses

One precaution to be observed is to keep the intensity control as low as possible when there is just a spot on the screen. Otherwise screen burn occurs and the phosphor is damaged. If possible, it is best to draw the spot into a line by having the time base running.

(i) Voltage measurements An oscilloscope has an input impedance of about 1 MΩ and can be used as a d.c./a.c. voltmeter by connecting the voltage to be measured across the Y-plates, with the time base off and the X-plates earthed. The spot is deflected vertically by d.c., Fig. 21.19(*a*); a.c. makes it move up and down and, if the motion is fast enough (e.g. at 50 Hz), a vertical line is obtained.

The gain control (marked VOLTS/DIV) on the Y-amplifier of a CRO is calibrated. If the gain control is on the '20 V' setting, a deflection of 1 division of the screen graticule would be given by a 20 V d.c. input; a line 1 division long would be produced by an a.c. input of 20 V peak-to-peak, i.e. peak voltage = 10 V and r.m.s. voltage = $0.7 \times$ peak voltage = 7 V. On the '5 mV' setting the gain is greatest and small voltages can be measured, 5 mV causing a deflection of 1 division.

While there are more accurate ways of measuring voltages, the CRO can measure alternating voltages at frequencies of several megahertz and is not damaged by overloading.

(ii) Time and frequency measurements These may be made if the CRO has a calibrated time base like that in Fig. 21.18. The time base control is marked TIME/DIV and gives several preset sweep speeds.

If the TIME/DIV control is on 10 ms, the spot takes 10 ms to move 1 division and so travels 10 divisions of the screen graticule in 100 ms (0.1 s).

(iii) Waveform display The a.c. signal, whose waveform is required is connected to the Y-input and the time base turned on. When the time base has the same frequency as the input, one complete wave is formed on the screen; if it is half that of the input, two waves are displayed. These and other effects are shown in Fig. 21.19(*b*).

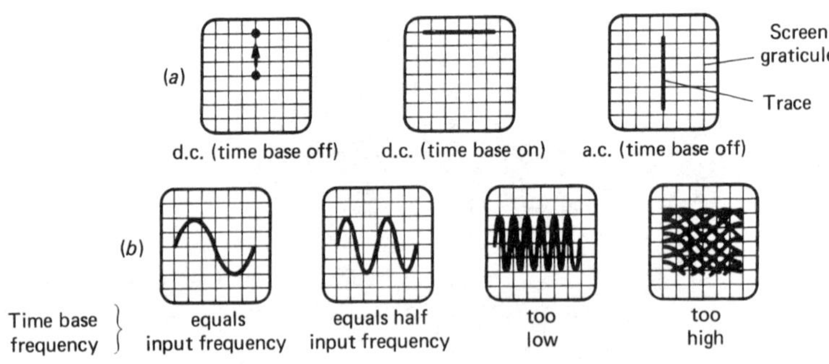

Fig. 21.19

21.9 Signal generators

A signal generator is an oscillator giving an output of known but variable frequency. There are two types.

(a) A.f. generators

These cover the range from 0.1 Hz to about 100 kHz. In modern instruments, sine, square and triangular wave outputs are produced by a waveform generator IC such as the 8038, described in section 17.5. To allow matching to different loads, they usually have a 'low' and a 'high' impedance output (e.g. values between 3 Ω and 100 Ω respectively). The former would be used for direct connection to a 4 Ω loudspeaker and the latter when feeding an oscilloscope. Most contain an *attenuator* to reduce the output voltage either by fixed steps (e.g. by a factor of 10 or 20 dB—see section 11.9) or continuously, when this is necessary.

A.f. generators are used to test audio circuits. One with a square wave output will give an idea of the frequency response of an a.f. amplifier if connected as in Fig. 21.20(*a*). A square wave contains many pure (sine wave) frequencies and if all are amplified equally, the square wave is not distorted. Types of response are shown in Fig. 21.20(*b*).

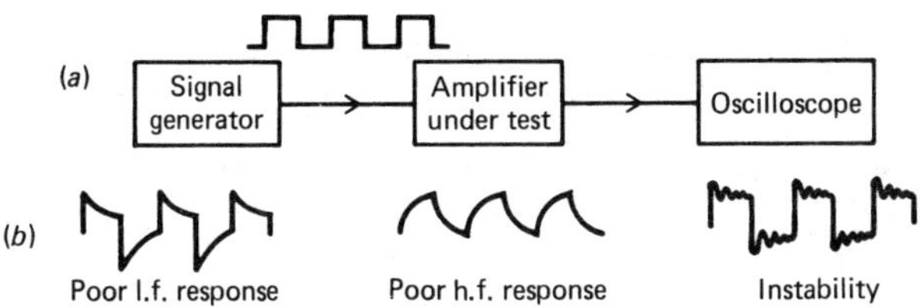

Fig. 21.20

(b) R.f. generators

The frequency coverage by this type is from 100 kHz to 300 MHz or higher. The quality of an instrument depends on the frequency stability of the oscillator. If it is an *LC* type, coarse adjustment of the frequency is obtained by switching in different coils and fine adjustment by a tuning capacitor. An attenuator allows the output, normally two or three hundred millivolts into a 75 Ω load, to be reduced if desired.

R.f. generators are used in radio work and their sine wave output can usually be both amplitude and frequency modulated.

21.10 Revision questions and Problems

1. a) What is meant by rectification?
b) Draw circuits for the following types of rectifier:
(i) half-wave,
(ii) centre-tap full-wave,
(iii) bridge full-wave.

2. Explain the statement that a heat sink has a *thermal resistance* of 4 °C W^{-1}.

3. In the reservoir capacitor smoothing circuit of Fig. 21.5(*a*):
a) what limits the maximum value of C_1?
b) if the r.m.s. output voltage of the transformer is 7 V what should be
(i) the maximum rating of C_1,
(ii) the maximum reverse voltage of D?

4. What is the frequency of the a.c. input to a full-wave rectifier if the frequency of the ripple on the output is 100 Hz?

5. a) What is meant by the regulation of a power supply?
b) Draw a block diagram for a stabilized power supply.
c) What is a voltage regulator?

6. a) Describe the action of
(i) a thyristor,
(ii) a triac,
(iii) a diac.
b) State two uses for a triac.

7. Name five types of battery used to power electronic equipment. Give one advantage and one disadvantage of each.

8. In the circuit of Fig. 21.21, V_1 and V_2 are two identical voltmeters with sensitivities of 10 kΩ V^{-1}. If both are used on their 10 V ranges, which one gives the more accurate reading? What are the readings?

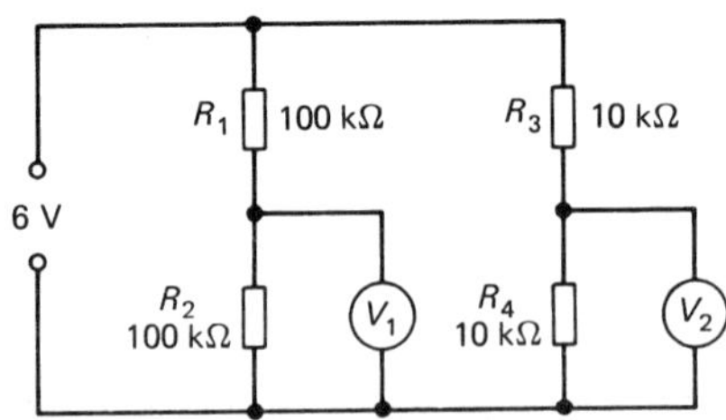

Fig. 21.21

9. a) Draw a block diagram for an oscilloscope.
b) When the Y-amplifier gain control on an oscilloscope is set on 2 V cm^{-1}, an a.c. input produces a 10 cm long vertical trace. What is
(i) the peak voltage,
(ii) the r.m.s. voltage, of the input?
c) What is the frequency of an alternating voltage which is applied to the Y-plates of an oscilloscope and produces five complete waves on a 10 cm length of the screen when the time base setting is 10 ms cm^{-1}?

APPENDIX

Pin connections of TTL and CMOS gates

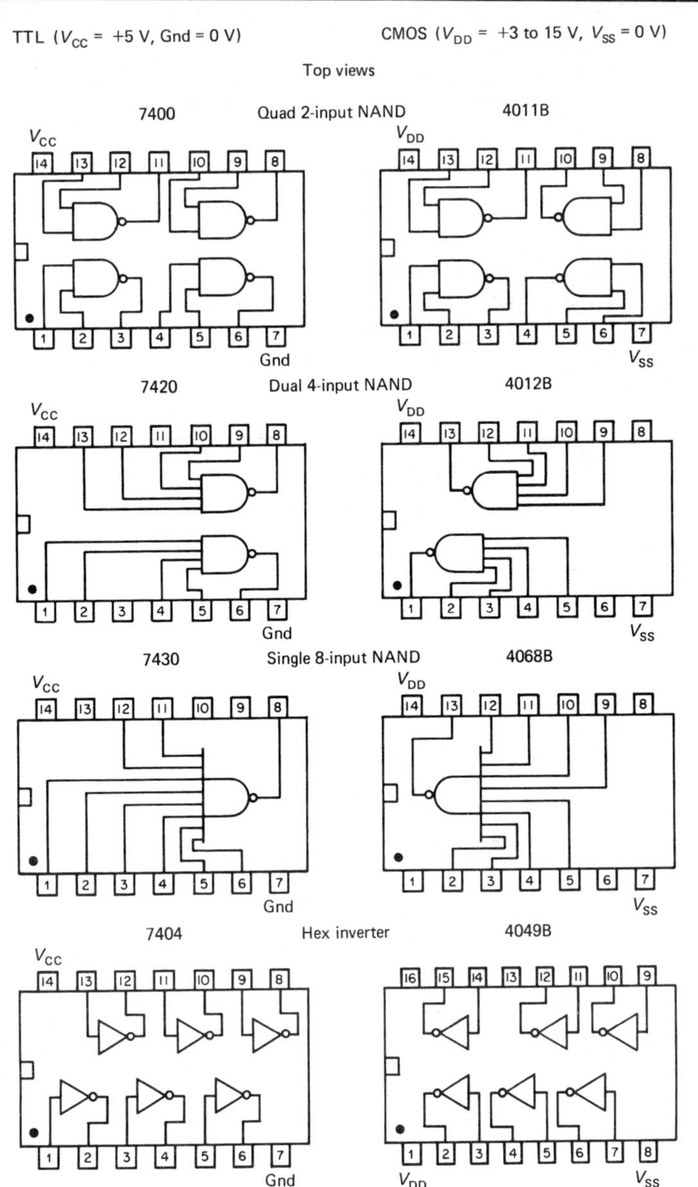

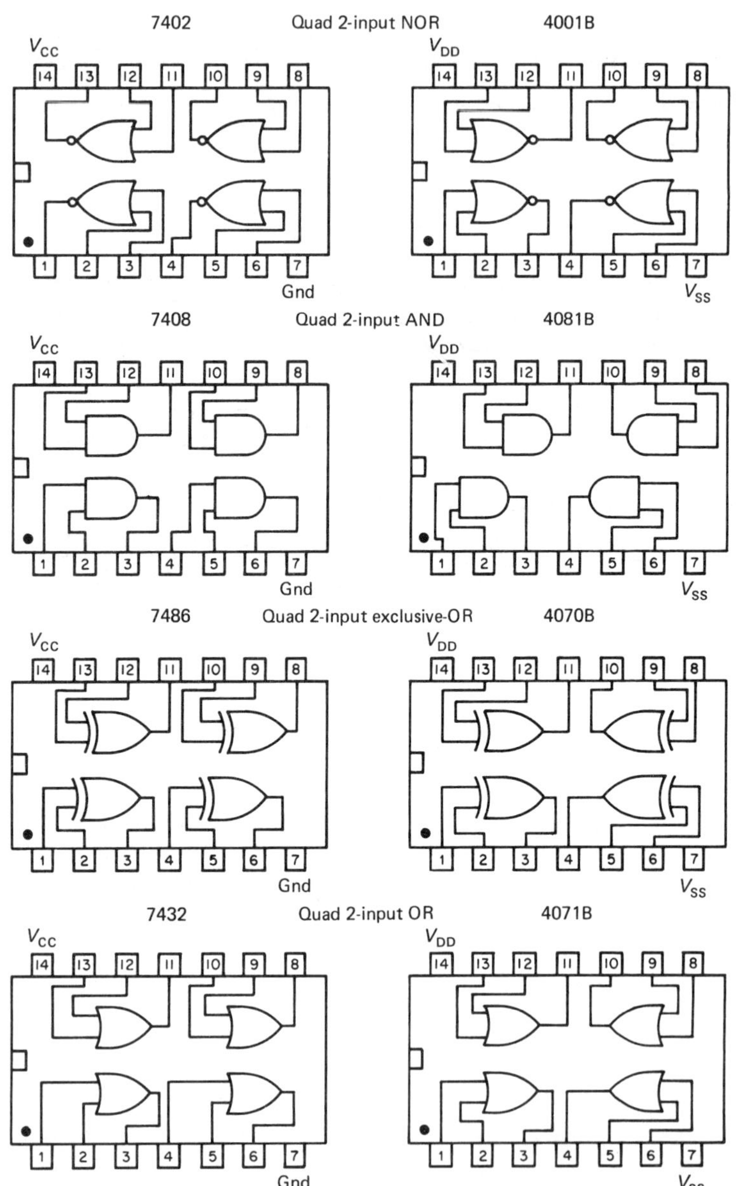
TTL
CMOS
Top views
7402
Quad 2-input NOR
4001B
V_{CC}
V_{DD}
14 13 12 11 10 9 8
1 2 3 4 5 6 7
Gnd
V_{SS}
7408
Quad 2-input AND
4081B
V_{CC}
V_{DD}
Gnd
V_{SS}
7486
Quad 2-input exclusive-OR
4070B
V_{CC}
V_{DD}
Gnd
V_{SS}
7432
Quad 2-input OR
4071B
V_{CC}
V_{DD}
Gnd
V_{SS}

Further reading

General

Jones, L.: *Basic Electronics for Tomorrow's World.* Cambridge University Press (Cambridge,1993).

Jones, M. H.: *A Practical Introduction to Electronic Circuits.* Cambridge University Press (Cambridge, 1985).

Horowitz, P. and Hill, W.: *The Art of Electronics.* Cambridge University Press (Cambridge, 1989).

Radio, Audio, Telephone and Television Systems

Green, D. C.: *Radio Systems for Technicians.* Pitman (London, 1985).

Hellyer, H. W. and Sinclair, I. R.: *Radio and Television.* Newnes Technical (London, 1976).

King, G. J.: *The Audio Handbook.* Newnes Technical (London, 1975).

Bigelow, S. J.: *Understanding Telephone Electronics.* Sams (Indiana, 1991).

Computers and Microprocessors

Fry, T. F.: *Beginner's Guide to Computers.* Newnes Technical (London, 1978).

Morgan, E.: *Microprocessors—a Short Introduction.* Department of Industry (London, 1980).

Coles, R.: *Microprocessors for Hobbyists.* Newnes Technical (London, 1979).

Ditch, W.: *Microelectronic Systems: a practical approach.* Edward Arnold (London, 1995).

Bolton, W.: *Programmable Logic Controllers: an introduction.* Butterworth-Heinemann (London, 1996).

Project Work

Sinclair, I. R.: *Practical Electronics Handbook.* Newnes Technical (London, 1992).

Bevis, G. and Trotter, M. eds.: *Microelectronics—Practical Approaches for Schools and Colleges.* B.P. Education Service (London, 1981).

Duncan, T.: *Adventures with Electronics.* John Murray (London, 1978).

Duncan, T.: *Adventures with Microelectronics.* John Murray (London, 1981).

Duncan, T.: *Adventures with Digital Electronics.* John Murray (London, 1982).

Social Implications

Jones, T. ed.: *Microelectronics and Society.* The Open University Press (Milton Keynes, 1980).

Mathematics

Bishop, O.: *Understanding Electrical and Electronic Maths.* Butterworth-Heinemann, (London, 1994).

Journals

Everyday with Practical Electronics

Electronics and Computing

Electronics Today International

Answers to revision questions and problems

Unit 2

Problems

1. a) 6 V; b) 1.5 V.

2. A_2 more; A_3 0.3 A; A_4 less (zero).

3. a) V = 18 V, V_2 = 2 V, V_1 = 8 V; b) equal.

4. (i) A = + 6 V, B = + 3 V, C = 0 V; (ii) A = 0 V, B = − 3 V, C = − 6 V; (iii) A = + 3 V, B = 0 V, C = − 3 V.

5. a) 3 Ω; b) 20 V; c) 2 A.

6. a) 0.66 V; b) 2 mA; c) 6 kΩ; d) 4 V.

7. a) (i) 5A, (ii) 0.5 A, (iii) 2 A; b) (i) 5 C, (ii) 50 C, (iii) 300 C.

8. a) 12 J; b) 60 J; c) 240 J.

9. a) 20 J; b) 100 J; c) 1 200 J.

10. a) 24 W; b) 3 J s^{-1}.

11. a) 8 V; b) 16 W.

12. a) 3 A; b) 4 Ω.

13. a) 3 V; b) 2.6 V; c) 0.4 V; d) 2 Ω; e) 13 Ω.

14. a) 6 V, 3 Ω; b) 3 Ω, 3 W.

Unit 3

Problems

1. a) 0.02 s = 1/50 s; b) 50 Hz; c) 10 V; d) 7 V.

2. a) 3:1; b) 1:4.

3. Steady direct voltage = 10 V; peak alternating voltage = 5 V.

4. a) 12 V; b) 12/0.7 = 17 V; c) 24 W.

Unit 4

Revision questions

3. a) 5.6 kΩ; b) 15 kΩ; c) 100 kΩ; d) 2.2 MΩ.

4. a) R_1:330 Ω ± 10%, R_2:10 kΩ ± 5%, R_3:560 kΩ ± 20%, R_4:2.2 MΩ ± 5%.; b) R_5:1 kΩ ± 1%, R_6:330Ω ± 5%, R_7:51 kΩ ± 2%, R_8:470 kΩ ± 10%.

5. a) (i) brown green brown brown; (ii) brown black black gold; (iii) yellow violet yellow; (iv) grey red red gold; (v) orange white orange red; (vi) brown black green silver.

b) (i) brown blue black black red; (ii) red yellow black brown gold; (iii) violet green black orange brown.

6. a) 2.2 kΩ ± 20%; b) 270 kΩ ± 5%; c) 1 MΩ ± 10%; d) 15 Ω ± 1%; e) 680 Ω ± 20%; f) 33 kΩ ± 5%.

7. a) 100RJ; b) 4K7G; c) 18KF; d) 390KM; e) 10MK; f) 68RJ.

11. a) 27 Ω; b) 4.7 kΩ; c) 68 kΩ; d) 330 kΩ.

12. a) 11 Ω, 9 Ω; b) 5.2 kΩ, 4.2 kΩ; c) 12 kΩ, 8 kΩ; d) 861 kΩ, 779 kΩ; e) 198 Ω, 162 Ω; f) 4.0 MΩ, 2.6 MΩ.

Problems

1. 0.25 W (250 mW).

2. 0.36 W (360 mW).

3. No; 1 W.

4. a) 12 kΩ; b) 15 kΩ.

5. a) 2 Ω; b) 1.5 Ω; c) 5 kΩ.

6. Combined resistance = 10 × 22/32 = 6.9 kΩ.

7. $80 = R \times 100/(R + 100)$, $R = 400$ kΩ, preferred value = 390 kΩ.

8. a) 4 Ω; b) 12 Ω; c) 3 Ω.

9. $\dfrac{(18 + 2) \times 10}{20 + 10} = 6.7$ kΩ, preferred value = 6.8 kΩ.

10. a) $\frac{10}{20} \times 6 = 3$ V; b) $\frac{150}{250} \times 5 = 3$ V; c) $\frac{100}{110} \times 9 = 8.2$ V.

11. a) 3 V; b) 4.5 V.

12. a) 4.5 V; b) 4.3 V; c) 3.0 V.

Unit 5

Revision questions

2. a) 0.1, 8.2, 10; b) 0.33, 1.0, 0.047; c) 820, 1000, 56 000, 100 000; d) 221, 823, 104.

3. a) 0.12 μF, 0.08 μF; b) 200 μF, 80 μF.

5. b) (i) $\frac{1}{C} = \frac{1}{C_1} + \frac{1}{C_2} + \frac{1}{C_3}$; (ii) $C = C_1 + C_2 + C_3$.

6. b) (i) Charge—after T, $V_C = 4$ V; after $2T$, $V_C = 5\frac{1}{3}$ V (since it rises by two-thirds of the voltage remaining, i.e. by $\frac{2}{3} \times (6 - 4) = 1\frac{1}{3}$ V);
(ii) Discharge—after T, $V_C = 2$ V; after $2T$, $V_C = \frac{2}{3}$ V (since it falls by two-thirds of the voltage across it at the start of the second time constant, i.e. by $\frac{2}{3} \times 2 = 1\frac{1}{3}$ V).

7. a) A; b) A.

8. d) (i) M_2; (ii) M_3.

9. B.

Problems

1. From $Q = V \times C$, a) 1 C; b) 450 μC.

2. a) $C = Q/V = 12\ \mu C/6\ V = 2\ \mu F$; b) $V = Q/C = 50\ \mu C/10\ \mu F = 5$ V.

3. a) $W = \frac{1}{2}Q \times V = \frac{1}{2} \times 2 \times 9 = 9$ J;
b) $Q = V \times C = 500 \times 1 = 500\ \mu C = 500 \times 10^{-6} C$ $W = \frac{1}{2}Q \times V = \frac{1}{2} \times 500 \times 10^{-6} \times 500 = 0.125$ J .

4. a) 6.9 μF; b) 0.32 μF; c) 50 μF.

5. 1 s, 1 000 μF, 0.033 s, 47 kΩ.

6. a) $I = V/R = 1.5/150$ A $= 1.5 \times 1000/150$ mA $= 10$ mA;
b) $Q = V \times C = 1.5 \times 1000 = 1500\ \mu C$;
c) Time constant $= C \times R = (1000 \times 10^{-6}) \times 150 = 0.15$ s = time to charge to $\frac{2}{3}$ of 1.5 V, i.e. to 1.0 V..

7. a) $X_C = 1/(2\ \pi fC) = 1/(2\pi \times 10^3 \times 10^{-6}) = 159\ \Omega$; b) 0.159 Ω.

8. $C = 1/(2\pi f X_C) = 1/(2\pi \times 700 \times 50)F = 4.55\ \mu F$.

9. Period of square wave input = 10 ms = 0.01 s.
a) Time constant $C \times R = 1\ \mu F \times 1\ k\Omega = 10^{-6} F \times 10^3\ \Omega = 0.001$ s. Time constant is 10 times *smaller* than period of input. Waveform same as Fig. 5.19(*c*) with spikes having 'peak' values of +1 V and −1 V.
b) Time constant $C \times R = 10 \times 10^{-6}\ F \times 10 \times 10^3\ \Omega = 0.1$ s. Time constant is 10 times *longer* than period of input. Waveform same as Fig. 5.19(*d*) with 'rough' square waves having 'peak' values of +0.5 V and −0.5 V.

Unit 6

Revision questions

5. a) magnetic field; b) magnet; c) changing, changing .
6. Resistance.
12. Greater, opposite polarity.

Problems

1. a) $X_L = 2\pi fL = 2\pi \times 100 \times 15\ \Omega = 9.4\ \text{k}\Omega$;
   b) $X_L = 2\pi \times 10^6 \times 10^{-3}\ \Omega = 6.3\ \text{k}\Omega$.
2. $L = X_L/(2\pi f) = 200/(2\pi \times 1000) = 0.032\ \text{H} = 32\ \text{mH}$.
3. a) 20 Ω; b) 0.5 A; c) $V_R = 6$ V, $V_L = 48$ V, $V_C = 40$ V.
4. a) $n_p/n_s = V_p/V_s = 240/12 = 20/1$; b) $n_p = 20n_s = 20 \times 80 = 1600$;
   c) $V_p \times I_p = V_s \times I_s$, $240 \times I_p = 12 \times 2$, $I_p = 0.1$ A $= 100$ mA.
5. a) $n_p/n_s = 15/1 = I_s/I_p = I_s/40$, $I_s = 600$ mA $= 0.6$ A

Unit 7

Problems

1. a) L_1 bright, L_2 Off; b) L_1 bright, L_2 off because the diode is forward biased and offers an easy path for the current to bypass L_2; c) L_1 and L_2 both dim because the reversed biased diode forces current through L_2 and there is then 4.5 V across each lamp—they need 6 V (or more) to be bright.
2. a) The diode is reverse biased if the supply is wrongly connected and offers a very high resistance so that very little current can flow to damage the circuit.

   b) If a large voltage is applied to the meter the diode conducts and acts as a bypass which diverts the current that would normally flow through M. A small voltage (less than 0.6 V for a silicon diode) does not turn on the diode.
3. a) Silicon; b) 22 Ω, 0.25 W.
4. Brightness gradually increases from 3 V to 6 V and is then 'tied' at that brightness by the Zener diode from 6 V to 9 V.
5. 0.5 A (500 mA).
6. (12 − 10) V/0.5 A = 4.0 Ω.

Unit 8

Revision questions

1. a) n-p-n; b) p-n-p.

2. Collector, base, emitter.

3. See section 8.2.

4. a) two; b) base-emitter; c) collector-base.

5. L lights in (*b*) and (*d*).

6. To allow as many as possible of the majority carriers to reach the collector from the emitter.

7. a) See section 8.2; b) See section 8.3.

8. a) about 0.6 V; b) about 0.1 V.

9. $h_{FE} = I_C/I_B$.

10. a) Graph of I_B against V_{BE}, $r_i = \Delta V_{BE}/\Delta I_B$;
b) graph of I_C against V_{CE}, $r_o = \Delta V_{CE}/\Delta I_C$;
c) graph of I_C against I_B, $h_{fe} = \Delta I_C/\Delta I_B$.

11. Leakage current (I_{CEO}) and d.c. current gain (h_{FE}).

12. a) (i) drain; (ii) source; (iii) gate;
b) FET has a much larger input resistance than a bipolar transistor and so draws less current from the 'driving' circuit.

14. Must be protected from static electricity.

Problems

1. a) 50; b) 102 mA.

2. 10 mA.

3. a) no; b) yes.

4. $V_{CC} = V_{R1} + V_{BE}$, $V_{R1} = 6.0 - 0.6 = 5.4$ V; $R_1 = V_{R1}/I_B = 5.4$ V/0.02 mA = 270 kΩ.

5. a) $V_{CC} = V_{R1} + V_{BE}$, $V_{R1} = 4.5 - 0.6 = 3.9$ V; $I_B = V_{R1}/R_1 = 3.9$ V/3.9 kΩ = 1 mA; b) $h_{FE} = I_C/I_B = 25/1 = 25$ (at 25 mA).

6. a) 5.4 V; b) 0.054 mA (54 μA); c) $I_C = h_{FE} \times I_B = 60 \times 0.054 = 3.2$ mA; d) $V_{R2} = I_C \times R_2 = 3.2$ mA × 1 kΩ = 3.2 V; e) $V_{CC} = V_{R2} + V_{CE}$, $V_{CE} = V_{CC} - V_{R2} = 6.0 - 3.2 = 2.8$ V.

Unit 9

Problems

1. 680 Ω.
2. a) $V = I \times R = \frac{10}{1\,000} \times 1\,000 = 10$ V; b) $I = \frac{V}{R} = \frac{6}{700}$A = 8.6 mA.
3. a) When S is switched on, *C* behaves like a very low resistor (i.e. a 'short circuit') in parallel with the relay coil and side-tracks the current from it. As *C* charges, it behaves like a larger resistor, less current flows into it and more goes through the relay coil. Eventually the coil current is large enugh to energize the relay. (*R* prevents *C* charging too rapidly at the start.)

   b) Immediately S is switched on, the maximum current flows because *C*, being uncharged, offers very little resistance at the start. This starting current is greater than the 'pull-in' current of the relay and its contacts close, lighting the bulb at once. As *C* charges the current falls, eventually to the 'drop-out' current of the relay.

Unit 10

Problems

1. a) 6.0 V; b) 15 μA (0.015 mA); c) 560 kΩ.
2. a) $V_{CC} = 10$ V, $R_L = 2$ kΩ; b) $V_{CE} = 5$ V, $I_C = 2.5$ mA, $I_B = 20$ μA; c) 220 kΩ.
3. b) $I_C = 2.5$ mA, $I_B = 30$ μA, $V_{CE} = 4.5$ V; c) Power $= I_C \times V_{CE} = 2.5 \times 4.5 =$ 11.3 mW; d) (i) 2.5 to 6.5 V = 4.5 V ± 2.0 V, (ii) ±2.0 V peak; e) ±20 mV;

   f) 2.0 V/20 mV = 100; g) $R_B = \frac{V_{CE} - V_{BE}}{I_B} = \frac{(4.5 - 0.6)\text{ V}}{30\ \mu\text{A}} = 130$ kΩ.

Unit 11

Problems

1. a) $V_B = (10 \times 9)/(47 + 10) = 1.6$ V; b) $V_E = V_B - V_{BE} = 1.6 - 0.6 = 1.0$ V ($V_{BE} = 0.6$ for silicon); c) $I_E = V_E/1$ kΩ = 1 V/1 kΩ = 1 mA; d) $V_C = V_{CC} - I_C \times 4.7$ kΩ $= 9 - 1 \times 4.7 = 4.3$ V (since $I_C \approx I_E$); e) $V_{CE} = V_C - V_E = 4.3 - 1.0 = 3.3$ V; f) $I_B = I_C/h_{FE} = 1$ mA/100 = 0.01 mA = 10 μA.
2. 40.
3. a) 1/100; b) 1/10.
4. 10:1.

5. Efficiency = (a.c. power output) × 100/(d.c. power from supply) = 60 mW × 100/ (50 mA × 6 V) = 20%.

6. Maximum a.c. power = $V_{CC}^2/8Z_L$ = 81/(8 × 8) = 1.3 W.

7. 7 dB.

8. At thc two frequencies a) the power is half and ; b) the voltage is 0.7, of its maximum value at frequencies mid-way between 10 Hz and 50 kHz.

9. a) 30 dB; b) 40 dB.

Unit 12

Revision questions

8. a) Z_1 – L_1, power supply, C_3; Z_2 – C_2, L_0, power supply, $\mathbf{C}_3$; Z_3 – C_1, C_2; b) Z_1 – C_1, C_3; Z_2 – C_2, C_0, C_3; Z_3 – L_1, C_2.

Problems

1. 2.25 MHz, 711.5 kHz.

2. 500 pF, 125 pF.

3. a) 1.6 MHz; b) $Z = R = 10\ \Omega$; c) 100.

4. a) 1.6 MHz; b) $Z = L/(CR) = 100\ \text{k}\Omega$; c) 100.

5. 3 MHz i.e. the frequency range over which the gain does not fall below 0.7 × 120 = 84 dB.

6. 1.6 kHz.

8. a) (i) 1/1; (ii) 1/2; b) (i) f = 100 kHz = 10^5 Hz = $1/T$ where T is the period, $T = 1/10^5$ s = 10 μs, (ii) 6 μs/(10 – 6) μs = 3/2, (iii) 6 μs/10 μs = 3/5.

Unit 13

Problems

1. ± 150 μV.

2. a) –2; b) 10 kΩ.

3. $V_o = -R_f(V_1 + V_2)/R_i = -2(3 + 2) = -10$.

4. a) +3; b) greater.

5. See Fig. A.01.

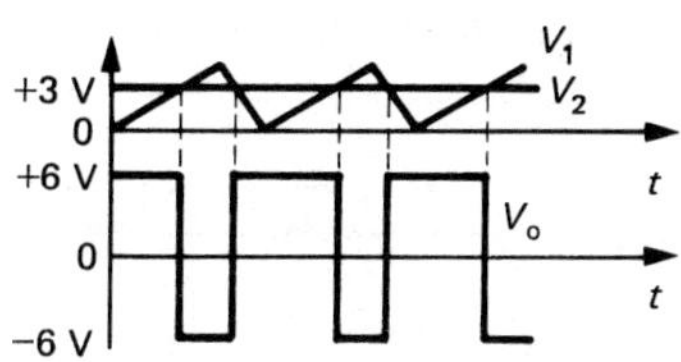

Fig. A.01

6. V_o switches from 'low' to 'high'.

7. a) $0 - 10^5$ Hz; b) $0 - 10^3$ Hz.

Unit 14

Revision questions

1. (a); (c); (d).

2. Four (red, red + amber, green, amber).

3. b) (i) zero, (ii) zero, (iii) zero; c) (i) very large, (ii) V_{CC}, (iii) zero; d) $I_C/I_B < h_{FE}/5$; e) Supply voltage and value of resistor in collector circuit; f) B since V_{CE} is smaller.

5. a) AND; b) OR.

Problems

1. a) +6 V; b) 0 V.

2. a) I_C (max) = supply voltage/R_L = 6/100 = 0.06 A = 60 mA;
b) (i) For hard bottoming $I_C/I_B = h_{FE}/5$, $60/I_B = 150/5$, $I_B = 2$ mA, (ii) $6 = V_{R_B} + V_{BE}$, $V_{R_B} = 5.4$ V, $R_B = V_{R_B}/I_B = 5.4$ V/2 mA = 2.7 kΩ.

3.

A	B	C	D	F
0	0	0	0	0
1	0	0	0	0
0	1	0	0	0
1	1	0	1	1
0	0	1	0	1
1	0	1	0	1
0	1	1	0	1
1	1	1	1	1

4. a) Output 'high' when both inputs 'high'—see Fig. A.02(*a*);
b) Output 'high' when both inputs 'low'—see Fig. A.02(*b*).

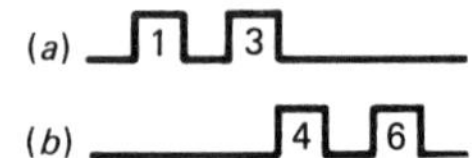

Fig. A.02

5. AND.

6. a) −6 V; b) +6V; c) −12 V.

7. 105 Hz.

8. a) See Fig. A.03; b) see Fig. A.04.

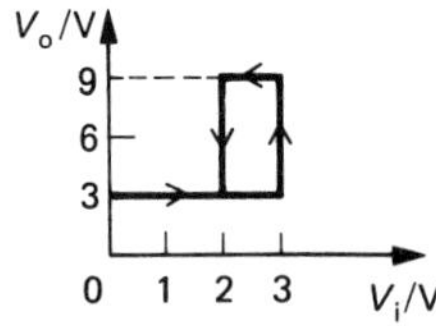

Fig. A.03

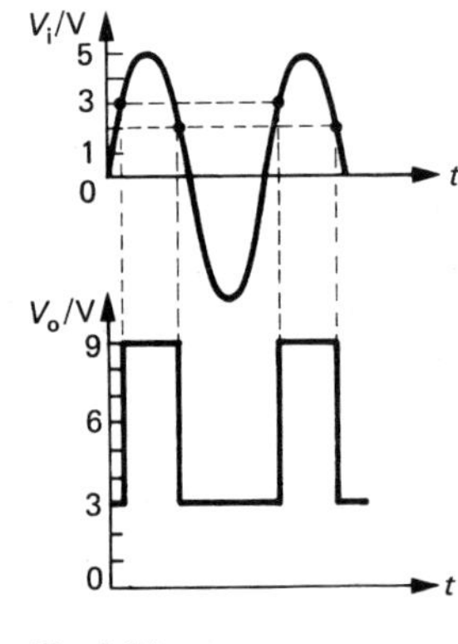

Fig. A.04

Unit 15

Revision questions and Problems

5. a) 8; b) 6; c) 8.

7. a) 100, 1101, 10101, 100110, 1000000; b) 7, 25, 42, 50; c) 6, B, F, 1F, 53, 12C; d) 3, 10, 13, 30, 421; e) 1001, 0001 0111, 0010 1000, 0011 0111 0000, 0110 0100 0101.

Unit 16

Revision questions and Problems

3. 50 kHz.

7. c) 8.

10. a) (i) $2^1 = 2$, (ii) $2^2 = 4$, (iii) $2^3 = 8$, (iv) $2^4 = 16$, (v) $2^5 = 32$; b) (i) 1, (ii) 3, (iii) 7, (iv) 15, (v) 31.

11. b) (i) 1, (ii) 3, (iii) 3, (iv) 4, (v) 5; c) $f/10$.

13. c) (i) 512, (ii) 64; d) $2^5 = 32$.

Unit 17

Revision questions and Problems

2. a) (i) R_1, R_2 and C_1 (ii) R_1, R_2, R_4 and C_1—the pitch increases when S_1 is 'off' because R_4 (1 kΩ) is now in parallel with R_1 (100 kΩ) and reduces its value to less than 1 kΩ. b) When C_3 has discharged sufficiently through R_3 for the voltage across it to be less than 0.7 V, the voltage at the reset pin is too low for astable operation. c) The second note would not be produced.

9. a) 50; b) 10.

10. a) (i) and (iii); b) (iii), because (i) requires the output of the TTL gate to act as a 10 mA source (i.e. provide the 10 mA required to light the LED); (iii) uses the TTL gate to sink the 10 mA which lights the LED and this it can easily do. c) 16/1.6 = 400/40 = 10. d) (i) and (iii).

Unit 18

Revision questions and Problems

3. a) 0.25 m; b) 150 MHz .

5. c) 9 kHz.

7. a) increased deviation; b) a very large number; c) narrower bandwidth.;

9. b) 0 to 5 MHz approx.

Unit 19

Revision questions and Problems

1. a) $Q = \overline{A} . \overline{B}$ (see Fig. A.05); b) $Q = \overline{A} . \overline{B} . \overline{C} + A . \overline{B} . C$ (see Fig. A.06).

2. a) $R = \overline{A} . \overline{B}$, $Y = \overline{A} . B$, $G = A . \overline{B}$, modulo-3 counter needed (see Fig. A.07); b) (i) $Q = A . \overline{B} + \overline{A} . B$ (see Fig. A.08), (ii) $Q = A . \overline{B} + \overline{A} . B$ (see Fig. A.09); c) $Q_0 = \overline{A} . \overline{B}$, $Q_1 = A . \overline{B}$, $Q_2 = \overline{A} . B$, $Q_3 = A . B$ (see Fig. A.10); d) $Q = \overline{A} . \overline{B} + A . B$ (see Fig. A.11).

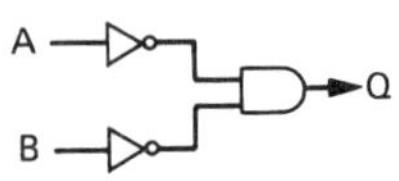

Fig. A.05

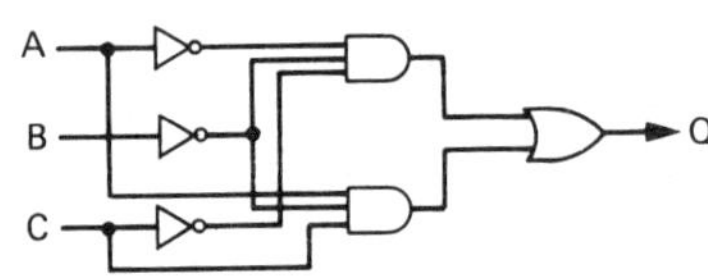

Fig. A.06

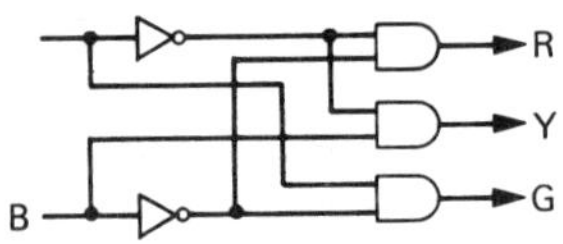

Fig. A.07

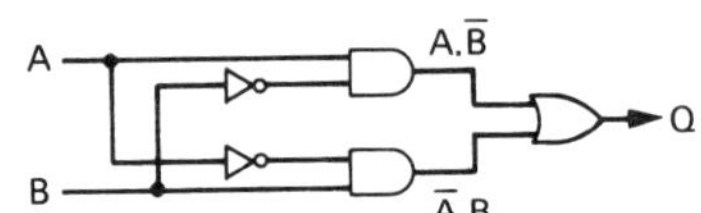

Fig. A.08

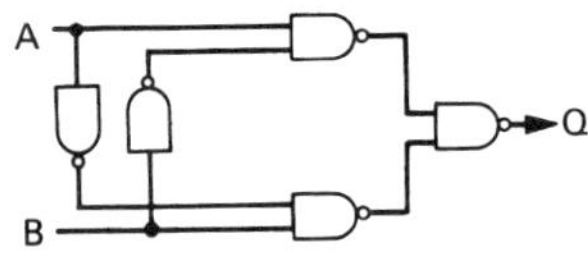

Fig. A.09

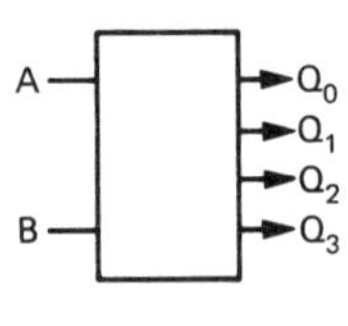

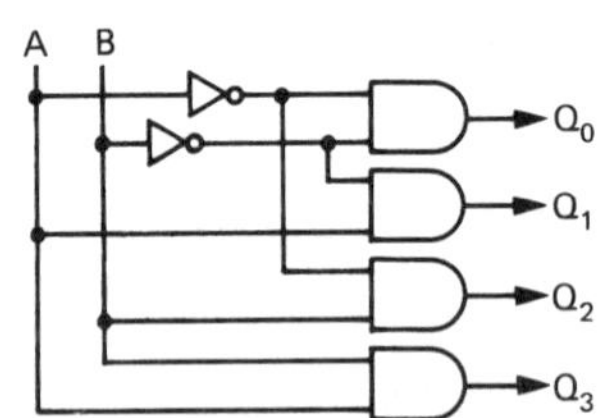

Fig. A.10

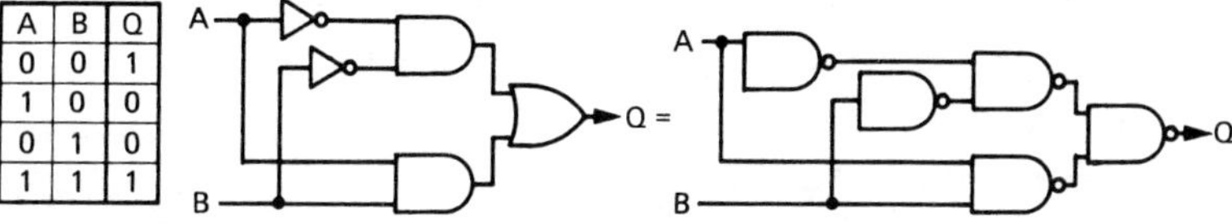

A	B	Q
0	0	1
1	0	0
0	1	0
1	1	1

Fig. A.11

5. a) A; b) A; c) 1; d) A; e) 1; f) A; g) 0; h) 0; i) 1; j) 0; k) 1; l) 1.

7. a) $A.(A+B) = A.A + A.B = A + A.B$ (since $A.A = A$)
$= A(1+B) = A$ (since $1 + B = 1$).

b) $(A+B).(B+C) = A.B + B.B + A.C + B.C$
$= A.B + B + A.C + B.C$ (since $B.B = B$)
$= B.(A+1) + C.(A+B)$ (since $A + 1 = 1$)
$= B + C.(A+B)$
$= B + A.C + B.C$
$= B(1+C) + A.C$
$= B + A.C$ (since $1 + C = 1$).

c) $\overline{A+B+C} = \overline{A+B}\,.\,\overline{C}$ (1st theorem)
$= \overline{A}.\overline{B}.\overline{C}$ (1st theorem).

d) $\overline{A}.\overline{B}.C + \overline{A}.B.C + A.C = \overline{A}.C(\overline{B} + B) + A.C$
$= \overline{A}.C + A.C$ (since $\overline{B} + B = 1$)
$= C(\overline{A} + A)$
$= C$ (since $\overline{A} + A = 1$).

e) $\overline{(A+B).(B+C)} = \overline{(A+B)} + \overline{(B+C)}$ (1st theorem)
$= \overline{A}.\overline{B} + \overline{B}.\overline{C}$ (1st theorem)
$= \overline{B}.(\overline{A} + \overline{C})$.

f) $\overline{\overline{A}.B+B.\overline{C}} = \overline{\overline{A}.B}\,.\,\overline{B.\overline{C}}$ (1st theorem)
$= (\overline{\overline{A}} + \overline{B}).(\overline{B} + \overline{\overline{C}})$ (1st theorem)
$= (A+\overline{B}).(\overline{B} + C)$ (since $\overline{\overline{A}} = A$ and $\overline{\overline{C}} = C$)
$= A.\overline{B} + \overline{B}.\overline{B} + A.C + \overline{B}.C$
$= A.\overline{B} + \overline{B} + A.C + \overline{B}.C$ (since $\overline{B}.\overline{B} = \overline{B}$)
$= \overline{B}.(A+1) + C.(A+\overline{B})$
$= \overline{B} + C.A + C.\overline{B}$ (since $A + 1 = 1$)
$= \overline{B}(1+C) + C.A$
$= \overline{B} + A.C$ (since $1 + C = 1$).

8. a) (i) $\overline{\overline{A.B}} = A.B$ (i.e. an AND gate), Q = 1 when A = 1 and B = 1;

(ii) $\overline{A+B}$ (i.e. a NOR gate), Q = 1 when A and B are both 0;

(iii) $\overline{A.B+C} = \overline{A.B}.\overline{C} = (\overline{A}+\overline{B}).\overline{C} = \overline{A}.\overline{C}+\overline{B}.\overline{C}$,

Q = 1 when A and C = 0 *or* B and C = 0;

(iv) $\overline{A}.\overline{B}.C$, Q = 1 when A = 0, B = 0 *and* C = 1;

(v) $\overline{\overline{A.\overline{B}}+\overline{C}} = \overline{\overline{A.\overline{B}}}.\overline{\overline{C}} = A.\overline{B}.C$, Q = 1 when A = 1, B = 0 *and* C = 1;

(vi) $\overline{A}.B+\overline{C}$, Q = 1 when A = 0 and B = 1 *or* when C = 0.

b) (iii)

A	B	C	A.B	Q	
0	0	0	0	1	← $\overline{A}.\overline{C}+\overline{B}.\overline{C}$ (A and C = 0 *or* B and C = 0)
0	0	1	0	0	
0	1	0	0	1	← $\overline{A}.\overline{C}$ (A and C = 0)
0	1	1	0	0	
1	0	0	0	1	← $\overline{B}.\overline{C}$ (B and C = 0)
1	0	1	0	0	
1	1	0	1	0	
1	1	1	1	0	

(iv)

A	B	C	$\overline{A}.\overline{B}$	Q	
0	0	0	1	0	
0	0	1	1	1	← $\overline{A}.\overline{B}.C$ (A = 0, B = 0 *and* C = 1)
0	1	0	0	0	
0	1	1	0	0	
1	0	0	0	0	
1	0	1	0	0	
1	1	0	0	0	
1	1	1	0	0	

9. a) To get an ex-NOR gate from an ex-OR gate the output from the former has to be inverted, hence if $Q = \overline{A}.B + A.\overline{B}$ is for an ex-OR gate, for an ex-NOR gate we can say, inverting the expression, that:

$$Q = \overline{\overline{A}.B+A.\overline{B}} = \overline{\overline{A}.B}.\overline{A.\overline{B}}$$
$$= (\overline{\overline{A}}+\overline{B}).(\overline{A}+\overline{\overline{B}}) = (A+\overline{B}).(\overline{A}+B)$$
$$= A.\overline{A}+\overline{A}.\overline{B}+A.B+\overline{B}.B$$
$$= \overline{A}.\overline{B}+A.B \quad (\text{since } A.\overline{A} = B.\overline{B} = 0)$$

b) In effect gates 1 and 2 each behave as an NOR gate (OR followed by NOT) to inputs A and B with an output $\overline{A+B}$; the other input to OR gate 4 is from an AND gate with output A . B. Therefore for OR gate 4 we can say:

$$Q = (\overline{A+B}) + A.B = \overline{A}.\overline{B} + A.B$$

which is the Boolean expression for an ex-NOR gate.

Unit 20

Revision questions and Problems

3. a) Falls at rate of 100 V s^{-1}; b) 100 kΩ.

Unit 21

Revision questions and Problems

3. b) (i) peak voltage = r.m.s. voltage/0.7 = 10 V, (ii) 4 × r.m.s. voltage = 28 V.

4. 50 Hz.

8. Both voltmeters have the same resistance of 10 × 10 = 100 kΩ but the resistance of V_2 is 10 times greater than that of R4 and gives the more accurate reading of 3 V. V_1 will read 2 V.

9. b) (i) 2 × 10/2 = 10 V, (ii) 0.7 × 10 = 7 V; c) 50 Hz.

Index